Green and Sustainable Infrastructure Construction Materials

Green and Sustainable Infrastructure Construction Materials

Editors

Jiaqing Wang
Shuaicheng Guo
Ruizhe Si
Chaochao Liu
Fangyuan Gong

MDPI

Basel • Beijing • Wuhan • Barcelona • Belgrade • Novi Sad • Cluj • Manchester

Editors
Jiaqing Wang
Nanjing Forestry University
Nanjing, China

Shuaicheng Guo
Hunan University
Changsha, China

Ruizhe Si
Southwest Jiaotong University
Chengdu, China

Chaochao Liu
Changsha University of Science & Technology
Changsha, China

Fangyuan Gong
Hebei University of Technology
Tianjin, China

Editorial Office
MDPI
St. Alban-Anlage 66
4052 Basel, Switzerland

This is a reprint of articles from the Special Issue published online in the open access journal *Materials* (ISSN 1996-1944) (available at: https://www.mdpi.com/journal/materials/special_issues/9245RHSO58).

For citation purposes, cite each article independently as indicated on the article page online and as indicated below:

Lastname, A.A.; Lastname, B.B. Article Title. *Journal Name* **Year**, *Volume Number*, Page Range.

ISBN 978-3-0365-9796-6 (Hbk)
ISBN 978-3-0365-9797-3 (PDF)
doi.org/10.3390/books978-3-0365-9797-3

Cover image courtesy of Jiaqing Wang

Contents

Preface

This reprint aims to discover cutting-edge techniques used on green and sustainable infrastructure construction materials. Due to the lack of natural resources and the rapid development of infrastructure construction in recent decades, the demand for the environmentally friendly and sustainable materials has increased. Therefore, multiple techniques have been introduced for their preparation, such as the reuse of waste resources via innovative pre-treatment methods and optimized material mixture designs, the application of sustainable biomass resource in the construction materials, and the utilization of composite structures to improve the serviceability of infrastructures.

Jiaqing Wang, Shuaicheng Guo, Ruizhe Si, Chaochao Liu, and Fangyuan Gong
Editors

materials

MDPI

Article

Parameter Study of Interfacial Capacities for FRP–Steel Bonded Joints Based on 3D FE Modeling

Jie Liu [1], Yu Yuan [1], Libin Wang [1,*], Zhongxiang Liu [2] and Jun Yang [3]

1 College of Civil Engineering, Nanjing Forestry University, Nanjing 210037, China
2 School of Transportation, Southeast University, Nanjing 210096, China
3 School of Civil Engineering, Suzhou University of Science and Technology, Suzhou 215129, China
* Correspondence: jhwlb@163.com

Abstract: This paper investigated the stress distribution of an adhesive layer for GFRP–steel bonded joints under 22.48 kN tensile loading using a three-dimensional numerical simulation. Firstly, a stress analysis of three paths was conducted, and after comparison, path II (through the middle layer of the bonding layer) was adopted as the analyzing path. Furthermore, a systemically parametric study of the effects of the FRP stiffness (i.e., elastic modulus and thickness), bonding length, adhesive thickness, and adhesive modulus was conducted. For the joints with different FRP elastic moduli, the minimum value of normal peeling stress was calculated as −3.80 MPa by the FRP for 10 GPa, showing a significantly severe stress concentration of FRP for 10 GPa. An analysis of the von Mises stresses proved that the increase in FRP stiffness could reduce the stress concentration of the adhesive layer effectively. The study of the effect of bonding lengths indicated that a more uniform peeling stress distribution could result from the longest bonding size; the largest peeling stress of 6.54 MPa was calculated for a bonding length of 30 mm. Further parameter analysis showed that the stress concentration of the adhesive layer could be influenced by the FRP thickness, bonding thickness, and elastic modulus of the adhesive layer.

Keywords: adhesive; equivalent stress; fiber-reinforced polymer–steel; finite element model; peeling stress

Citation: Liu, J.; Yuan, Y.; Wang, L.; Liu, Z.; Yang, J. Parameter Study of Interfacial Capacities for FRP–Steel Bonded Joints Based on 3D FE Modeling. *Materials* **2022**, *15*, 7787. https://doi.org/10.3390/ma15217787

Academic Editors: Jiaqing Wang, Fangyuan Gong, Shuaicheng Guo, Ruizhe Si and Chaochao Liu

Received: 30 September 2022
Accepted: 2 November 2022
Published: 4 November 2022

1. Introduction

A large number of steel structures such as buildings, bridges, offshore platforms, and large mining equipment (wind turbines for instance) require repairing/retrofitting due to deterioration under loading and environmental effects [1,2]. The conventional repairing (or retrofitting) method for steel structures is to cut out and replace plating or attach (or weld) external steel plates. In recent years, with the successful use of artificial fiber-reinforced polymer (FRP) in the aerospace industry, FRP reinforcement and its repairing technique have attracted extensive attention in engineering steel structures due to FRP's superior characteristics such as a high strength-to-weight ratio, easy fabrication, superior corrosion/fatigue resistance, and good durability [3–6]. Many researchers [7–9] have mentioned that FRP reinforcement could effectively improve/repair the bearing capacity of engineering steel structures, including the static bearing capacity, fatigue resistance, etc.

For composite structures (i.e., structures composed of different component materials), the bonding layer is often the weak connection area [10,11]; debonding of FRP from a steel substrate is one of the main failure modes [3]. Therefore, to achieve a superior capacity of reinforced steel structures, the bonding behavior between a steel matrix and FRP needs to be studied in depth.

In 2001, Miller et al. [12] studied the bonding performance of carbon-fiber-reinforced polymer (CFRP) on a steel plate and found that the force transfer length was a function of the geometric and material properties of the steel substrate, CFRP reinforcement, and adhesive. Fawzia et al. [13] studied the bonding behaviors of FRP–steel double-strap bonded joints and mentioned that the bonding capacities and failure modes could be affected by the elastic modulus and bonding length significantly. Yu et al. [14] discussed the performance of CFRP-to-steel bonded interfaces based on an experimental study and concluded that nonlinear adhesives with a lower elastic modulus but a larger strain capacity were more beneficial than linear adhesives with a similar or even higher tensile strength. Neto et al. [15] conducted a parametric study of adhesive joints with composites and discussed the failure in adhesive joints with different bonding parameters. Furthermore, in 2016, Wang et al. [16] presented an experimental study on the behavior of CFRP-to-steel bonded joints with a ductile adhesive by testing single-strap pull joints using an approximately trapezoidal shape for bonded joints. He et al. [3] carried out numerical modeling of bond behavior between steel and CFRP laminates with a ductile adhesive and proposed a model for the effective bond length for steel–CFRP bonded joints.

The above researchers studied the performance change law of FRP–steel bonded structures based on macro performance analysis methods. However, when considering the complexity and difficulty in the quantitative detection of the research on the interface characteristics of the bonding structure (such as the stress distribution inside the bonding layer, the internal trend of failure, etc.) [17–21], the current research on the failure characteristics and change trends of micro details of the bonding layer is still relatively limited. To fill this gap, this paper presents a study of the interfacial stress for GFRP–steel double-strap bonded joints via the finite element (FE) method. Based on this study, the peeling stress and von Mises stress of the adhesive layer under different bonding parameters are analyzed and the failure mechanism of the steel–FRP bonding interface is discussed and expounded upon rationally.

2. Bonding Materials

2.1. Material Properties of GFRP and Steel

For this study, a GFRP plate fabricated using glass fibers and unsaturated polyester resin was adopted. The steel substrates were composed of Q345b steel. Based on the tensile tests of steel coupons, an average yield strength and Young's modulus of 291.3 Mpa and 204.0 Gpa were measured, respectively. The main mechanical properties of the GFRP, adhesive, and steel are listed in Table 1.

Table 1. Mechanical properties of the GFRP, adhesive, and steel.

Mechanical Parameter	GFRP	Adhesive *	Steel
Young's modulus, Mpa	15,400 (longitudinal direction) 6850 (transverse direction) 7630 (thickness direction)	213.24 * (Secant Young's modulus)	204,000
Strength, Mpa	291.1 (longitudinal direction) 125.3 (transverse direction) 144.6 (thickness direction)	18.85 *	291.3 (Yield stress)
Poisson's ratio	0.37	0.40 (According to product manual)	0.3

* I mechanical properties were determined by a tensile test.

2.2. Tensile Test of the Adhesive Specimen

The methacrylate adhesive material was the super commercial two-part structural adhesive PLEXUS MA 310 (with a fixture cure time of 55 min at 25 °C and a specified operating temperature range of 55 °C to 121 °C.), which was manufactured by ITW Performance Polymer (Wujiang) Co., Ltd. in Wujiang, China. Taking into account that the adhesive used in FRP–steel bonded joints is significantly affected by the environment, the adhesive's properties in a field environment were worth testing specifically. A unidirectional quasistatic tensile test of the adhesive specimen was conducted in the authors' previous study [2]. The uniform specimens of the adhesive were formed into a dog-bone shape with a thickness and testing length of 15 mm and 100 mm, respectively. According to this tensile test, the mechanical behaviors of the adhesive could be adopted as shown in Figure 1. Based on the tensile test of the adhesive specimens, we concluded that the tensile strength of the adhesive was 18.85 MPa. With the increase in the specimen length, the tensile force increased nonlinearly until it fractured. To simulate the adhesive material, a secant elastic modulus (213.24 MPa) of the adhesive material was adopted.

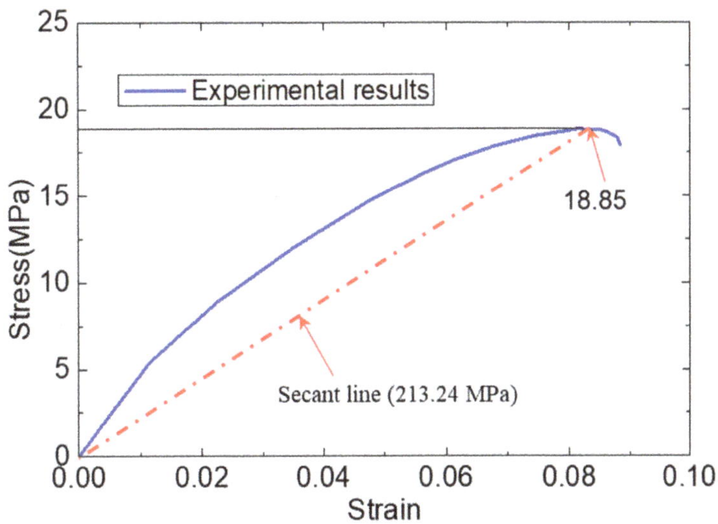

Figure 1. Mechanical behavior of adhesive in tensile experiment and finite element model.

3. Tensile Property of GFRP–Steel Specimen

To investigate the bonding performance of the GFRP–steel composite structure, GFRP–steel double-strap joints were manufactured and tensile-tested using a servo-hydraulic test machine (MTS Landmark with a maximum load capacity of 50 kN, Eden Prairie, MN, USA) as shown in Figure 2 and presented in the authors' previous study [22]. The thickness of the steel plate and GFRP for the standard specimen was 5 mm, and the thickness of the adhesive (bonding layer) was 1 mm, as Figure 2a shows. During the static tensile testing, the load was applied through the two ends of the specimen with the displacement-control mode set at a rate of 1 mm/min and a tensile capacity of 22.48 kN [22].

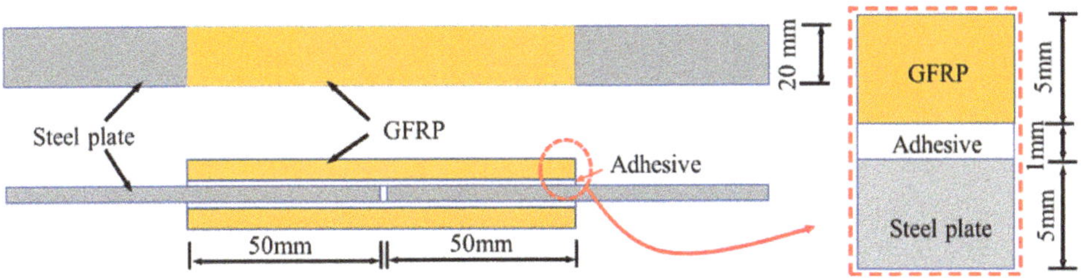

(**a**)

Figure 2. *Cont.*

(**b**)

Figure 2. The GFRP–steel double-strap joints and the tensile test graph. (**a**) Dimensions of the GFRP–steel double-strap joints; (**b**) raw materials and test loading graph of the double-strap joint.

4. Finite Element Model

4.1. The Geometry of the Finite Elements

Although experimental research can obtain more intuitive and reliable conclusions, it is difficult to monitor the internal mechanical characteristics and change mechanisms of a component. Benefiting from its convenience and economy, the finite element analysis method makes up for the shortage in experimental research methods [10,11,23–27], so it is widely used in the engineering research field.

To analyze the interfacial stress distribution of the GFRP–steel joints with different bonding parameters, three-dimensional finite element models were developed by using the FE software package ANSYS 14.5, as shown in Figure 3. For all finite element analyses, an implicit calculation method was performed. While taking into account the symmetry of the specimen $\frac{1}{4}$ structural models were established. The GFRP plates, adhesive layers, and steel plates were modeled using a 3D eight-node solid element (i.e., Solid 185 in ANSYS) with three degrees of freedom (DOFs) at each node (translations in the *x*, *y*, and *z* directions) and meshed with a size of $0.5 \times 0.5 \times 0.5$ mm. Under this meshing strategy, the bonding layer was simulated by a two-layer hexahedral element.

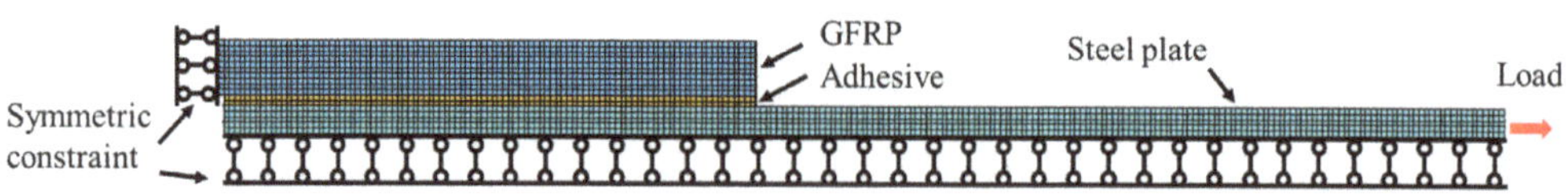

Figure 3. Finite element model of GFRP–steel bonded joint.

The steel, adhesive, and GFRPs, which were assumed to be homogeneous and linearly elastic, were defined according to the values given in Table 1. To simulate the bonding effects between different components of the specimen, glue operation was used in the models. Constraints were applied to the nodes of the left end faces for the GFRP plates, and the tensile loading was simulated by applying an external load to the right end faces of the steel plate models, as Figure 3 shows.

4.2. Validation of the FE Model

To validate the effectiveness of the FE model, a tensile test result for a standard GFRP–steel bonded joint cured in air conditions was adopted and compared. The tensile test of the GFRP–steel bonded joint was reported in the authors' previous study [22]. According to the tensile test, the load–displacement curve and ultimate tensile strength

(22.48 kN) were adopted. Furthermore, based on the tensile test, a typical cohesive failure mode was observed for the bonded joint, which meant that the fracture was inside the adhesive material rather than at the adhesive–structure interface [22]. Based on the results of the quasistatic tensile test, we observed that the fracture was initiated at the adhesive near the gap area of the specimen; that is to say, as the load increased, the cracking propagated toward the free end of the GFRP along the bonding layers and eventually resulted in a complete disengagement of the GFRP plate from the steel substrate. Therefore, we considered that for the proposed bonded joint, the composite specimen began to fail only when the stress of the adhesive layer reached its failure strength.

While taking into account that the tensile capacity of the GFRP–steel bonded joint depended on the adhesive layers (for the cohesive failure mode) and an ultimate tensile strength of 22.48 kN was obtained in the tensile test, the FE model of the standard joint was loaded with 22.48 kN and calculated. Figure 4 shows the stress contours of the adhesive layer; it can be seen that the numerical results agreed with the experimental results well. By applying a 22.48 kN tensile loading, the maximum stress of the adhesive layer for the FE model was 18.80 MPa, which was very close to the failure strength (18.85 MPa) of the adhesive material shown in Figures 1 and 2. Therefore, we believe that the model strategy could simulate the tensile properties of the GFRP–steel double-strap bonded joints effectively.

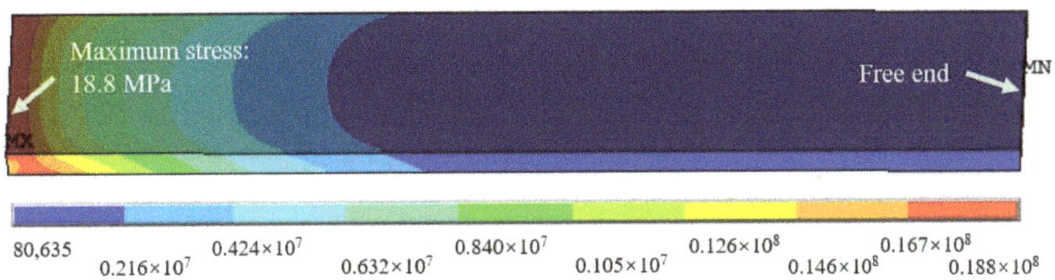

Figure 4. The von Mises stresses of the adhesive layer for the tensile joint.

For the double-strap bonded specimen, the force transmission route under tension was divided into two paths and transmitted to the two FRP plates through the shear stress of the adhesive layer. Considering that the shear path transmitted by the adhesive layer did not coincide with the central axis of the FRP plate, therefore, a normal stress perpendicular to the interface (called the peeling stress) appeared in the strap joint area when the bonding structure was stretched; this peeling stress was the key reason for the interlaminar failure of the bonding specimen [28,29]. To analyze the distribution of the interfacial peeling stress, the normal stress of the bonding zone was extracted using path analysis technology, as shown in Figure 5: three extraction paths were respectively defined to extract the normal (Z-direction) stress at the center interface center of the FRP–adhesive layer, the Z direction stress at the middle layer of the adhesive layer, and the normal (Z direction) stress at the center interface of the steel–adhesive layer.

Figure 6 shows the normal (Z-direction) peeling stress distribution curve extracted from the three paths of the bonding zone. Based on Figure 6, we observed that the peeling stress extracted from the three paths was very close. These maximum values were 2.77; the three minimum values stresses were −3.17, −3.10, and −3.01 MPa, respectively; therefore, to facilitate the comparative analysis, path II was used as the analysis path in the subsequent path analysis. Furthermore, when analyzing the stresses of the node for a different location, the peeling stress near the left side (the center line of the test specimen) and right side were found to be much greater than the others.

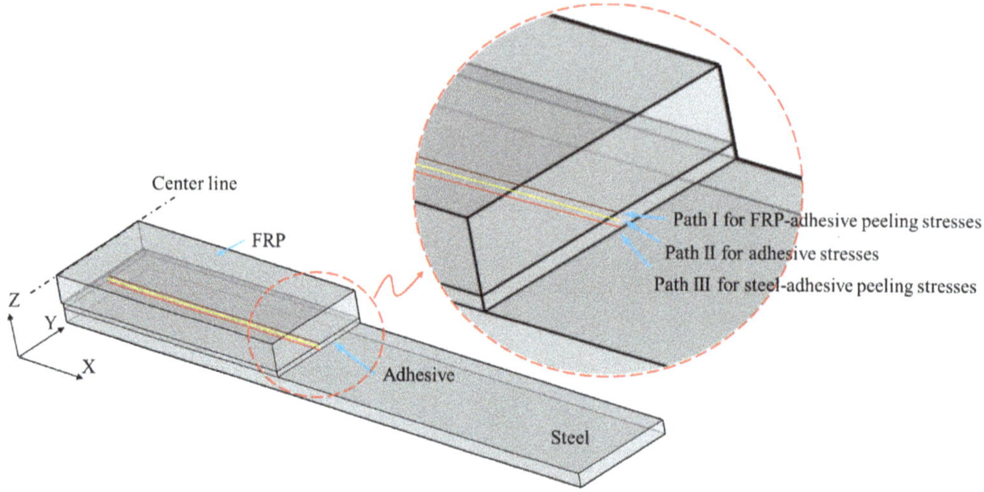

Figure 5. Paths for calculated adhesive stresses.

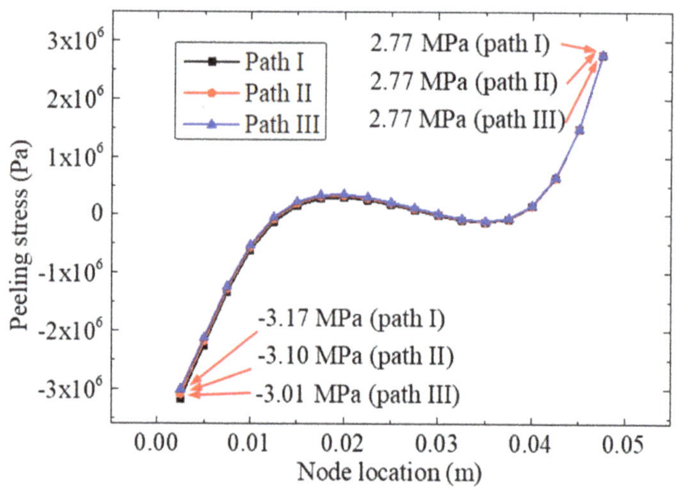

Figure 6. Peeling stresses distribution curve for the three paths.

Figure 7 shows the von Mises stresses of nodes on the three paths. Compared to the peeling stress distribution curve, a different change law was found according to the von Mises stress curve; the stress value decreased with an increase in the coordinate value. The maximum stress (14.71 Mpa) of the bonding layer was calculated near the centerline. This phenomenon indicated that the stress concentration at the left end of the adhesive was much greater than that of the others, which explained the failure mechanism of the bonded specimen observed previously [22].

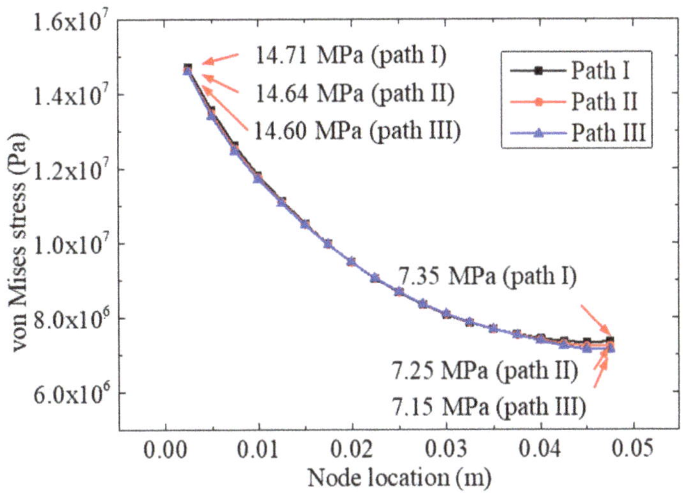

Figure 7. The von Mises stresses of nodes on path II.

5. Parametric Studies

According to previous studies by researchers, it was found that the tensile capacities of the double-strap GFRP–steel bonded joints were influenced by the FRP stiffness, bonding length, and other adhesive bonding parameters [1,30,31]. Therefore, the proposed FRP model was applied to investigate the effects of these parameters on the double-strap GFRP–steel bonded joints. The workflow of the parametric studies is shown in Figure 8.

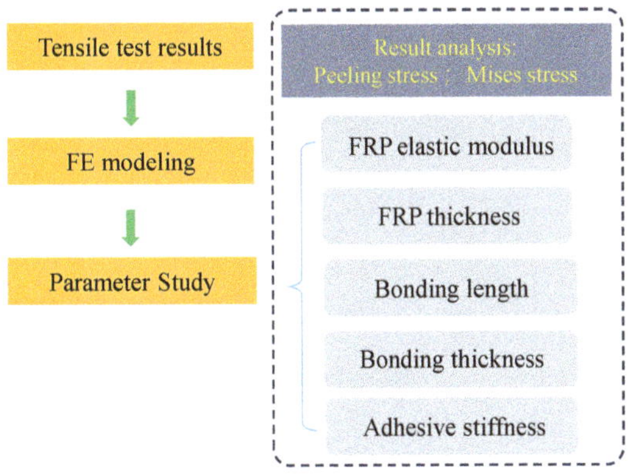

Figure 8. Workflow of the parametric studies.

5.1. Effect of FRP Elastic Modulus

Based on the producing materials, FRP can be divided into four categories: glass-fiber-reinforced polymer (GFRP), carbon-fiber-reinforced polymer (CFRP), basalt-fiber-reinforced polymer (BFRP), and aramid-fiber-reinforced polymer (AFRP). Different material properties are shown by these composites, among which the impact of the material stiffness (including elastic modulus and thickness of the FRP) on the performance of the bonded structure has received widespread attention [32,33]. To investigate the effect of the FRP's elastic modulus on the FRP–steel bonded joints, the FRP moduli E_f = 10 Gpa, 50 Gpa, 100 Gpa, 200 Gpa, and 400 Gpa were adopted in this study; the other parameters of the model were set as shown in Table 1 and Figure 3a. The results of the numerical simulation analysis are listed in Table 2 and Figures 9 and 10.

Table 2. Stress analysis results for five elastic moduli.

Elastic Modulus of FRP Plates/Gpa	Ultimate Peeling Stresses for Path II/Mpa	Von Mises Stresses for Path II/Mpa
10	−3.80	16.96 (left)/6.15 (right)
50	−1.95	11.02 (left)/9.25 (right)
100	−1.61	10.09 (left)/9.84 (right)
200	−1.41	9.59 (left)/10.18 (right)
500	−1.28	9.29 (left)/10.41 (right)

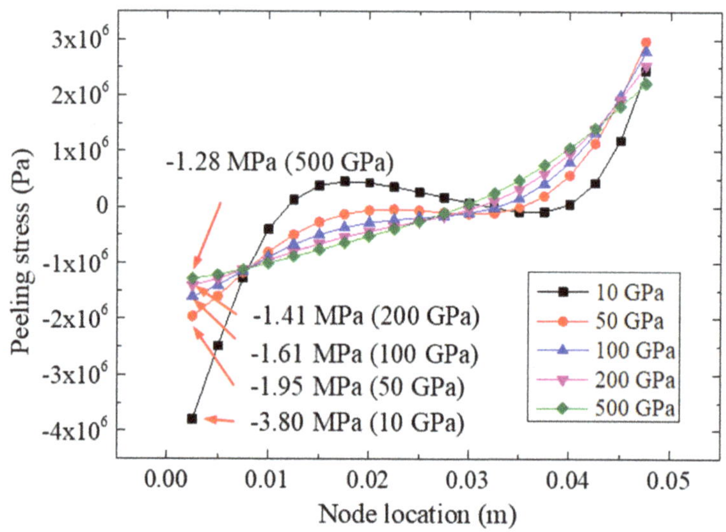

Figure 9. Peeling stresses on path II for different FRP elastic moduli.

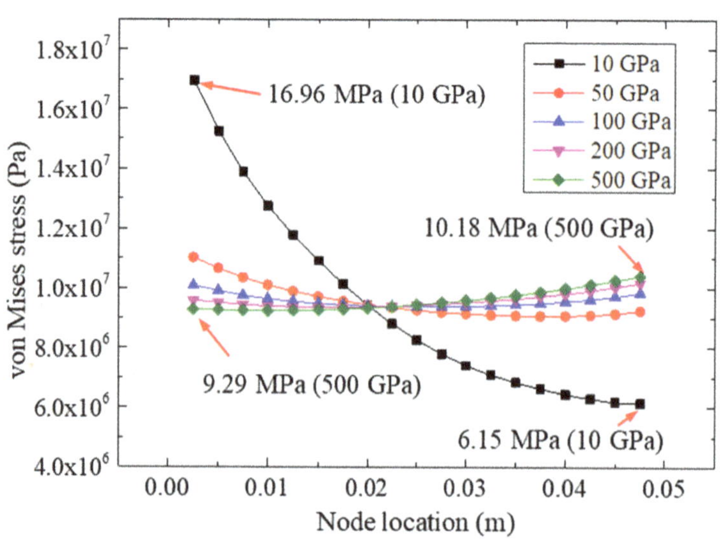

Figure 10. The von Mises stresses on path II for different FRP elastic modulus.

Figure 9 shows the peeling stresses of the adhesive for the FRP–steel bonded joints with different FRP elastic moduli (i.e., E_f = 10 Gpa, 50 Gpa, 100 Gpa, 200 Gpa, and 400 Gpa) under a 22.48 kN tensile loading. According to Figure 9, it was found that the elastic modulus of the FRP could significantly affect the normal peeling stress distribution of the bonding layer. The normal peeling stress distribution of the bonding layer was more uniform for the specimens bonded with FRP materials with a higher elastic modulus. The minimum value of the normal peeling stress was calculated as −3.80 Mpa for the FRP with an elastic modulus of 10 Gpa; as the elastic modulus of the FRP increased, the ultimate value of the normal peeling stress increased gradually. The changes in the extreme normal peeling stress showed that the stress concentration at the end of the bonding layer decreased with an increase in the FRP plate stiffness and that the bonding strength of the joint could be improved accordingly [14]. Furthermore, for the FRP with 10 Gpa, the variation range of the normal stresses of the adhesive layer was −3.80~2.44 Mpa, whereas the variation range of the bonding condition for 500 Gpa was −1.28~2.22 Mpa.

Figure 10 plots the von Mises stresses on path II for different FRP elastic moduli under a 22.48 kN tensile loading. With the increase in FRP stiffness, the von Mises stresses on the left side of the adhesive layer decreased gradually, while the stresses on the right side increased gradually. The extreme value of the von Mises stress for the left-side was 16.96 Mpa (for 10 Gpa), which was much higher than that of the specimen for 500 Gpa, and the excess ratio was about 83%. However, for the right side, the extreme value (10.18 Mpa) of the von Mises stress was calculated for the specimen with 500 Gpa, which was much higher than that of the specimen with 10 Gpa. This phenomenon showed that the change in the elastic modulus of the FRP could improve the distribution of the equivalent stress (i.e., von Mises stress) in the bonding area. In other words, an FRP with a higher elastic modulus could make the equivalent stress more uniform, which may have effectively reduced the stress concentration in the bonding end zone.

5.2. Effect of FRP Thickness

This section investigates the effect of the FRP's thickness on the tensile behaviors of the bonded structures. FRP–steel double shear specimens with different FRP thicknesses (i.e., 3 mm, 4 mm, 5 mm, 6 mm, and 7 mm) were simulated and analyzed. The elastic moduli and dimensions (except for FRP thickness) of the bonding joints were set according to Table 1 and Figure 2a.

Figure 11 presents the peeling stresses on path II for different FRP thicknesses under a 22.48 kN tensile loading. Based on Figure 11, we observed that the peeling stress could be influenced by FRP thickness. As the bonding length increased, the extreme peeling stress on the right side increased accordingly. The largest peeling stress was calculated for the FRP thickness of 7 mm, which had an extreme stress of 3.50 MPa on its right side. Furthermore, for the left side, the smallest value (−3.58 MPa) of the peeling stress was calculated for the 3 mm bonding specimen, while the largest extreme peeling stress on the left side for the 4 mm, 5 mm, 6 mm, and 7 mm specimens was −3.31, −3.10, −2.96, and −2.87 MPa, respectively, showing that the change in FRP thickness likely may not have affected the left-side stress remarkably.

Figure 12 plots the von Mises stresses of the adhesive layer for different FRP thicknesses under a 22.48 kN tensile loading. We observed that the FRP thickness may have affected the extreme equivalent stress of the adhesive layer. The extreme values of the five FRP plates were 17.40 (3 mm), 15.72 (4 mm), 14.64 (5 mm), 13.89 (6 mm), and 13.34 MPa (7 mm). Compared with the other specimens, the largest extreme stress was adopted in the model of the 3 mm FRP, which indicated that the increase in the FRP thickness would reduce the extreme equivalent stress of the bonding layer. By comparing Figures 10 and 12, similar trends of stress curves could be found; for example, the stress distribution of specimens made with the more rigid FRP plates (e.g., 7 mm FRP and 500 GPa) was more uniform than that of the more flexible FRP plates (e.g., 3 mm FRP and 10 GPa).

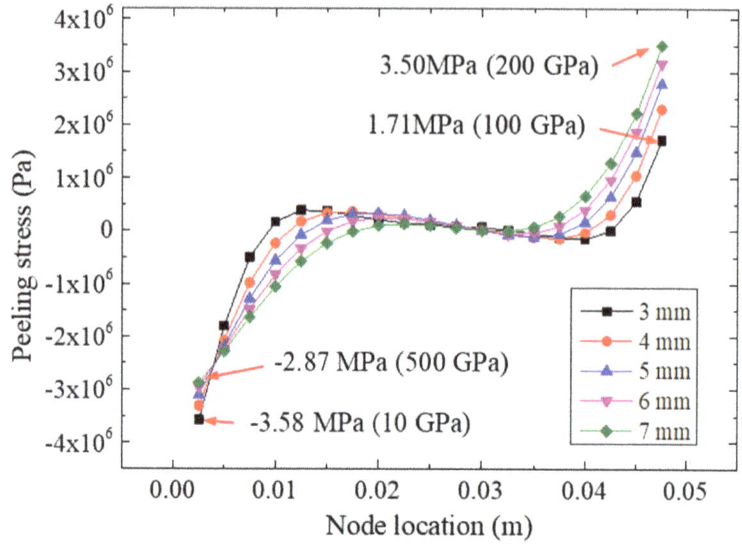

Figure 11. Peeling stresses on path II for different FRP thicknesses.

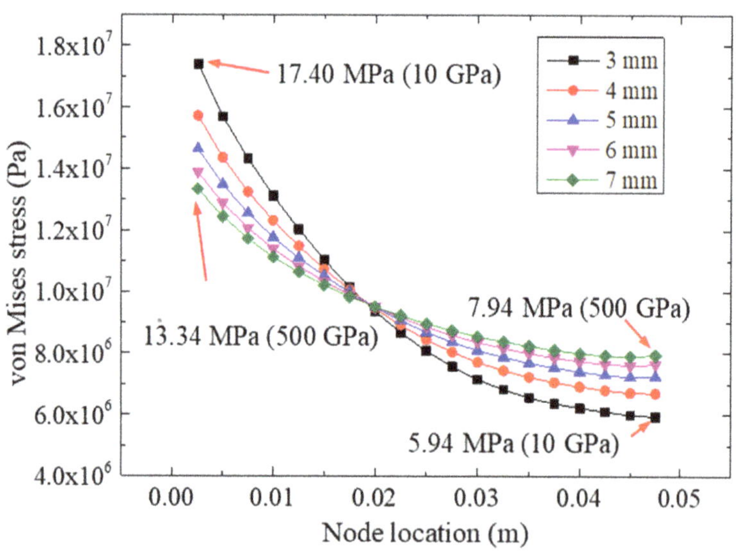

Figure 12. The von Mises stresses for different FRP thicknesses.

5.3. Effect of Bonding Length

The bonding parameter (for example, the bonding length) is another important factor that may affect the strength of bonded structures [2]. Therefore, to study the influence of bonding length on the tensile performance of the bonded joints, FRP–steel double-strap specimens with different bonding lengths (i.e., 30 mm, 60 mm, 90 mm, 120 mm, and 150 mm) were simulated and analyzed. The elastic moduli and dimensions (except for bonding lengths) were set according to Table 1 and Figure 2a.

Figure 13 plots the peeling stresses of the adhesive layer for different bonding lengths under a 22.48 kN tensile loading. We found that the bonding length may have influenced the peeling stress of the adhesive layer significantly. The largest peeling stress was calculated for a bonding length of 30 mm, which had an extreme stress of 6.54 MPa on its right side. As the bonding length increased, the extreme peeling stress decreased accordingly. Furthermore, for the left side, the smallest value (−4.50 Mpa) of the peeling stress was calculated for the 30 mm bonding specimen, which showed that a longer bonding size would lead to a more uniform peeling stress. The homogenization of the peeling stress could effectively improve the strength of specimens; similar conclusions were reached and reported in [34]. Therefore, to reduce the extreme value of the peeling stress for the adhesive layer, the bonding length should be appropriately increased. In addition, for the specimens with a bonding length greater than 90 mm, the increase in the bonding length did not have such a remarkable effect on the bonding stresses; for example, the extreme peeling stress of the 150 mm bonding specimen (0.12 Mpa) was only 0.60 Mpa less than that of the 90 mm bonding specimen (0.72 Mpa).

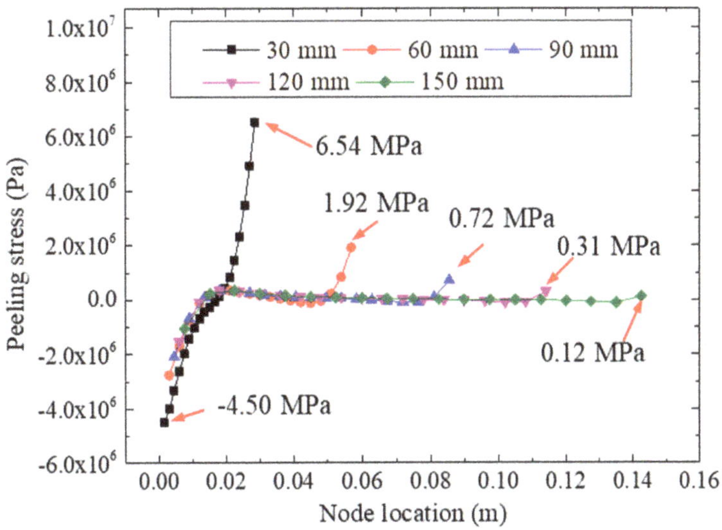

Figure 13. Peeling stresses for different bonding lengths.

Figure 14 shows the von Mises stresses of the adhesive layer for different bonding lengths under a 22.48 kN tensile loading. According to the distribution of the von Mises stresses, we observed that a larger bonding length could reduce the equivalent stress of the adhesive layer significantly. The stresses of the specimen with a 30 mm bonding length were much greater than those of the other specimens. For instance, for the specimen with a 30 mm bonding length, the extreme stresses were 19.53 Mpa and 13.90 Mpa, whereas extreme stresses of 10.63 Mpa and 0.93 Mpa were calculated for the 150 mm bonding specimen, which indicated that the increase in the bonding length would reduce the equivalent stress of the bonding layer. In addition, the extreme stresses for the 90 mm bonding specimen were calculated as 12.07 Mpa and 2.85 Mpa, and quite low stress values were found. Therefore, although a longer bonding length would lead to a more appropriate and uniform adhesive stress, a bonding length for the double-strap specimen of 90 mm is recommended when considering the manufacturing costs.

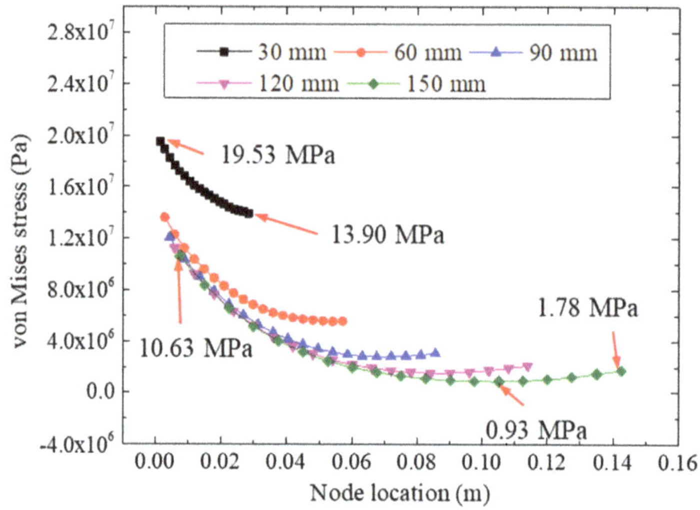

Figure 14. The von Mises stresses for different bonding lengths.

5.4. Effect of Bonding Thickness

To study the influence of the bonding thickness on the tensile performance, FRP–steel double-strap bonded joints with different thicknesses (0.5, 1, 1.5, 2, and 2.5 mm) of the adhesive layer were modeled and analyzed. The elastic moduli and dimensions (except for the bonding thickness) were set according to Table 1 and Figure 2a.

Figure 15 plots the peeling stresses of the adhesive layer for different bonding thicknesses under a 22.48 kN tensile loading. For different bonding thicknesses (i.e., 0.5, 1.0, 1.5, 2.0, and 2.5 mm), the minimum peeling stresses of the five models were −3.71, −3.10, −2.86, −2.77, and −2.74 MPa respectively; the maximum peeling stresses of the five models were 2.08, 2.77, 3.09, 3.30, and 3.44 Mpa, respectively. The extreme difference in peeling stress for each model was 5.79, 5.87, 5.95, 6.07, and 6.18 Mpa, respectively, indicating that the difference in bonding thickness likely may not have affected the nonuniformity of the interface peeling stress. By analyzing the stress values of the different bonding thicknesses, we found that the change in the bonding thickness could adjust the peeling stress of the interface to a certain extent. This was specifically reflected by the fact that the thinner the bonding layer was, the more prominent the extreme value of peeling stress on the left side (near the centerline) was, and conversely, the thicker the bonding layer was, the more prominent the extreme value on the right side was.

Figure 16 shows the von Mises stresses of the adhesive layer for different bonding thicknesses under a 22.48 kN tensile loading. It can be seen from the curves that for each specimen with a specific bonding thickness, with an increase in the node coordinates, the equivalent stress at the interface showed a decreasing trend, showing that under the current specimen parameters (i.e., bonding length and stiffness), the early failure of the left bonding zone was the main cause of the failure of the composite structures. Furthermore, according to the distribution of the von Mises stresses for the different bonding thicknesses, we observed that the change in the bonding thicknesses could affect the equivalent stresses remarkably. The largest value of the equivalent stress was 18.14 Mpa (in the 0.5 mm bonded model); the four extreme equivalent stresses for bonding thicknesses of 1.0, 1.5, 2.0, and 2.5 mm were 14.64, 13.26, 12.60, and 12.31 Mpa, respectively, all of which were calculated on the left side of the adhesively bonded models. We concluded that with an increase in the bonding thickness (from 0.5 mm to 2.5 mm), the extreme value of the bonding layer stress decreased gradually, indicating that the increase in the bonding layer thickness could effectively reduce the stress concentration in the bonding area and increase the strength-bearing capacity of the specimen accordingly.

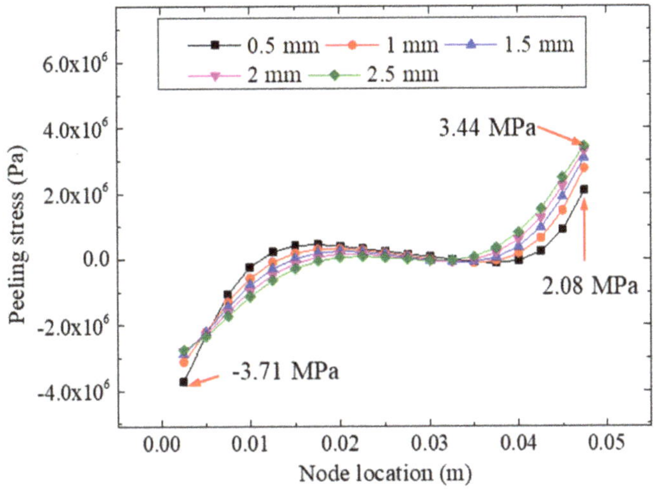

Figure 15. Peeling stresses for different bonding thicknesses.

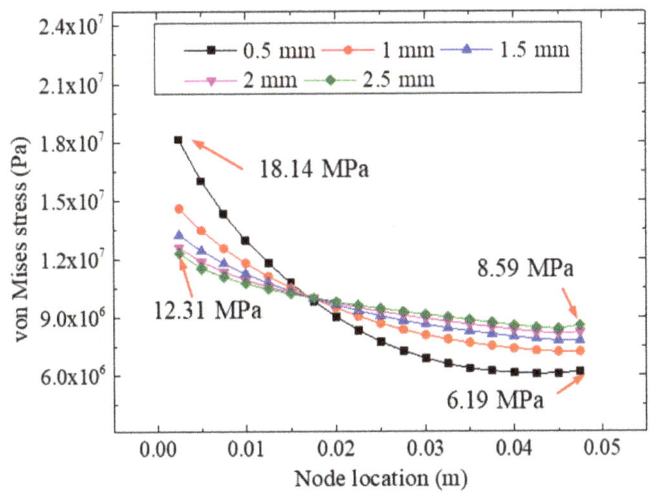

Figure 16. The von Mises stresses for different bonding thicknesses.

5.5. Effect of Adhesive Stiffness

In this section, the effects of the stiffness of the adhesive material are discussed based on the bonded specimens. The model strategies and parameters (except for the elastic modulus of the adhesive layer) followed were according to Table 1 and Figure 2a. Five elastic moduli (200, 500, 800, 1100, and 1400 MPa) of the adhesive were adopted; the analyzed results of the peeling and von Mise stresses of the adhesive layers are presented in Figures 17 and 18.

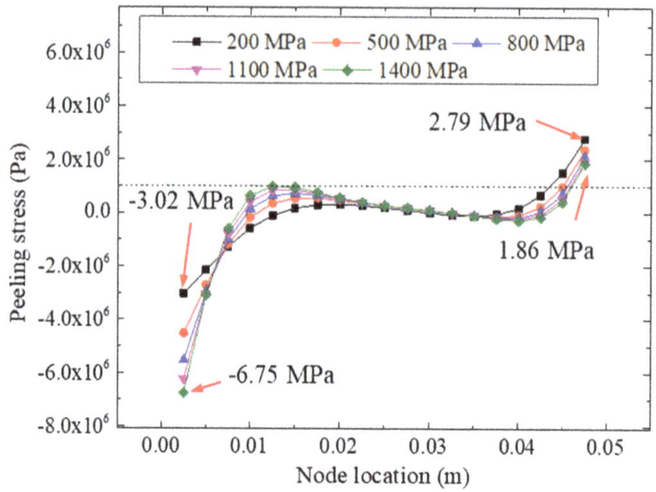

Figure 17. Peeling stresses for different adhesive modulus.

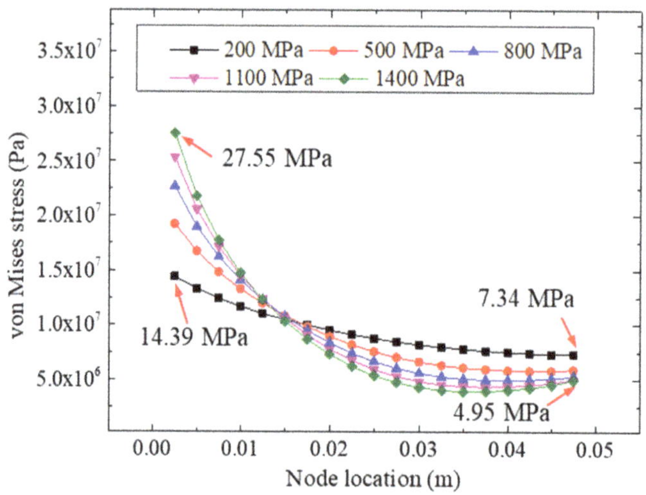

Figure 18. The von Mises stresses for different adhesive modulus.

Figure 17 plots the peeling stresses of the adhesive layer for different adhesive moduli under a 22.48 kN tensile loading. We found that the elastic modulus of the adhesive layer had a significant influence on the peeling stress of the bonded specimen. For different adhesive moduli (i.e., 200, 500, 800, 1100, and 1400 MPa), the maximum peeling stresses of the right side were 2.79, 2.38, 2.12, 1.96, and 1.86 MPa, respectively, showing that the elastic modulus of the adhesive layer had little influence on the free end (i.e., the right side of the bonded specimens). However, for the left side, the minimum peeling stresses were −3.02 (200 MPa), −4.53 (500 MPa), −5.53 (800 MPa), −6.24 (1100 MPa), and −6.75 (1400 MPa) MPa, respectively, indicating that the peeling stress on the left side of the bonding layer was more sensitive to the stiffness of the adhesives.

Figure 18 presents the comparison of the von Mises stresses of the adhesive layers for different elastic moduli under a 22.48 kN tensile loading. Similar to the results under other conditions, the difference in the stresses on the right side of the adhesive layer was quite small, whereas a remarkable stress concentration was observed on the left side of the adhesive layers, showing that the stress concentration of the left bonding layer was the main cause of the failure of the bonding zone of the specimen. The maximum equivalent stresses of the five models were 14.39 (200 MPa), 19.22 (500 Mpa), 22.68 (800 Mpa), 25.36 (1100 Mpa), and 27.55 (1400 Mpa) Mpa, respectively, indicating that the bonding materials with a higher elastic modulus were more likely to cause stress concentration in the bonding zone. Therefore, we concluded that with the same bonding parameters and conditions, the more flexible bonding material was more beneficial to the stress distribution in the bonding zone.

6. Conclusions

This study investigated the effects of different bonding parameters on the bonding behaviors of FRP–steel double-strap bonded structures. During this study, the effects of FRP stiffness, bonding length, adhesive thickness, and adhesive modulus were discussed, and the following conclusions were drawn:

1. When taking into account that the thickness of the bonding layer of the bonded specimen was relatively small, the stress results (for the interface peeling stress and the von Mises stress) of different calculating paths of the bonding material were very close. Therefore, path II (through the middle layer of the bonding layer) was used as the subsequent stress analysis path.
2. The study of different FRP stiffnesses (elastic modulus and thickness) showed that the normal peeling stress and von Mises stress distributions in the bonding layer were more uniform in the specimens bonded with more rigid FRP materials, while the FRP with a higher stiffness was more conducive to eliminating the stress concentration in the adhesive layer.
3. An increase in the bonding length could effectively reduce the stress concentration in the adhesive layer. When considering the manufacturing costs, we recommend a bonding length of 90 mm for double-strap bonded specimens.
4. A difference in bonding thickness likely may not affect the nonuniformity of the interface peeling stress. However, a change in the bonding thickness can affect the equivalent stress remarkably; with an increase in the bonding thickness (from 0.5 mm to 2.5 mm), the extreme value of the bonding layer stress decreased gradually.
5. The elastic modulus of the adhesive layer had a significant influence on the peeling stress of the bonded specimen. The peeling stress on the left side of the bonding layer was more sensitive to the stiffness of the adhesives; bonding materials with a higher elastic modulus were more likely to cause stress concentration in the bonding zone. This research included a detailed analysis of the influence of various bonding parameters on the tensile behaviors of FRP–steel double-strap bonded structures; however, when considering the relatively complex failure mechanisms of bonded composite structures, a more profound experimental analysis (including different bonding parameters and conditions) is required. It is worth noting that this study was aimed at the bonding stresses of FRP–steel double-strap bonded joints based on a linear elastic assumption and cohesive failure mode; therefore, the failure mechanisms of FRP–steel composite specimens composed of nonlinear materials with other failure modes (except for cohesive failure) require further research.

Author Contributions: Conceptualization, L.W.; Data curation, Y.Y.; Formal analysis, J.L.; Methodology, J.L.; Visualization, J.Y.; Writing—original draft, J.L.; Writing—review & editing, Z.L. All authors have read and agreed to the published version of the manuscript.

Funding: This research was funded by the National Natural Science Foundation of China (Grant Nos. 52108151 and 52108272).

Institutional Review Board Statement: Not applicable.

Informed Consent Statement: Not applicable.

Data Availability Statement: Not applicable.

Acknowledgments: The authors gratefully acknowledge the financial support provided by the National Natural Science Foundation of China (Grant Nos. 52108151 and 52108272).

Conflicts of Interest: The authors declare no conflict of interest.

References

1. Zhao, X.L.; Zhang, L. State-of-the-art review on FRP strengthened steel structures. *Eng. Struct.* **2007**, *29*, 1808–1823. [CrossRef]
2. Liu, J.; Guo, T.; Feng, D.M.; Liu, Z.X. Fatigue performance of rib-to-deck joints strengthened with FRP angles. *J. Bridg. Eng.* **2018**, *23*, 04018060. [CrossRef]
3. He, J.; Xian, G.J.; Zhang, Y.X. Numerical modelling of bond behaviour between steel and CFRP laminates with a ductile adhesive. *Int. J. Adhes. Adhes.* **2021**, *104*, 102753. [CrossRef]
4. Wei, Y.; Bai, J.W.; Zhang, Y.R.; Miao, K.T.; Zheng, K.Q. Compressive performance of high-strength seawater and sea sand concrete filled circular FRP-steel composite tube columns. *Eng. Struct.* **2021**, *240*, 112357. [CrossRef]
5. Tang, Y.; Sun, Z.; Wu, G.; Wei, Y. Compressive Behavior of Steel-FRP Composite Bars Confined with Low Elastic Modulus FRP Spirals in Concrete Columns. *J. Compos. Constr.* **2022**, *26*, 04022058. [CrossRef]
6. Tang, Y.; Sun, Z.; Wei, Y.; Zou, X. Compressive behavior and design method of BFRP bars constrained with a BFRP spiral with different spacings in concrete members. *Eng. Struct.* **2022**, *268*, 114757. [CrossRef]
7. Guo, T.; Liu, J.; Deng, Y.; Zhang, Z.L. Fatigue performance of orthotropic steel decks with FRP angles: Field measurement and numerical analysis. *J. Perform. Constr. Facil.* **2019**, *33*, 04019042. [CrossRef]
8. Zheng, Z.H.; Du, Y.S.; Chen, Z.H.; Li, S.Y.; Niu, J.Q. Experimental and theoretical studies of FRP-Steel composite plate under static tensile loading. *Constr. Build. Mater.* **2021**, *271*, 121501. [CrossRef]
9. Harries, K.A.; Peck, A.J.; Abraham, E.J. Enhancing stability of structural steel sections using FRP. *Thin Wall Struct.* **2009**, *47*, 1092–1101. [CrossRef]
10. Su, P.; Dai, Q.; Li, M.; Ma, Y.; Wang, J. Investigation of the mechanical and shrinkage properties of plastic-rubber compound modified cement mortar with recycled tire steel fiber. *Constr. Build. Mater.* **2022**, *334*, 127391. [CrossRef]
11. Wang, J.; Li, Q.; Song, G.; Luo, S.; Ge, D. Investigation on the comprehensive durability and interface properties of coloured ultra-thin pavement overlay. *Case Stud. Constr. Mater.* **2022**, *17*, 01341. [CrossRef]
12. Miller, T.C.; Chajes, M.J.; Mertz, D.R.; Hastings, J.N. Strengthening of a steel bridge girder using CFRP plates. *J. Bridg. Eng.* **2001**, *6*, 514–522. [CrossRef]
13. Fawzia, S.; Zhao, X.L.; Al-Mahaidi, R.; Rizkalla, S. Bond characteristics between CFRP and steel plates in double strap joints. *Adv. Steel. Constr.* **2005**, *1*, 17–27.
14. Yu, T.; Fernando, D.; Teng, J.G.; Zhao, X.L. Experimental study on CFRP-to-steel bonded interfaces. *Compos. Part B Eng.* **2012**, *43*, 2279–2289. [CrossRef]
15. Neto, J.A.B.P.; Campilho, R.D.S.G.; da Silva, L.F.M. Parametric study of adhesive joints with composites. *Int. J. Adhes. Adhes.* **2012**, *37*, 96–101. [CrossRef]
16. Wang, H.T.; Wu, G.; Dai, Y.T.; He, X.Y. Experimental study on bond behavior between CFRP plates and steel substrates using digital image correlation. *J. Compos. Constr.* **2016**, *20*, 04016054. [CrossRef]
17. Duc, N.D.; Trinh, T.D.; Do, T.V.; Doan, D.H. On the Buckling Behavior of Multi-Cracked FGM Plates. In *International Conference on Advances in Computational Mechanics*; Springer: Singapore, 2017; pp. 29–45.
18. Nam, V.H.; Nam, N.H.; Vinh, P.V.; Khoa, D.N.; Thom, D.V.; Minh, P.V. A new efficient modified first-order shear model for static bending and vibration behaviors of two-layer composite plate. *Adv. Civ. Eng.* **2019**, *2019*, 6814367.
19. Nguyen, H.N.; Nguyen, T.Y.; Tran, K.V.; Tran, T.T.; Nguyen, T.T.; Phan, V.D.; Do, T.V. A finite element model for dynamic analysis of triple-layer composite plates with layers connected by shear connectors subjected to moving load. *Materials* **2019**, *12*, 598. [CrossRef]
20. Castagnetti, D.; Spaggiari, A.; Dragoni, E. Assessment of the Cohesive Contact method for the analysis of thin-walled bonded structures. *Int. J. Adhes. Adhes.* **2012**, *37*, 112–120. [CrossRef]
21. Castagnetti, D.; Dragoni, E.; Spaggiari, A. Efficient post-elastic analysis of bonded joints by standard finite element techniques. *J. Adhes. Sci. Technol.* **2009**, *23*, 1459–1476. [CrossRef]
22. Liu, J.; Guo, T.; Hebdon, M.H.; Liu, Z.X.; Wang, L.B. Bonding Behaviors of GFRP/Steel Bonded Joints after Wet–Dry Cyclic and Hygrothermal Curing. *Appl. Sci.* **2020**, *10*, 5411. [CrossRef]
23. Ammarullah, M.I.; Santoso, G.; Sugiharto, S.; Supriyono, T.; Kurdi, O.; Tauviqirrahman, M.; Winarni, T.I.; Jamari, J. Tresca stress study of CoCrMo-on-CoCrMo bearings based on body mass index using 2D computational model. *J. Tribol.* **2022**, *33*, 31–38.
24. Wang, J.; Dai, Q.; Lautala, P.; Yao, H.; Si, R. Rail Sample Laboratory Evaluation of Eddy Current Rail Inspection Sustainable System. *Sustainability* **2022**, *14*, 11568. [CrossRef]
25. Chen, S.; Wang, J.; Li, Q.; Zhang, W.; Yan, C. The Investigation of Volatile Organic Compounds (VOCs) Emissions in Environmentally Friendly Modified Asphalt. *Polymers* **2022**, *14*, 3459. [CrossRef]
26. Yang, J.Q.; Smith, S.T.; Feng, P. Effect of FRP-to-steel bonded joint configuration on interfacial stresses: Finite element investigation. *Thin Wall. Struct.* **2012**, *62*, 215–228. [CrossRef]

27. Ammarullah, M.I.; Santoso, G.; Sugiharto, S.; Supriyono, T.; Wibowo, D.B.; Kurdi, O.; Tauviqirrahman, M.; Jamari, J. Minimizing Risk of Failure from Ceramic-on-Ceramic Total Hip Prosthesis by Selecting Ceramic Materials Based on Tresca Stress. *Sustainability* **2022**, *14*, 13413. [CrossRef]
28. Mukhtar, F.M.; Faysal, R.M. A review of test methods for studying the FRP-concrete interfacial bond behavior. *Constr. Build. Mater.* **2018**, *169*, 877–887. [CrossRef]
29. Razaqpur, A.G.; Lamberti, M.; Ascione, F. A nonlinear semi-analytical model for predicting debonding of FRP laminates from RC beams subjected to uniform or concentrated load. *Constr. Build. Mater.* **2020**, *233*, 117838. [CrossRef]
30. Chiew, S.P.; Yu, Y.; Lee, C.K. Bond failure of steel beams strengthened with FRP laminates–Part 1: Model development. *Compos. Part B Eng.* **2011**, *42*, 1114–1121. [CrossRef]
31. Robinson, M.J.; Adams, T.C. Performance of FRP composite strap joints utilizing fiber tow steering. *Compos. Part B Eng.* **2020**, *190*, 107910. [CrossRef]
32. Wei, Y.; Zhu, C.; Miao, K.T.; Chai, J.L.; Zheng, K.Q. Compressive behavior of rectangular concrete-filled fiber-reinforced polymer and steel composite tube columns with stress-release grooves. *Compos. Struct.* **2021**, *281*, 114984. [CrossRef]
33. Tang, Y.; Sun, Z.; Wu, G.; Wei, Y. Experimental Study on Cyclic Behavior of SFCBs with Different Slenderness Ratios. *J. Mater Civ. Eng.* **2021**, *33*, 04021204. [CrossRef]
34. Hu, B.; Li, Y.; Jiang, Y.T.; Tang, H.Z. Bond behavior of hybrid FRP-to-steel joints. *Compos. Struct.* **2020**, *237*, 111936. [CrossRef]

materials

Article

Analysis on Effects of Joint Spacing on the Performance of Jointed Plain Concrete Pavements Based on Long-Term Pavement Performance Database

Jiaqing Wang [1], Xiaojuan Luo [1], Xin Huang [1,*], Yao Ye [1] and Sihan Ruan [2,3,*]

1 College of Civil Engineering, Nanjing Forestry University, Nanjing 210037, China
2 Key Laboratory of RC&PC Structures of Ministry of Education, Southeast University, Nanjing 210096, China
3 School of Civil Engineering, Southeast University, Nanjing 210096, China
* Correspondence: xinhuang@njfu.edu.cn (X.H.); sihanruan@seu.edu.cn (S.R.)

Abstract: With the day–night temperature and moisture levels changing every day, expansion and shrinkage of concrete slabs is always occurring; therefore, joints provide extra room for concrete slab deformation. The joint spacing in jointed plain concrete pavement (JPCP) is continuously affecting long-term pavement behaviors. In this study, data from the Long-Term Pavement Performance (LTPP) program were analyzed, and the behaviors of JPCP with different joint spacings were compared to discover the joint spacing effects. Since LTPP has an enormous database, three representative sections located in different states were selected for analysis, where the variable factors such as temperature, moisture, and average annual daily truck traffic (AADTT) were almost the same between the three sections. Three different joint spacings, including 15 ft (4.5 m), 20 ft (6 m), and 25 ft (7.5 m), were compared based on the collected LTPP data. The involved long-term pavement performances, such as average transverse cracking (count), average JPCP faulting, international roughness index (IRI), and falling weight deflectometer (FWD) deflections were compared between JPCP with different joint spacings. Based on the comparative analysis, the JPCP constructed with a 15 ft joint spacing demonstrated the best long-term performance. It showed no transverse cracking, the lowest average JPCP faulting, the best IRI value, and the smallest FWD deflection during the entire in-service period. With proper joint spacing, the cost of road maintenance throughout the life cycle could be significantly reduced due to there being less distress. Therefore, it is recommended to optimize the joint spacing to about 15 ft in JPCP in future applications.

Keywords: jointed plain concrete pavement (JPCP); long-term pavement performance (LTPP); joint spacing; pavement distresses; optimum design

Citation: Wang, J.; Luo, X.; Huang, X.; Ye, Y.; Ruan, S. Analysis on Effects of Joint Spacing on the Performance of Jointed Plain Concrete Pavements Based on Long-Term Pavement Performance Database. *Materials* **2022**, *15*, 8132. https://doi.org/10.3390/ma15228132

Academic Editor: Simon Hesp

Received: 19 October 2022
Accepted: 12 November 2022
Published: 16 November 2022

1. Introduction

As one of the most frequently used construction materials, concrete is widely used in the construction and civil engineering industries. The first concrete road was built in Bellefontaine, USA in 1893 and is still in use. In response to the national green energy-saving policy, the civil engineering industry has been improving the performance of concrete in recent years to achieve the purpose of both reducing the amount of concrete used and meeting the requirements of building use. Concrete is made of portland cement, aggregates, and water, which are restricted to rigid pavements. There are five main types of concrete pavement: jointed plain concrete pavement (JPCP), jointed reinforced concrete pavement (JRCP), continuously reinforced pavement (CRCP), permeable concrete pavement, and compacted concrete pavement (RCC). Transverse cracks are cracks that are perpendicular to roadway alignment. These are generally due to temperature changes with shrinkage during curing. If there is no room for concrete to expand or contract, the concrete will crack. In fact, all transverse cracking in concrete pavements is due to tension within the slabs, and it can be very difficult to predict where the cracking will occur. In order to avoid transverse cracks, contraction joints have been constructed. JPCP is the original type of rigid pavement, constructed of closely spaced plain concrete without reinforcement, divided by joints to resist the development of cracks. It is the most common type of concrete pavement in the U.S. and Canadian highway systems. Rigid pavements transfer vehicle loads generated during vehicle travel through joints, and relative joint depths of less than 25% can affect joint load efficiency [1]. The load transfer efficiency (LTE) is provided by the joint, the dowel bar, or the aggregate interlocking structure that can be used for load transfer through the joint, where the aggregate interlocking structure is supported by the joint, the dowel bar, the base, and the concrete shoulder. Lack of aggregate interlocking and reduced shear resistance along the joints can lead to reduced LTE, while LTE is also affected by erosion in the dowel bar [2]. Sadeghi and Hesami [3] found that increasing the modulus of elasticity or thickness of the concrete slab base can increase the efficiency of load transfer, reduce the deflection of joints, and reduce damage to pavement joints. Dowel bars can provide uniform deflection of the slab when subjected to moving axial loads [4]. H.B. Sii et al. [5] analyzed dowel bar loosening on JPCP based on a three-dimensional finite element and demonstrated that the presence of gaps between dowel bars and slabs leads to increased stresses in the subgrade and a significant reduction in LTE. The use of force transfer bars is an effective method to control joint fracture, and Davids et al. [6] proposed a force transfer bar modeling technique that allows explicit modeling of force transfer bar loosening. This technique found that the principal tensile stresses in the concrete slab increased significantly and the vertical stresses in the subgrade increased under wheel and temperature loads when the distance between the dowel bar and the concrete slab was lower than 0.24 mm. Without joints, transverse and longitudinal cracking would form in natural directions and locations, and once those cracks formed, the concrete pavement would be damaged quickly under repeated load, temperature, and moisture changes. Joints have the most considerable effects on the performance of plain concrete pavements. In addition to considering the single joint performance, some scholars have also studied the long-term performance of the JPCP structures.

The Long-Term Pavement Performance (LTPP) program is the largest pavement investigation conducted to date and is an important source of monitoring the road performance information for the pavement system. The program monitors more than 2500 test sections of operational highways in the United States and Canada and collects data through a partnership between highway agencies and LTPP program organizations. Malla et al. [7] constructed a resilient modulus prediction model using data extracted from the LTPP database, and Dong et al. [8] analyzed the effect of different factors on asphalt pavement cracking based on LTPP data. Gong et al. [9] illustrated that the LTPP database is vital and can be used to analyze the structural performance of a wide range of pavements. The JPCP performance was also studied based on LTPP data. For traffic safety and to ensure pavement performance, JPCP needs to be maintained and rehabilitated during its service life. JPCP health assessments are based on conditions such as plate cracking direction, length, and width, which can lead to more accurate maintenance decisions by referring to LTPP data. Wang et al. [10] analyzed the LTPP cracking data to explore the relationship between fatigue failure points and various influencing factors, and the results demonstrated that an increase in traffic volume and precipitation intensity, and the number of freeze–thaw cycles, will reduce the fatigue life of the JPCP. Additionally, a survival model that can be used to predict the pavement condition was also proposed based on the collected LTPP data. Other research also found that short-slab JPCP can be used to reduce crack width, enhance aggregate interlocking, and reduce pavement deterioration [11]. The main form of JPCP structural damage is fatigue cracking, and Kim et al. [12] believed that fatigue cracking spreading from the bottom to the top is not the only formation possible; in a critical condition, cracking could also develop from the top to bottom. Having optimum joints in the pavement can guide the cracking direction based on the research findings. Meanwhile, other factors such as temperature, humidity, slab thickness, and subgrade design also affect the long-term performance of infrastructures [13,14]. Buckling and warping can affect the structural behavior of the concrete pavement, and higher temperatures can affect the connection at the joints and cause the concrete to warp. Without joints, transverse and longitudinal cracks will form in the deformation of concrete slabs. Once these cracks are formed, concrete pavements can be damaged rapidly under repeated wheel loads and changes in the environment, while lower humidity can also cause excessive upward warping of the slab [15]. With changing temperatures, thermal stresses are generated inside the concrete slabs, while the heat of cement hydration and evaporation from the pavement surface also greatly influences the internal temperature and humidity [16]. Asbahan et al. [17] found experimentally that temperature and humidity in concrete slabs caused the slab to curl upward and produced cracks that developed from top to bottom. The total curvature of the restrained slab was also higher than that of the unrestrained slab. Shafiee et al. [18] found that the combination of high-temperature weather, high wind speed, and low relative humidity produced a large internal temperature gradient in concrete slabs, which reduced the strength and durability of JPCP. Besides the surface layer of JPCP, granular and stabilized subgrades are often used in the design of JPCP structures, and the stabilized subgrade is used more frequently. Mu et al. [15] evaluated the AASHTO pavement M–E design guidelines and found that the M–E design overestimated the stresses in the unbonded subgrade while underestimating the stresses in bonded subgrade when the environmental effect is obvious. It also underestimated transverse cracking in JPCP with stabilized subgrade while overestimating that with the aggregate subgrade. In addition to the structure design effects, different types of concrete materials also show different effects on the concrete structures [19,20]. Shi et al. [21,22] studied the utilization of RAP instead of coarse aggregate in concrete pavement materials. The results showed that the application of RAP decreased the tensile strength of JPCP, increased the stress concentration, and led to an increased chance of transverse cracking. Conventional plain concrete (PC) leads to a large design thickness of concrete slabs, and different fiber-reinforced concrete (FRC) materials are used in JPCP to reduce the slab thickness. Ali et al. [23] found that JPCP constructed using FRC with glass fiber and polypropylene (PP) fiber was more economical and environmentally friendly than the JPCP produced conventionally. Colley et al. [24] found that when the load, slab thickness, and subgrade were kept constant, the joint effectiveness was reduced with the increased crack width. With a constant crack width, the joint effectiveness decreased with the application of load, but with the increase in the number of loading cycles, the rate of decrease became slower. It can be summarized that no matter the changes in structure designs and material types, the joints are the most important influencing factor in the performance of JPCP. Based on the current research, it was found that few studies have been conducted on the effect of joint spacing on JPCP performance, which could also be a dominant factor that affects the long-term performance. The joint spacing is also a construction related issue when designing a JPCP structure. The presence of joints cannot be dismissed with regard to the limitations of natural cracks and the expansion/shrinkage of concrete materials [25]. The range of joint spacing is usually between 15 ft (4.5 m) and 30 ft (9 m), but the optimum joint spacing is not clear when considering the long-term performance.

This research is based on the LTPP database. In order to find the effects of joint spacing on the JPCP performance, three different joint spacings (15 ft (4.5 m), 20 ft (6 m), and 25 ft (7.5 m)) were selected and compared in terms of the extracted LTPP data collection. Not only annual average daily truck traffic (AADTT) was chosen as a constant, with a level of 3000~6000, but also the climate situation was controlled to make sure the only variable value was the joint spacing. Based on the extracted LTPP data, the transverse cracking, faulting, IRI, and FWD deflection during the service periods were compared to the long-term behaviors of JPCP with various joint spacings. The findings from this study could contribute to the future application of JPCP with performance-optimized joint spacing.

2. Collection and Extraction of the LTPP Database

It is generally known that road performance is significantly affected by climate changes despite pavement types. To investigate the joint spacing effects in JPCP, three different survey sections with similar weather conditions were chosen, thus the weather influences could be generally eliminated. In the LTPP database, the road surface type was selected as JPCP. The climatic zone was set as Wet, Non-Freeze. In addition, the AADTT range was between 3000~6000. Based on the database, 46 sections were found that meet those defined conditions. After manually analyzing all 46 sections, 3 different sections with various critical joint spacings were determined. The number of lanes was 2 in each of the three sections. The sections were in Georgia, Tennessee, and Arkansas, respectively, and their corresponding joint spacings are summarized in Figure 1. The basic information for the three sections is also listed in Table 1.

Figure 1. Locations of three extracted sections in terms of the LTPP database.

Table 1. The locations and joint spacings of selected projects.

State Code	13	47	05
Section ID	3016	0601	0219
State/Province	Georgia	Tennessee	Arkansas
County	Haralson	Madison	Saline
Joint spacing (ft)	20 (6 m)	25 (7.5 m)	15 (4.5 m)

Based on the aforementioned considerations, the three sections including 13-3016, 47-0601, and 05-0219 were then compare as shown in Table 2. The three sections had similar concrete slab thickness, functional class, number of lanes, and climate zone. All three sections had the LTPP data collected for more than 15 years, which is enough to accurately represent the long-term performance of the JPCP. In addition, dowel bars were involved in all of these sections at the joints.

Table 2. The circumstances and physical conditions of selected sections.

Sections	13-3016	47-0601	05-0219
Route, Direction	Interstate–1, East Bound	Interstate–1, West Bound	Interstate–1, West Bound
GPS Lat, Long	33.68136, 85.29316	35.71699, 88.6383	34.51776, 92.68929
Functional Class	Rural Principal Arterial-Interstate	Rural Principal Arterial–Interstate	Rural Principal Arterial-Interstate
Number of Lanes	2	2	2
Concrete Slab Thickness (in.)	11.10 (28 cm)	9.00 (23 cm)	11.10 (28 cm)
Climatic Zone	Wet, Non-Freeze	Wet, Non-Freeze	Wet, Non-Freeze
Date of Construction	1 December 1977	1 June 1964	1 October 1995
Date Included in LTPP	1 January 1987	1 January 1987	1 September 1993
LTPP Monitoring Status (Data Inactive)	Out–of–study (15 June 2007)	Out–of–study (1 May 2004)	Out–of–study (20 August 2013)

The pavement performance was largely affected by the traffic volume. To eliminate the effect of traffic volume, based on the LTPP data, the changes in AADTT in the three different sections were compared as shown in Figure 2, which includes the estimated and monitored traffic data throughout the entire survey period. Obviously, the AADTT volumes increased during the service life, with the rapid development of economics and vehicles, and were all around 3000 to 6000 throughout the in-service period. The AADTT volume between 3000 and 6000 is usually defined as the medium traffic level, which can represent most of the traffic conditions in the field application of JPCP. Thus, the traffic volume for all selected sections was at the same level.

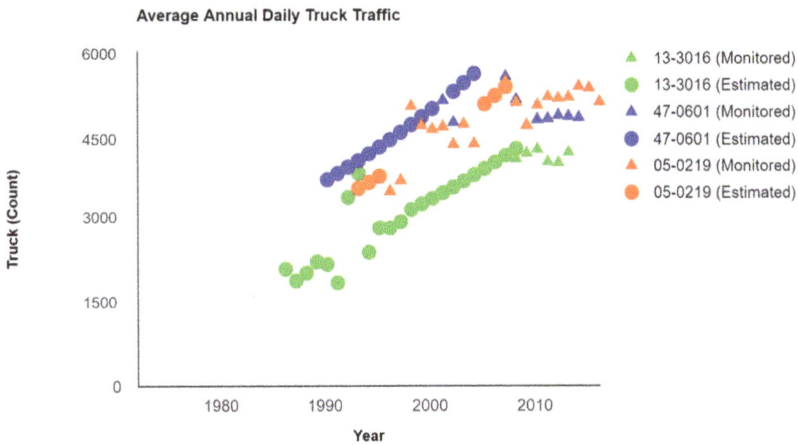

Figure 2. The change in average annual daily truck traffic (AADTT) in different sections.

The average annual precipitation at the three different locations was also compared as illustrated in Figure 3. It can be observed that the average annual precipitation in these three sections was mostly in the range of 1000 to 1600 mm during the survey periods. The maximum average annual precipitation of about 2100 mm was found in the 05-0219 section in the year 2010, while the lowest average annual precipitation of about 800 mm was found in the 13-3016 section in the year 2008. Overall, the precipitation conditions in these selected sections were kept consistent at a certain level, therefore the precipitation effect could be dismissed.

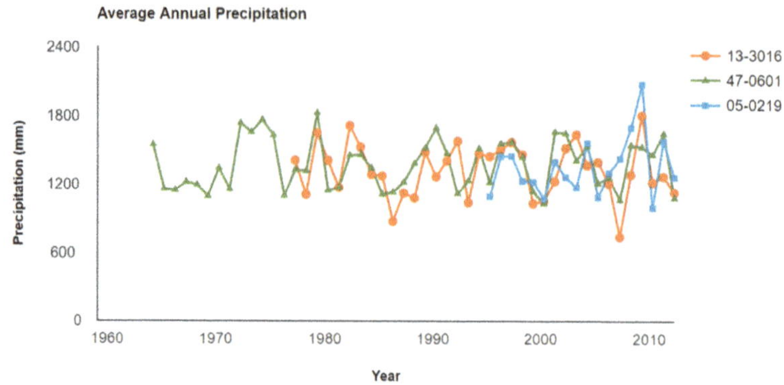

Figure 3. The change in precipitation in different sections.

As mentioned before, the temperature also influences the structure behaviors of JPCP to some extent. Therefore, not only the AADTT and precipitation have been considered, but also the temperature changes in these three sections have been compared. Figure 4 shows the average temperature changes through 12 months in sections 13-3016, 47-0601, and 05-0219, respectively. The temperature changes in the three different states showed the same trend and values, the lowest temperature was about 30 °F (−1.1 °C) and the highest temperature was about 90 °F (32.2 °C).

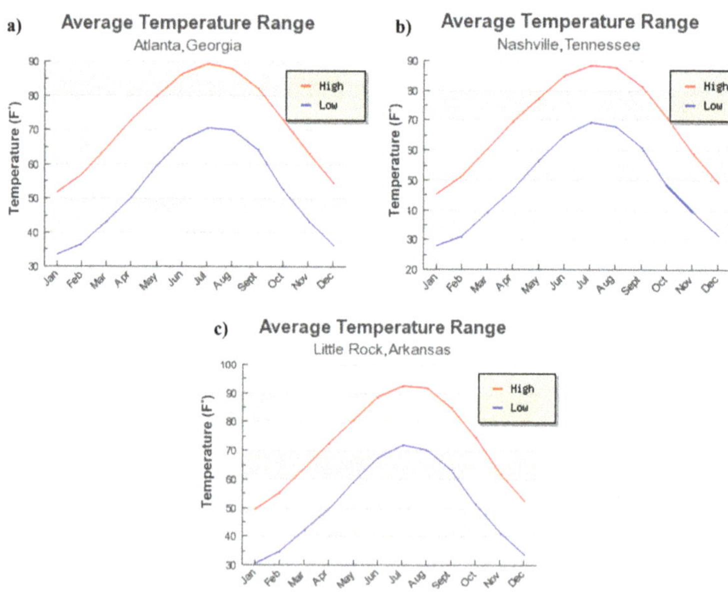

Figure 4. Average temperature changes: (**a**) Changes in section 13-3016; (**b**) Changes in section 47-0601; (**c**) Changes in section 05-0219.

After comparing the basic structure design and environmental conditions in these sections, it can be predicted that the long-term performance of JPCP in these sections was mainly influenced by joint spacing changes. The transverse crack (count), average JPCP faulting, average IRI, and average FWD deflection in these three sections were compared to analyze the long-term pavement performance.

3. Results and Discussion

3.1. JPCP Transverse Cracking

The JPCP transverse crack count was compared between three different joint spacings, as illustrated in Figure 5. From the results, it is obvious that section 0219 with the 15 ft joint spacing did not have any JPCP transverse cracks during the observation period. For sections 0601 (25 ft joint spacing) and 3016 (20 ft joint spacing), the transverse cracking volume was similar at first, but section 0601 had a higher number after 20 years of use. The total JPCP transverse cracking number was 13.6 in section 3016 and 18.2 in section 0601. The transverse cracking number increased with the increase in the in-service period. However, section 0601 had a higher accumulated number of cracks than section 3016. In terms of transverse cracking, the longer joint spacing was associated with a higher cracking volume.

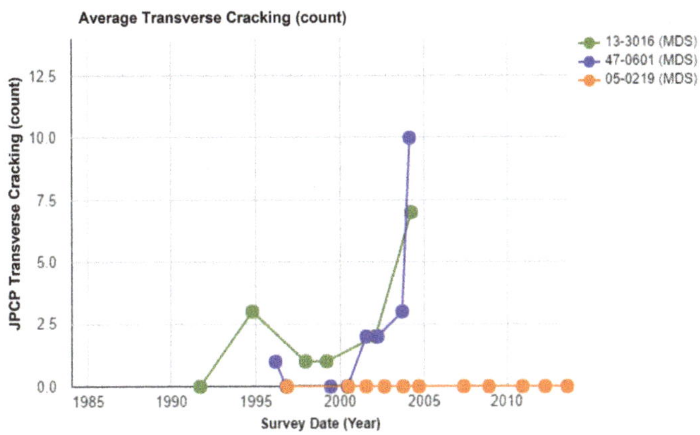

Figure 5. The comparison of the transverse cracking (count) with different joint spacings.

Based on the mechanistic–empirical pavement design method, the maximum joint spacing can be calculated by Equation (1):

$$l = \sqrt[4]{\frac{Eh^3}{12(1-\mu^2)k}} \tag{1}$$

where, l = radius of relative stiffness, E = modulus of elasticity of the concrete, h = slab thickness, k = modulus of subgrade reaction, and μ = Poisson's ratio for concrete material.

With the determination of the radius of relative stiffness, the slab length is generally recommended to be in the range of $4l$ to $5l$. By using plain concrete materials, the calculated spacing of joints is about 15 ft (4.5 m) for plain concrete. The LTPP extracted results were consistent with the traditional mechanistic–empirical calculation in terms of the optimum joint spacing for resistance to transverse cracking.

From the long-term performance analysis based on the LTPP data, it can be noticed that joint spacing is a considerable factor when designing a JPCP. To avoid transverse cracks and keep the pavement in good performance, joint contractions must be placed. Sometimes, dowel bars are also used to transfer load between concrete slabs. From the data comparison of different pavement performances, the behaviors of JPCP do show a relationship with the concrete slab joint spacing. According to the comparison, lower joint spacing (less than 20 ft) will support better long-term performance of the pavement than joint spacing which is higher than 25 ft.

3.2. Average JPCP Faulting

Faulting is a common distress type in JPCP and is defined as the difference in elevation across a transverse joint or crack. Faulting can result from a combination of factors such as inefficient load transfer at joints, slab pumping, slab settlements, curling, warping, and inadequate base support conditions. Faulting plays a predominant role in pavement surface roughness over time, affecting ride comfort and driver safety. Moreover, significant joint faulting has an adverse impact on pavement lifecycle costs for maintenance and rehabilitation as well as vehicle operating costs.

According to the data from the LTPP database, the average JPCP faulting in these three sections showed significant differences. It is universally known that faulting is related to repeated load. From Figure 6, it is obvious that section 0601 (with 25 ft joint spacing) had the highest average JPCP faulting since the data began being collected. In contrast to this, section 0219 (with 15 ft joint spacing) had the lowest average JPCP faulting in the LTPP research period. Section 3016 (with 20 ft joint spacing) had a medium average JPCP faulting value. The average JPCP faulting depth in section 0601 (with 25 ft joint spacing) showed a reduction from the year 2000 to the year 2004. This could be attributed to differences and deviations in test methods. A previous study found that the JPCP faulting increased from 2 mm to 3 mm when the joint spacing increased from 15 ft to 20 ft [15], which is consistent with the results found in this study.

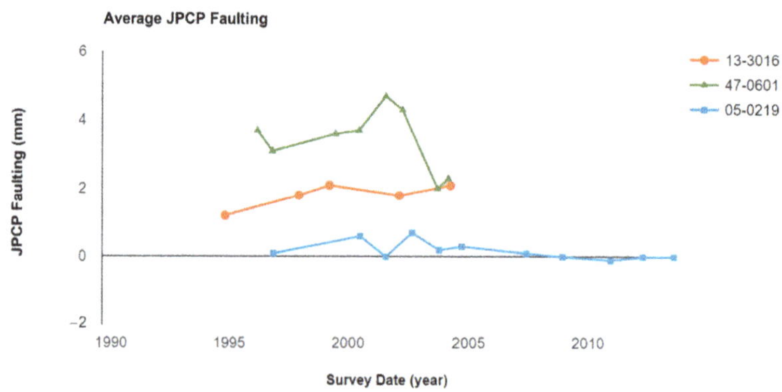

Figure 6. The comparison of the average JPCP faulting with different joint spacings.

A Georgia fault meter (GFM) was used to collect joint and crack faulting data for the LTPP Program. The GFM is a manual faulting measurement (MFM) instrument that is time-consuming and inconvenient. The manual GFM uses a dial gauge to determine the positive or negative difference at a joint or crack, and the automated GFM uses the linear variable differential transformer (LVDT) to determine positive or negative faulting at a joint or crack. The legs of the GFM's base are set on the slab in the direction of traffic on the left side of the joint. The joint must be centered between the guidelines shown on the side of the meter. The measuring probe contacts the slab on the approach side. Vertical movement of this probe is transmitted to an LVDT to measure joint faulting. Therefore, MFM data collection may contain some possible mistakes, due to equipment inaccuracy and operational error. In a word, even though LTPP data collection is a good data source, there are still some deviations within those data. In LTPP data collection methods, the faulting is determined by MFM, which is not accurate. Some more up-to-date ways can be used to detect faulting at a more accurate level.

However, the average JPCP faulting in these three different sections is different and has stratification. Section 0219, with 15 ft joint spacing, demonstrated the best performance among these three sections, suggesting that in the long-term service period, a lower joint spacing will contribute to lower JPCP faulting.

3.3. Average International Roughness Index

The international roughness index (IRI) is a scale for roughness based on the simulated response of a generic motor vehicle to the roughness in a single-wheel path of the road surface [26]. IRI is related to pavement flatness and driving comfort [27]. Ride comfort depends on human response to vibration, vehicle response to the road, and road roughness. The only factor which influences the ride comfort that is controlled by pavement is the road surface roughness. Therefore, IRI is a very important factor in considering long-term pavement performance.

From LTPP data, the average IRI was collected. As can be seen in Figure 7, section 0219 (with 15 ft joint spacing) had the lowest IRI value consistently throughout the LTPP research period. Section 0219 had almost the same IRI value after construction but had some minor increases in the later years. For section 0601 (with 25 ft joint spacing), the IRI value was almost the same at first but then increased dramatically in the latter years, which meant the surface smoothness was significantly diminished during the increasing traffic volume and reduced durability. The IRI value showed a constant trend and was mostly lower than 1.5 m/km in section 3016 and section 0219. The lowest joint spacing of 15 ft contributed to the lowest and the most consistent IRI value (1.3 m/km) among the three different joint spacings, which was also in agreement with the findings in terms of the JPCP faulting

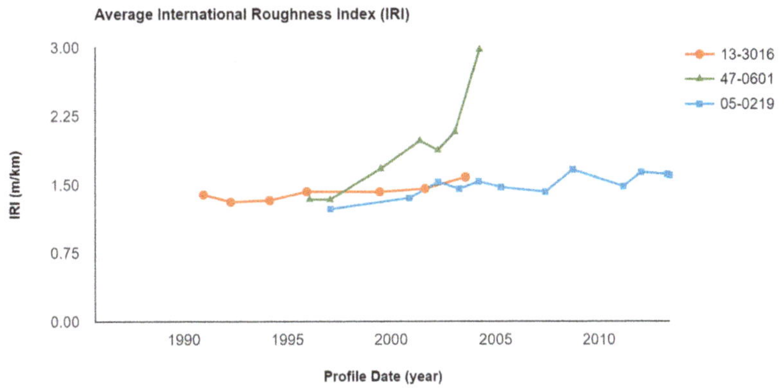

Figure 7. The comparison of the IRI with different joint spacings.

There are IRI categories given by the Federal Highway Administration (FHWA), shown in Table 3. An IRI value that is lower than 1.5 m/km means the pavement roughness is in good condition, and the acceptable IRI value should be lower than 2.7 m/km to maintain a workable surface roughness. Section 3016 (with 20 ft joint spacing) and section 0219 (with 15 ft joint spacing) fell into the good roughness category by this standard. However, section 0601 which had the highest joint spacing (25 ft) had an unacceptable value after being in service for ten years. According to those comparisons, joint spacing which is lower than 20 ft will contribute to good performance in long-term pavement smoothness and lower IRI values. Another study found that the IRI increased when the joint spacing changed from 12 ft to 20 ft, with a slab thickness that was greater than 7 in [28], which is also in agreement with the findings in this investigation.

Table 3. FHWA IRI Categories.

Roughness Category	IRI Value (m/km)
Good	<1.5
Acceptable	<2.7

Throughout the whole life cycle of the JPCP, maintenance is a considerable part of the economic cost during the reduction of IRI. The pavement with a higher IRI will result in a higher cost of maintenance. Therefore, it is important to determine an optimum joint spacing when designing a JPCP to diminish the maintenance costs.

3.4. Average FWD Deflection

The FWD deflection can be used to investigate the dynamic deformation behavior of pavement structures. During the service period, the average FWD deflections changed continuously in the selected sections, as shown in Figure 8.

Figure 8. Average FWD deflection.

A comparison of the FWD deflections among the three sections shows that the FWD deflection was always higher in the 47-0601 section with a joint spacing of 25 ft, while the lowest FWD deflection was generally found in the 05-0219 section with a joint spacing of 15 ft. It is known that the lower the FWD deflection, the better the dynamic stability of the pavement structures. However, the trend of FWD deflection change was not constant in all sections. For instance, the FWD deflection first increased and then decreased in section 05-0219, in which an even higher FWD deflection measurement was found compared to that of section 13-3016 (with a joint spacing of 20 ft) around the year 2002. Consequently, even if the FWD deflection measurements can reflect the dynamic stability of the JPCP, it is not recommended to directly determine the status of the pavement with a single-year measurement since the FWD deflection measurements are considerably affected by other unexpected factors. It would be better to evaluate the FWD deflection with a long-term investigation, such as the LTPP method used in this study.

In this study, load transfer between adjacent concrete slabs was not considered. Aggregate interlocking between two slabs is the simplest way to transfer load. However, cracks will occur once the joints have been constructed, thus aggregate interlock performance within two slabs will be considerably reduced since dowel bars are placed in the joint contractions to help transfer load from one concrete slab to another. The long-term behaviors of dowel bars will also have an impact on the performance of JPCP. Moreover, sealed and unsealed joints are also factors that contribute to the JPCP long-term performance. In future studies, those factors could be investigated further to obtain a more all-around optimized JPCP performance-based design.

4. Conclusions

In this study, the effects of joint spacings on the long-term pavement performance in JPCP were investigated using the extracted LTPP data, collected for more than 15 years. With the increase in joint spacing, the development of transverse cracks was significantly increased, especially in the largest joint spacing of 25 ft, which showed 10 transverse cracks in later service life. With the change in joint spacings in JPCP, the dimension of the faulting was also changed, and decreased joint spacings could effectively control the development of faulting. The longer the joint spacing, the larger the IRI values in later service life, demonstrating worse surface roughness. The changing trend of FWD deflection was not consistent during the survey period, which is not critical to justify the long-term pavement performance of JPCP when compared with other involved performances.

In the future application of JPCP, the joint spacing should be designed to be lower than 20 ft to minimize the reduction in durability during the service life. An optimized joint spacing, considering the long-term pavement performance, is recommended to be 15 ft in most cases.

Author Contributions: Conceptualization, J.W.; methodology, J.W.; formal analysis, J.W. and X.L.; investigation, X.L.; resources, J.W. and X.H.; data curation, S.R. and Y.Y.; writing—original draft preparation, J.W.; writing—review and editing, J.W. and X.L.; supervision, X.H. All authors have read and agreed to the published version of the manuscript.

Funding: This research was funded by the National Natural Science Foundation of China (Grant No. 52108408), and the Natural Science Foundation of Jiangsu Province (Grant No. BK20210617).

Institutional Review Board Statement: Not applicable.

Informed Consent Statement: Not applicable.

Data Availability Statement: The data can be provided if needed by the corresponding author.

Acknowledgments: The first author would like to acknowledge the support from the National Natural Science Foundation of China (Grant No. 52108408), and the support from the Natural Science Foundation of Jiangsu Province (Grant No. BK20210617).

Conflicts of Interest: The authors declare no conflict of interest.

References

1. Pradena, M.; Houben, L. Sustainable Pavements: Influence of the Saw-Cutting Method on the Performance of JPCP. In Proceedings of the 14th International Multidisciplinary Scientific GeoConference SGEM 2014, Albena, Bulgaria, 17–26 June 2014.
2. Su Jung, Y.; Zollinger, D.G.; Won, M.; Wimsatt, A.J. *Subbase and Subgrade Performance Investigation for Concrete Pavement*; Technical Rep No FHWA/TX-09/0-6037-1; Texas Transportation Institute, Texas A&M University: College Station, TX, USA, 2009.
3. Sadeghi, V.; Hesami, S. Investigation of load transfer efficiency in jointed plain concrete pavements (JPCP) using FEM. *Int. J. Pavement Res. Technol.* **2018**, *11*, 245–252. [CrossRef]
4. Hansen, W.; Liu, Z. *Improved Performance of JPCP Overlays*; Bureau of Field Services, Michigan Department of Transportation: Lansing, MI, USA, 2013.
5. Sii, H.B.; Chai, G.W.; van Staden, R.; Guan, H. Effect of dowel looseness on response of jointed concrete pavements using three-dimensional finite element analysis. *Adv. Mater. Res.* **2014**, *900*, 435–444. [CrossRef]
6. Davids, W.G. Effect of dowel looseness on response of jointed concrete pavements. *J. Transp. Eng.* **2000**, *126*, 50–57. [CrossRef]
7. Malla, R.B.; Joshi, S. Resilient modulus prediction models based on analysis of LTPP data for subgrade soils and experimental verification. *J. Transp. Eng.* **2007**, *133*, 491–504. [CrossRef]
8. Dong, Q.; Huang, B. Evaluation of effectiveness and cost-effectiveness of asphalt pavement rehabilitations utilizing LTPP data. *J. Transp. Eng.* **2012**, *138*, 681–689. [CrossRef]
9. Gong, H.; Huang, B.; Shu, X. Field performance evaluation of asphalt mixtures containing high percentage of RAP using LTPP data. *Constr. Build. Mater.* **2018**, *176*, 118–128. [CrossRef]
10. Wang, Y.; Mahboub, K.C.; Hancher, D.E. Survival analysis of fatigue cracking for flexible pavements based on long-term pavement performance data. *J. Transp. Eng.* **2005**, *131*, 608–616. [CrossRef]
11. Pradena, M.; Houben, L. Sustainable pavements: An analysis of the cracks width in jointed plain concrete pavements with short slabs. In Proceedings of the 14th International Multidisciplinary Scientific GeoConference SGEM 2014, Albena, Bulgaria, 17–26 June 2014.
12. Kim, H.B. Top-Down Cracking of Jointed Plain Concrete Pavements. *J. East. Asia Soc. Transp. Stud.* **2010**, *8*, 1529–1541.
13. Smiley, D.; Hansen, W. *Investigation of Early Cracking on Selected JPCP Projects*; Construction and Technology Division, Michigan Department of Transportation: Lansing, MI, USA, 2007.
14. Qiudong, W.; Libin, W.; Bohai, J.; Zhongqiu, F. Modified effective notch stress method for fatigue evaluation of rib-deck welds integrating the critical distance approach. *J. Constr. Steel Res.* **2022**, *196*, 107373. [CrossRef]
15. Mu, F.; Vandenbossche, J.; Gatti, K.; Sherwood, J. An evaluation of JPCP faulting and transverse cracking models of the mechanistic-empirical pavement design guide. *Road Mater. Pavement Des.* **2012**, *13*, 128–141. [CrossRef]

16. Huang, K.; Zollinger, D.G.; Shi, X.; Sun, P. A developed method of analyzing temperature and moisture profiles in rigid pavement slabs. *Constr. Build. Mater.* **2017**, *151*, 782–788. [CrossRef]
17. Asbahan, R.E.; Vandenbossche, J.M. Effects of temperature and moisture gradients on slab deformation for jointed plain concrete pavements. *J. Transp. Eng.* **2011**, *137*, 563–570. [CrossRef]
18. Shafiee, M.; Maadani, O.; Shirkhani, H. Evaluation of climate impacts on jointed plain concrete pavement structures. In Proceedings of the 2019 TAC-ITS Canada Joint Conference & Exhibition, Halifax, NS, Canada, 22–25 September 2019.
19. Zhang, Y.; Wei, Y.; Li, B.; Wang, G.; Huang, L. A novel seawater and sea sand concrete-filled CFRP-carbon steel composite tube column: Seismic behavior and finite element analysis. *Eng. Struct.* **2022**, *270*, 114872. [CrossRef]
20. Wei, Y.; Zhu, C.; Miao, K.; Chai, J.; Zheng, K. Compressive behavior of rectangular concrete-filled fiber-reinforced polymer and steel composite tube columns with stress-release grooves. *Compos. Struct.* **2022**, *281*, 114984. [CrossRef]
21. Shi, X.; Mukhopadhyay, A.; Zollinger, D.; Huang, K. Performance evaluation of jointed plain concrete pavement made with portland cement concrete containing reclaimed asphalt pavement. *Road Mater. Pavement Des.* **2021**, *22*, 59–81. [CrossRef]
22. Shi, X.; Zollinger, D.G.; Mukhopadhyay, A.K. Punchout study for continuously reinforced concrete pavement containing reclaimed asphalt pavement using pavement ME models. *Int. J. Pavement Eng.* **2020**, *21*, 1199–1212. [CrossRef]
23. Ali, B.; Qureshi, L.A.; Kurda, R. Environmental and economic benefits of steel, glass, and polypropylene fiber reinforced cement composite application in jointed plain concrete pavement. *Compos. Commun.* **2020**, *22*, 100437. [CrossRef]
24. Colley, B.; Humphrey, H. *Aggregate Interlock at Joints in Concrete Pavements*; Portland Cement Association: Skokie, IL, USA, 1967.
25. Carroll, J.C.; Helminger, N. Fresh and Hardened Properties of Fiber-Reinforced Rubber Concrete. *J. Mater. Civ. Eng.* **2016**, *28*, 49–69. [CrossRef]
26. Múčka, P. International Roughness Index specifications around the world. *Road Mater. Pavement Des.* **2017**, *18*, 929–965. [CrossRef]
27. Abdelaziz, N.; Abd El-Hakim, R.T.; El-Badawy, S.M.; Afify, H.A. International Roughness Index prediction model for flexible pavements. *Int. J. Pavement Eng.* **2020**, *21*, 88–99. [CrossRef]
28. Chen, Y.-A.; Taylor, P.C.; Ceylan, H.; Kim, S.; Wang, X. Effect of joint spacing and pavement thickness on concrete overlay performance. *Int. J. Pavement Res. Technol.* **2019**, *12*, 64–69. [CrossRef]

materials

MDPI

Article

Mixture Design and Mechanical Properties of Recycled Mortar and Fully Recycled Aggregate Concrete Incorporated with Fly Ash

Lijuan Zhang [1], Dong Ding [1], Jun Zhao [1], Guosen Zhou [2] and Zhi Wang [1,*]

1 School of Mechanics and Safety Engineering, Zhengzhou University, No. 100 Science Avenue, Zhengzhou 450001, China
2 Forth Co., Ltd. of China Construction Fifth Division, 69 Shihua Road, Guancheng District, Zhengzhou 450004, China
* Correspondence: wangzhi@zzu.edu.cn; Tel.: +86-138-3852-7945

Abstract: Recycled aggregate concrete (RAC) is a sort of green, low carbon, environmental protection building material, its application is of great significance to the low carbonization of the construction industry. The performance and strength of RAC are much lower than natural aggregate concrete (NAC), which are the key factors restricting its application. Class F fly ash is a cementitious material that is considered environmentally hazardous. In this paper, appropriate water-binder (w/b) ratios were found through a mortar expansion test at first. The compressive strength of recycled mortar incorporated with class F fly ash was further studied. On this basis, the mechanical properties of nine groups of fully recycled aggregate concrete (FRAC) with a w/b ratio of 0.3, 0.35, and 0.4, and fly ash replacement ratios of 0, 20%, and 40%, were studied. The influence of the w/b ratio and fly ash replacement ratio on mechanical properties was analyzed and compared with previous research results. In addition, the conversion formulas between the splitting tensile strength, flexural strength, and compressive strength of FRAC were fitted and established. The research results have a certain guiding significance for the mixture design of FRAC and further application of class F fly ash.

Keywords: recycled mortar; recycled concrete; mixture proportion method; mechanical properties; fly ash

Citation: Zhang, L.; Ding, D.; Zhao, J.; Zhou, G.; Wang, Z. Mixture Design and Mechanical Properties of Recycled Mortar and Fully Recycled Aggregate Concrete Incorporated with Fly Ash. *Materials* **2022**, *15*, 8143. https://doi.org/10.3390/ma15228143

Academic Editor: Danuta Barnat-Hunek

Received: 28 September 2022
Accepted: 12 November 2022
Published: 17 November 2022

1. Introduction

Concrete is essential in building construction and infrastructure, and the global demand for concrete is constantly on the rise. The main constituent materials of concrete, river sand and gravel, are non-renewable resources and are extracted most, even more than fossil fuels [1]. Approximately 30–50 billion tons of sand are consumed annually worldwide, and the current rate of extraction of river sand and natural aggregates far exceeds the rate of natural renewal, but cannot fully meet the demands of engineering construction [2]. Meanwhile, a massive amount of construction and demolition (C&D) waste is produced as concrete structures are demolished and disposed of in landfills [3]. An important way to recycle concrete materials is fabricating C&D waste as recycled aggregate (RA) to manufacture recycled mortar and recycled aggregate concrete (RAC). The application of RA can not only consume C&D waste and solve the urban environmental issue contributing to C&D waste, but also reduce the mining and use of natural sand and gravel, protect the natural environment, and solve the current natural resources dilemma. However, the performance of RA changes dramatically due to complex sources and diverse reconstruction techniques of C&D waste.

Currently, concrete materials that use exclusively recycled coarse aggregate (RCA) and recycled fine aggregate (RFA) are referred to as fully recycled aggregate concrete (FRAC). Generally speaking, the surface of RCA is wrapped with a layer of mortar, with a high content of needles and flakes, while the crushing index, the water absorption rate, and the void ratios are larger than for natural coarse aggregate (NCA), and the apparent density is smaller [4]. The chemical composition of RCA and NCA is also different. The content of SiO_2 in RCA can hit 50% to 60%, while the chemical composition of calcium (CaO) in RCA is mostly below 20%. The NCA is mostly composed of CaO (49.16%) and SiO_2 (5.46%) [5,6]. RCA, as a siliceous aggregate, shows clearly inadequate water resistance [7]. Thus, recycled aggregate concrete (RAC) generally performs worse than natural aggregate concrete (NAC), with workability [8], density, and mechanical properties [9] decreasing as the replacement ratio of RCA increases. RFA is determined as having a higher water absorption and a more adherent mortar in surface compared to natural fine aggregate (NFA), resulting in in the application of RFA being greatly limit [10]. RFA is firmly excluded from the production of concrete and mortar in most countries and territories [11]. The chemical composition of RFA is similar to that of RCA, with SiO_2 content being between 60% and 80%, and CaO content being between 5% and 20%. Besides, RFA also contains a certain amount of Al_2O_3 (below 10%) and Fe_2O_3 (below 10%) [12]. Previous studies in our laboratory have shown that the main mineral compositions of RFA are quartz (SiO_2), calcite ($CaCO_3$), albite ($Na_2O{\cdot}Al_2O_3{\cdot}6SiO_2$), and anorthite ($CaO{\cdot}Al_2O_3{\cdot}2SiO_2$). $CaCO_3$ in calcite can react with C_3A in cement to form calcium calcium aluminate ($C_3A{\cdot}CaCO_3{\cdot}11H_2O$), thus hindering the formation of calcium hydroxide ($Ca(OH)_2$) and hydrated calcium sulphoaluminate (Aft), which improves the compactness of the interfacial transition zone (ITZ) and enhances the bonding property of the interface between the aggregate and cement paste, thus increases the strength of the mortar [13]. Besides, the rough irregular surface of RFA effectively enhances the mechanical interlocking of ITZ between RFA and the cement matrix, which improves the splitting tensile strength [14]. The $Ca(OH)_2$ groups in RFA are found to react with fly ash to form secondary hydration compounds that increase the mechanical strength of the mortar [12,15]. Based on the above results, RFA can be more fully applied by an appropriate mixture design method. At present, the main mixture design method of RCA is the additional water consumption method. However, the study in our laboratory showed that, when RFA is used, the mechanical properties of RAC decrease with an increasing RFA replacement ratio and additional water consumption. When the RFA replacement ratio is 100%, the elastic modulus is reduced by up to 20% [14]. Therefore, a new mixture design method should be proposed to obtain FRAC with better mechanical properties. The mechanical properties of concrete are closely related to its workability, especially slump. For FRAC, the slump of fresh FRAC is more difficult to control due to the high water absorption of RCA and RFA. The fluidity of mortar directly affects the slump of concrete, which is mainly affected by the water-binder ratio and sand-cement ratio. Therefore, the fluidity of recycled mortar can be used as a main reference in the mixture design of FRAC.

The water-binder (w/b) ratio is an important parameter affecting the workability and mechanical properties of concrete. For RAC, RA will absorb a large amount of water so that the effective water in the mixing process will be significantly reduced; using the unilateral water consumption of NAC to calculate the mix proportion of RAC will lead to a significant decrease of RAC slump [16]. Therefore, additional water and admixtures are needed to optimize the performance of the RAC. Wagih et al. [17] reported that a 13% increase in water consumption is required for RAC to achieve a similar performance as NAC. Wang et al. [18] suggested that, when calculating the mix proportion of RAC, in addition to the water consumption designed in accordance with the NAC ratio, he considered the additional water consumption required to make RA reach the saturation state of water absorption, called the additional water consumption method. Before pouring RA, soak RA to make it reach the saturation state of water absorption [19]. However, in the actual pouring process, both methods are not very feasible. For FRAC, since the fine aggregate used is RFA, its unilateral water consumption is even higher than plain RAC, thus the w/b ratio of FRAC deserves more attention.

Fly ash, one of the main waste products of coal-fired power plants, is considered as being deleterious material due to the significant variability of its leaching characteristics regarding heavy metals [20]. However, fly ash has a good adsorption capacity for NO_2, SO_2, organic compounds, and mercury [21], so it is usually used as a supplementary cementitious material in the construction industry [22]; itis concluded that it can improve the workability, mechanical properties, microstructure, and durability of concrete, and meanwhile reduce the heat of hydration and the cracking tendency [23]. In addition, adding liberal fly ash at a low w/b ratio is an effective method for fabricating high performance concrete [24,25]. The hydration of fly ash and $Ca(OH)_2$ groups in RFA ameliorates the adverse effect of RFA on mechanical properties [12,15]. The effect of fly ash on concrete is greatly affected by its characteristics, especially the content of calcium oxide and fineness. The American Society for Testing Materials (ASTM) specification C618-19 [26] defines two classes of fly ash (classes C and F). As the total sum of SiO_2, Al_2O_3, Fe_2O_3, and SO_3 exceeds 70% of the total volume, it is considered as class F, otherwise it is class C. The Chinese standard GB/T 1596-2017 [27] is more intuitive for defining the class of fly ash; fly ash is defined as class F if the CaO content is less than 10%, otherwise it is class C. China is the world's largest producer of fly ash, with an annual output of 600 million tons, but only 70% of fly ash is fully utilized. Moreover, the CaO content of most fly ash produced in China is between 1.1% and 7% [28], belonging to class F fly ash. Thus, the application of class F fly ash should be further studied.

Sand ratio also influences the mechanical properties and workability of concrete. In the design of an ordinary concrete ratio, it is considered that the essence of determining the best sand ratio is to determine the best sand-cement ratio. As the surface of RCA generally adheres to a layer of old mortar, if the sand ratio of RAC is calculated according to the calculation method of NAC, it will make its content of mortar much higher than NAC, which is one of the major reasons for the deterioration of the performance of RAC. In addition, the particle morphology of RFA and NFA is also very different; if the outer surface of RFA is rough, it will further affect the workability of RAC.

At present, the strength of FRAC can be improved by reducing the water-binder ratio, however, there is no effective method to control the workability of FRAC. The workability not only affects the application of FRAC in practical engineering, but also plays a decisive role in its strength, compactness, and durability after hardening. Therefore, how to determine the mix design parameters of FRAC to ensure the workability and strength of RC is the focus of this paper. A new mixture design method of FRAC is provided. The workability of FRAC can be significantly improved by an optimal sand-cement ratio and sand ratio, and the effects of the water-binder ratio and fly ash replacement ratio on FRAC made by this method were studied, which provides the experimental basis and theoretical guidance for the mixture design of FRAC and the application of class F fly ash.

2. Materials and Methods

2.1. Mixture Design Method

In order to ensure the workability of FRAC, a new mixture design method was conducted based on the fluidity of recycled mortar. The determination process of the main parameter mixture design is as follows:

Firstly, the optimal sand-cement ratio corresponding to different water-binder ratios was obtained through the expansion test of recycled mortar. Secondly, the sand ratio was calculated through the "Mortar abundant coefficient method". This method considers that, in addition to filling the void between the CA, FA also needs to have a certain amount of surplus to form the cement mortar and ensure that the fresh concrete mix has enough flow performance. Finally, the water content, cementitious materials, and RCA and RFA of FRAC under different water-binder ratios can be directly obtained by the "Absolute volume method".

On this basis, the influence of different water-binder ratios and fly ash replacement ratios on recycled mortar and FRAC made by this method was studied. The main research steps of the paper are shown in Figure 1. Based on this mixture design method, not only FRAC, but also recycled mortar with good workability can be prepared.

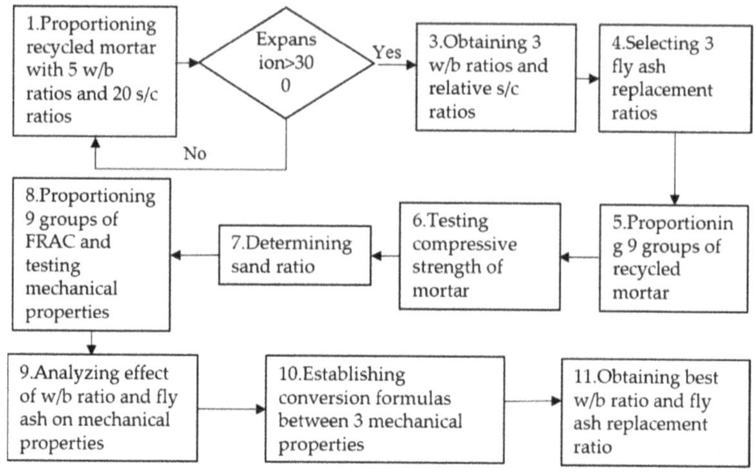

Figure 1. Research steps.

2.2. Materials

Grade 42.5 ordinary Portland cement and fly ash were used as binder materials. The physical properties and main chemical composition of cement and fly ash are listed in Tables 1 and 2, respectively. The particle size distribution of both binder materials are listed in Table 3 and Figure 2. The particle size of fly ash is larger than that of cement. The CaO content of fly ash is 3.36% and the total sum of SiO_2, Al_2O_3, Fe_2O_3, and SO_3 is 91.303%. Therefore, the fly ash used in the paper belongs to class F, according to C618-19 [26] and GB/T 1596-2017 [27]. The fly ash replacement ratio, which is the ratio of the weight of fly ash to the weight of total cementitious material, is taken as 0, 20%, and 40%. RCA and RFA are derived from discarded laboratory concrete specimens. The waste concrete was crushed and screed to prepare as RCA (5–20 mm) and RFA (0–4.75 mm). The material properties of RCA and RFA are tested by the Chinese standard JGJ 52-2006 [29] and shown in Table 4. The crush index of RCA is 14.1%, which is higher than the crush index of NCA. The fineness modulus of RFA is 3.15, which is larger than 3, and therefore belongs to coarse sand. The surface of RFA is rougher than the NFA. The water-reducing rate of polycarboxylic acid water-reducing agent used in the experiment is 27%.

Table 1. The physical properties of binder materials.

	Density (kg/m^3)	Packing Density (kg/m^3)	Specific Surface Area (m^2/kg)	Water Consumption at Standard Consistency (%)	28 d Compressive Strength (MPa)
Cement	3126	1550	356	27	44.8
Fly ash	2550	1120			

Table 2. Main chemical composition of binder materials.

Chemical Composition (%)	SiO_2	CaO	Al_2O_3	Fe_2O_3	MgO	SO_3	K_2O	TiO_2
Cement	27.73	46.31	13.54	3.09	3.09	2.82	0.984	0.688
Fly ash	54.74	3.36	33.33	2.30	0.867	0.933	2.21	1.01

Table 3. Particle size distribution of cement and fly ash.

Particle Size (μm)	Surface Weighted Mean D	Vol. Weighted Mean	d (0.1)	d (0.5)	d (0.9)
Cement	5.194	19.999	2.691	12.673	46.709
Fly ash	5.692	24.373	3.307	16.502	57.635

Note: d (0.1), d (0.5), and d (0.9) represent the diameters corresponding to 10%, 50%, and 90% of the cumulative size distribution (0 to 100%), respectively.

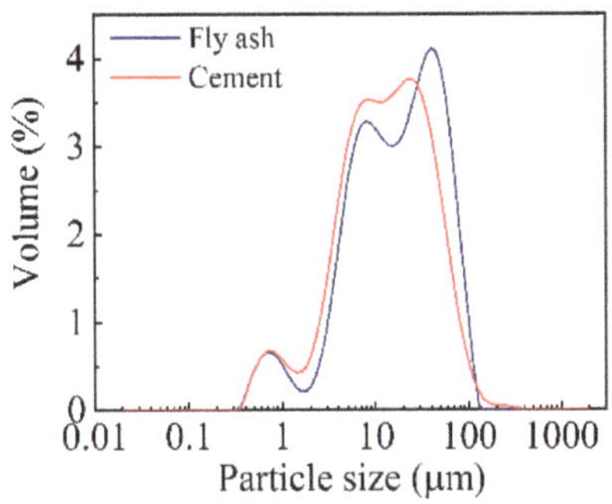

Figure 2. Particle size distribution of binder materials.

Table 4. Properties of RCA and RFA.

Aggregate	Apparent Density (kg/m^3)	Water Absorption (%)	Crushing Index (%)	Fineness Modulus	Loose Stack Density (kg/m^3)	Compact Stack Density (kg/m^3)	Void Ratio (%)
RCA	2734.8	3.56	14.1	-	1291.8	1378.7	52.8
RFA	2594.7	7.6	-	3.15	1486		

2.3. Specimen Preparation

Owing to the large water absorption of RCA and RFA, the workability of FRAC will be seriously affected if made using the mix design of ordinary concrete. In order to ensure that the slump of the FRAC can reach more than 50 mm, the expansion test of the recycled mortar should be carried out first. The expansion test was carried out by the Chinese standard GB 50119-2013 [30]. The detailed method is to pour the mortar into the mortar expansion cylinder twice (each time the amount is 1/2 of the cylinder height) and use the pounding rod along the edge to the center clockwise to evenly pound 15 times. After pounding and scraping, the cylinder is slowly lifted and the maximum diameter in two directions perpendicular to each other is measured with a steel ruler after 10 s, and the average value is taken as the mortar expansion. The w/b ratios of mortar were selected as 0.3, 0.35, 0.4, 0.45, and 0.5 and adjusted the sand-ash ratio to make the expansion of the mortar reach more than 300 mm. In addition, on this basis, the fly ash replacement ratio was 0, 20%, and 40%.

The detailed mix design of FRAC needs to be determined by the properties of the recycled mortar. The mixing process of FRAC is as follows: firstly, the cement and RFA were mixed without water for 120 s. In the second step, the water-reducing agent and half of the water were added and mixed for two 120 s so that the cement and water-reducing agent could fully react. Point three, the RCA and the remaining half of water were added and mixed for 120 s. The slump was tested first, then the concrete was loaded into the molds and vibrated on a vibrating table for 90 s, cured indoors for 24 h, then demolded and placed in a standard curing room. After 28 days of curing, the mechanical properties of FRAC were tested.

2.4. Test Methods

Standard mortar test blocks with a side length of 70.7 mm were used to test the compressive strength of recycled mortar, three test pieces were in each group and the loading speed was 0.5 KN/s. The compressive strength and splitting tensile strength of FRAC were tested with 100 mm^3 test blocks, with three test pieces in each group, and were performed by a digital display pressure tester. The loading speed of the compressive strength test was 0.5–0.8 MPa/s, and the loading speed of the splitting tensile strength test was 0.08 MPa/s. As the size of the specimens used is smaller than the standard specimen block with a side length of 150 mm, the compressive strength and splitting tensile strength are multiplied by the discount factor of 0.95 and 0.85, respectively. Beams with a side length of 100 mm × 100 mm × 400 mm were tested for flexural strength with a loading speed of 0.1 mm/min. Mortar specimens and FRAC specimens were maintained in a standard maintenance room for 28 d before testing. The mechanical properties testing equipment and methods of mortar specimens and FRAC specimens meet the requirements of the Chinese standard JGJ/T70-2009 [31] and GB/T50081-2019 [32], respectively.

3. Properties of Recycled Mortar

3.1. Effect of Sand-Cement Ratio

According to Table 5, the variation relationship between the expansion of mortar and sand-cement ratio under various w/b ratios is shown in Figure 3. Figure 3 shows that the mortar's sand-cement ratio should decrease as the w/b ratio decreases.

Table 5. The relationship between expansion and sand-cement ratio under different w/b ratio.

Mortars NO	W/B	S/C	Expansion (mm)
M-0.5	0.5	2.5	170
		2.3	200
		2.1	230
		2	420
M-0.45	0.45	2	170
		1.8	210
		1.7	230
		1.6	450
M-0.4	0.4	1.6	170
		1.5	220
		1.4	260
		1.3	430
M-0.35	0.35	1.4	170
		1.3	190
		1.2	240
		1.1	340
M-0.3	0.3	1.25	150
		1.1	160
		1.05	210
		1	300

The expansion of the highest sand-cement ratio for each group of tests is only 150–170 mm, which is attributed to the high water absorption of RFA. Therefore, a large amount of water will be absorbed so that the actual amount of water involved in the hydration reaction of the cement is much smaller, resulting in a greatly reduced amount of cement paste, and the water-reducing agent also does not fully react with cement. With the reduction of the sand-cement ratio, it is obvious to observe the air bubbles generated by the reaction of the water-reducing agent and cement when mixing the mortar, which shows that in the production of recycled mortar, it is important to ensure an effective amount of water and a sufficient mixing time in order to make the water-reducing agent take effect and ensure the fluidity of mortar.

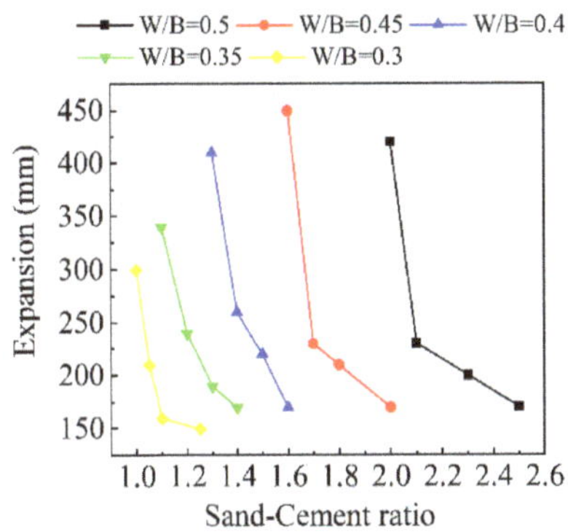

Figure 3. Variation of expansion with sand-cement ratio under various w/b ratio.

The expansion of mortar increases with the decrease of the sand-cement ratio, with slow changes in the early stage. As the sand-cement ratio continues to become small, the curves of all five groups will have an inflection point, after which the expansion will suddenly become larger. Large expansion will reduce the strength of both mortar and concrete. Meanwhile, the sand-cement ratio decreases and the content of cement in the mortar increases, which will increase the economic cost. In an effort to ensure the strength and workability of FRAC, the sand-cement ratio should be maximized after satisfying the mortar fluidity, while ensuring optimal economic efficiency. Therefore, the maximum sand-cement ratio of expansion over 300 mm is selected as the test optimal sand-cement ratio.

In this test, three w/b ratios of 0.3, 0.35, and 0.4 were selected, and the optimal sand-cement ratios of 1, 1.1, and 1.3 were selected under the three w/b ratios. The fly ash replacement ratio was taken as 0, 20%, and 40%. Subsequently, an expansion test and a compressive strength test of the recycled mortar were executed. The detailed mixture design and results are shown in Table 6. From Table 6, the compressive strength of plain recycled mortar under three water-binder ratios is 92.0%, 85.7%, and 90.0% of standard mortar produced with high quality standard sand, respectively, which proves that using this mixture design method can produce recycled mortar with high strength.

Table 6. Mixture design and compressive strength of recycled mortar.

Mortars NO	Cement (g)	Fly Ash (g)	Water (g)	Water-Reducing Agent (g)	Sand (g)	S/C	Expansion (mm)	f_{cu} (MPa)
M0.30F0	1100	0	330	33	1100	1	300	41.2
M0.30F20	880	220	330	27	1100	1	340	40.4
M0.30F40	660	440	330	22	1100	1	400	39.4
M0.35F0	1100	0	385	33	1210	1.1	340	38.4
M0.35F20	880	220	385	27	1210	1.1	370	35.2
M0.35F40	660	440	385	22	1210	1.1	420	31.8
M0.40F0	1100	0	440	33	1430	1.3	410	40.3
M0.40F20	880	220	440	27	1430	1.3	430	31.9
M0.40F40	660	440	440	22	1430	1.3	440	25.3

3.2. Effect of Fly Ash Replacement Ratio

The variation of expansion with the fly ash replacement ratio is shown in Figure 4. As the fly ash replacement ratio increases, the expansion of the recycled mortar increases continuously, which shows that fly ash can effectively increase the fluidity of recycled mortar. This is because the shape of fly ash is mostly spherical [33], and the particle size of fly ash is also larger than cement. The addition of fly ash helps to improve the particle size distribution and the fluidity of mortar. In addition, fly ash reduces the flocculation of cement particles and plays a role in dilution [34]. However, when the w/b ratio is high, the increase of expansion is decreasing with the increase of the fly ash replacement ratio, which shows that the improvement effect of fly ash on the workability of mortar is more obvious when the w/b ratio is small. When the w/b ratio is 0.3, the increase of expansion from 300 mm to 400 mm is the most obvious as the fly ash replacement ratio increases from 0 to 40%.

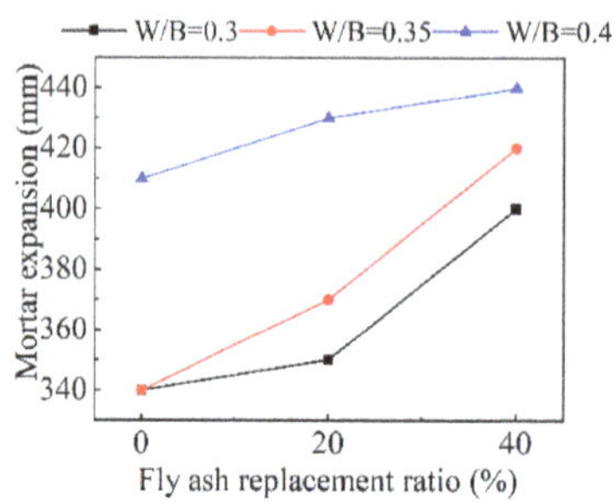

Figure 4. Variation of expansion with fly ash replacement ratio.

The variation of the compressive strength of recycled mortar with the fly ash replacement ratio under different w/b ratios is shown in Figure 5. The strength of recycled mortar decreases continuously with the increase of the fly ash replacement ratio. Since the active substances in fly ash that can participate in the hydration reaction are much smaller than those in cement, the increase of fly ash admixture will reduce the degree of the hydration reaction of the whole cementitious material, thus reducing the compressive strength of mortar. Moreover, with the increase of the w/b ratio, the decreasing effect of fly ash on compressive strength is more obvious. When the w/b ratio is 0.3, the compressive strength only decreases by 1.9%. When the w/b ratio increases to 0.4, the compressive strength is reduced by 34.1%. It can be seen from Figure 6 that the compressive strength of recycled mortar generally shows a downward trend as the w/b ratio increases, and the decreasing amplitude of compressive strength increases with the increase of the fly ash replacement ratio.

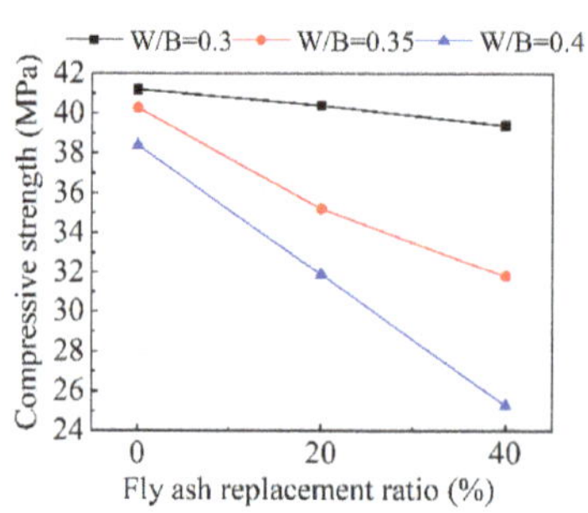

Figure 5. Variation of recycled mortar compressive strength with fly ash replacement ratio.

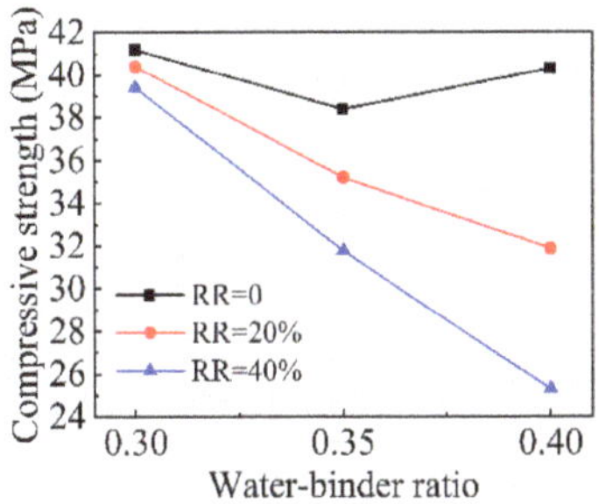

Figure 6. Variation of recycled mortar compressive strength with w/b ratio. RR represents replacement ratio of fly ash.

In summary, although fly ash can ameliorate the fluidity of mortar, it will reduce the compressive strength. Especially for mortar with a larger w/b ratio, the effect of fly ash to ameliorate the fluidity of mortar is reduced, and the reduction of compressive strength is increased. Thus, for mortar with a high w/b ratio, it is not suitable to mix a large magnitude of fly ash. In the meanwhile, as the fly ash replacement ratio is high, the compressive strength of recycled mortar also decreases more quickly as the w/b ratio increases.

4. Mix Design and Properties of FRAC

4.1. Determination of Sand Ratio

Generally speaking, the fine aggregate needs to fill the void between the coarse aggregate first, and then form mortar to wrap the surface of the coarse aggregate, so the sand ratio of FRAC can be calculated by the following two formulas.

$$V_s = \gamma \times V_g \times P_g \tag{1}$$

$$\beta_s = \frac{m_s}{m_s + m_g} \tag{2}$$

where, V_g, V_s are the volumes of the coarse aggregate and fine aggregate per cubic concrete, respectively. P_g is the void ratio of the RCA, which is 52.8% in this test. γ is the mortar rich coefficient, which is the ratio of the volume of mortar to the void volume of coarse aggregate; γ = 1.1–1.4 for ordinary concrete [28]. In the test, the value of γ is taken as 1.3.

β_s is the sand ratio. m_g, m_s are the mass of RCA and RFA per cubic FRAC. The values of m_g and m_s are equal to the product of the volume and apparent density of RCA and RFA per cubic FRAC. From Table 4, the apparent densities of RCA and RFA are 2734.8 kg/m^3 and 2594.7 kg/m^3. Putting these data into the Equation (2), β_s = 39.4%. Therefore, the sand ratio in this test is taken as 40%. Then, based on the w/b ratio and sand-cement ratio obtained from the performance test of mortar, the mix design of FRAC is calculated.

4.2. Mixture Design of FRAC

According to the optimum sand-cement ratio determined in the recycled mortar test, the sand ratio was 40%, the water consumption per cube was 200 kg, and the water-reducing agent dosage was 3% of the mass of cement. The test scheme of FRAC with a strength class of 30 MPa was designed with the w/b ratio and fly ash replacement ratio as the main test variables. The w/b ratios were 0.3, 0.35, and 0.40, and the fly ash replacement ratios were 0, 20%, and 40% for a total of nine groups of specimens. The slump of these nine groups FRAC were between 50–80 mm. The mixture design and test results of the cube compressive strength, splitting tensile strength, and flexural strength of these nine groups FRAC are shown in Table 7.

Table 7. Mixture design (kg/m^3) and strength (MPa) of recycled concrete.

NO	Specimen ID	Water	Cement	Fly Ash	S/C	RFA	RCA	WRA	f_{cu}	$f_{t,sp}$	f_{tf}
1	C0.30F0	200	600	0	1	600	900	18	42.0	2.88	5.7
2	C0.30F20	200	480	120	1	600	900	14.4	42.4	2.73	4.8
3	C0.30F40	200	360	240	1	600	900	10.8	33.2	2.59	4.0
4	C0.35F0	200	571	0	1.1	628	942	17.1	46.7	3.49	5.6
5	C0.35F20	200	457	114	1.1	628	942	13.7	43.1	3.25	5.2
6	C0.35F40	200	343	228	1.1	628	942	10.3	34.0	2.74	4.7
7	C0.40F0	200	500	0	1.3	650	975	15	35	3.05	4.2
8	C0.40F20	200	400	100	1.3	650	975	12	30.7	2.63	3.9
9	C0.40F40	200	300	200	1.3	650	975	9	26.6	2.27	3.7

Note: S/C represents sand-cement ratio, WRA represents water-reducing agent, f_{cu} represents cube compressive strength, $f_{t,sp}$ represents splitting tensile strength, f_{tf} represents flexural strength.

4.3. Compressive Strength

4.3.1. Effect of w/b Ratio

The variation of the compressive strength of FRAC with the w/b ratio is shown in Figure 7. When the w/b ratio increases, the variation trend of compressive strength is consistent. The compressive strength of FRAC increases first as the w/b ratio increases from 0.3 to 0.35, and subsequently decreases as the w/b ratio grows to 0.4. The increase in range was the highest when fly ash was not incorporated. The compressive strength of FRAC reached a climax when the w/b ratio was 0.35 and without adding fly ash. It is generally believed that the compressive strength of concrete will decrease with the increase of the w/b ratio. In this paper, the compressive strength of FRAC with a w/b ratio of 0.35 is higher than the other two types, while the compressive strength of recycled mortar with a water-binder ratio of 0.35 is lower than that of recycled mortar with a water-binder ratio of 0.3. The reason is that, when the w/b ratio is low, mortar is not sufficient to fully coat all RCA and the bonding effect between mortar and RCA is poor, thus compressive damage is mainly caused by the damage of mortar, the compressive strength of mortar, and is close to FRAC, even slightly higher than FRAC.

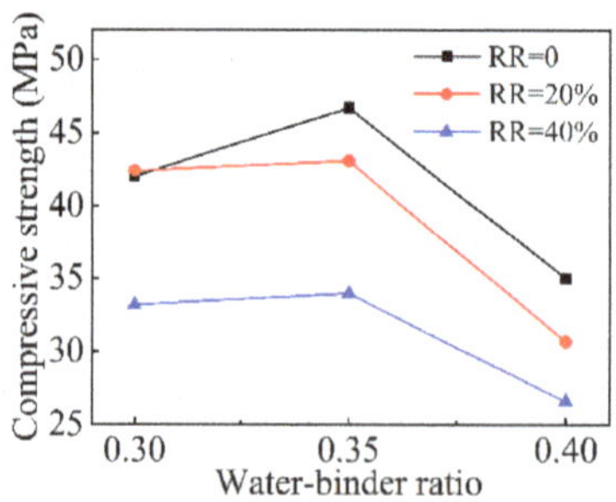

Figure 7. Variation of FRAC compressive strength with w/b ratio.

4.3.2. Effect of Fly Ash

According to the data in Table 7, the effect of the fly ash replacement ratio on the compressive strength of FRAC is shown in Figure 8. The compressive strength of FRAC decreases with the increase of the fly ash replacement ratio, except when the w/b ratio is 0.3; as the replacement ratio increases from 0 to 20%, the compressive strength of FRAC hardly changes. However, the compressive strength of FRAC with 40% fly ash decreases by 20.9%. When the w/b ratio is 0.35, the compressive strength reduces by 7.7% as the fly ash replacement ratio increases to 20%. the compressive strength of FRAC with 40% fly ash further decreases, by comparison with the FRAC without fly ash, and the compressive strength is reduced by 27.2%. When the w/b ratio increases to 0.4, the compressive strength decreases by 12.3% when the fly ash replacement ratio increases from 0 to 20%. When the replacement ratio increases to 40%, the compressive strength decreases by 24%, compared with FRAC without fly ash. Past studies have indicated that the compressive strength of concrete mixed with fly ash is quite different from that of plain concrete at 28 days, and the difference is gradually decreased at 90 days. The study by Saha et al. [35] indicated that the compressive strength of class F fly ash (calcium content 0.6%) concrete at 28 days decreased, and decreased sharply with the increase of the fly ash replacement ratio. The compressive strength of concrete with 20% and 40% fly ash at 28 days was 68.4% and 52.6% of the compressive strength of plain concrete. As the curing day increases to 90 days, the compressive strength development was 79% and 72.1% of the plain concrete compressive strength for the concrete with 20% and 40% fly ash. For FRAC, Corinaldesi et al. [36] studied the compressive strength of FRAC using 20% fly ash (calcium content 3.08%) and the cement at a 0.40 w/b ratio is 14.0% higher than plain FRAC. An investigation by Saravanakumar et al. [37] showed that the compressive strengths of FRAC mixed with 40%, 50%, and 60% fly ash (calcium content 1.07%) are 81.1%, 79.2%, and 65.8% of plain FRAC. Kurda et al. [38] calculated and fitted the compressive strength of 665 concrete mixed with fly ash and recycled aggregate. Fitting results showed that, when the replacement rate of coarse and fine recycled aggregate is 100%, the compressive strengths are 112.7% and 76.2% of the plain concrete compressive strength for the concrete with 20% and 40% fly ash. In this paper, only when the w/b ratio is 0.3, is the compressive strength of FRAC with 20% fly ash higher than that of plain FRAC. The ratio of the compressive strength of 40% fly ash FRAC to the compressive strength of plain FRAC is close to fitting the results of Kurda. The compressive strength ratios of 40% fly ash FRAC to plain FRAC under three w/b ratios are 79.0%, 72.8%, and 76.0%, respectively.

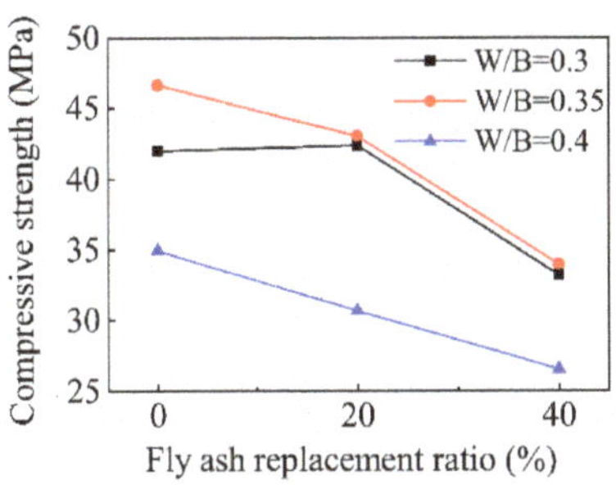

Figure 8. Variation of FRAC compressive strength with fly ash replacement ratio.

The effect of fly ash on the mechanical properties of FRAC is influenced to a large extent by its fineness and content of calcium-containing compounds. In general, high-calcium fly ash reacts faster and provides better early-strength, while low-calcium fly ash reacts slower in the early ages of the hydration reaction because of the presence of more chemically inert crystalline phases. The study by Hemalatha et al. [22] has shown that there is still unreacted fly ash and there are incompletely hydrated particles in concrete mixed with low-calcium fly ash after curing 28 days. The fineness of fly ash used in concrete also is a factor which influences the properties of concrete [39]. Chindaprasirt et al. [40] suggested that the packing and nucleation effects in the hydration reaction of fly ash mortar are mainly determined by the fineness of fly ash. Their research also showed that fine fly ash can enhance compressive strength and reduce shrinkage. On the other hand, coarse fly ash is less reactive, requires more water, and creates more pores inside the mortar, causing a decrease in the strength of mortar and concrete [41]. In this experiment, the fly ash used has a lower content (3.36%) of calcium-containing compounds and a coarser fineness, which leads to a decrease in the compressive strength of FRAC. Furthermore, the reduction of compressive strength is relatively low as the w/b ratio is 0.4, because the water is enough at this time and the reaction of the fly ash is better, so the reduction is smaller.

4.3.3. The Relationship of Compressive Strength between FRAC and Recycled Mortar

The relationship between the compressive strength of FRAC and mortar is shown in Table 8 and Figure 9. It is obvious that the compressive strength of most of the FRAC is higher than that of the recycled mortar, which illustrated that the RCA plays a good role in supporting the compressive damage process. This also shows that FRAC made by this mixture design method has good compressive strength. For recycled mortar, the compressive strength decreases as the w/b ratio increases, whereas for FRAC, the compressive strength is at a maximum when the w/b ratio is 0.35. The ratio of the compressive strength of FRAC to the recycled mortar is the largest at this w/b ratio; this is because the damage of FRAC is caused by the damage of RAC rather than mortar when the w/b ratio is 0.3.

Table 8. Ratio of compressive strength (MPa) between FRAC and mortar.

NO	Specimen ID	Compressive Strength of FRAC	Compressive Strength of Mortar	Ratio
1	C0.30F0	42.0	41.2	1.02
2	C0.30F20	42.4	40.4	1.05
3	C0.30F40	33.2	39.4	0.84
4	C0.35F0	46.7	40.3	1.16
5	C0.35F20	43.1	35.2	1.22
6	C0.35F40	34.0	31.8	1.07
7	C0.40F0	35	38.4	0.91
8	C0.40F20	30.7	31.9	0.96
9	C0.40F40	26.6	25.3	1.05

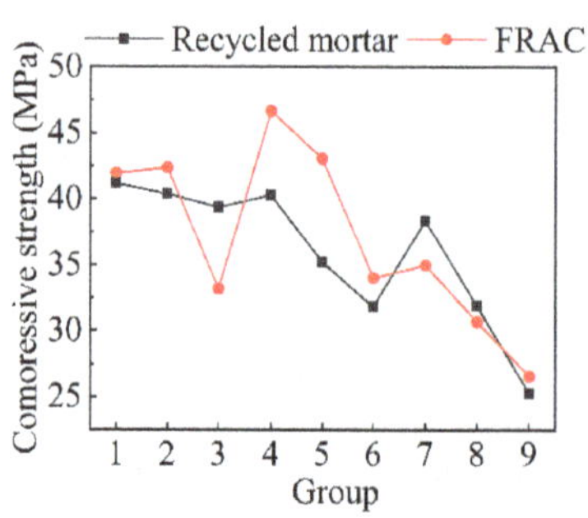

Figure 9. The relationship between mortar strength and compressive strength of FRAC.

4.4. *Splitting Tensile Strength*

4.4.1. Effect of w/b Ratio

From Figure 10, the variation between the splitting tensile strength and the w/b ratio shows a tendency of first increasing and, succeeding that, decreasing as the w/b ratio increases. When the w/b ratio increases from 0.3 to 0.35, the corresponding increasing rates of splitting tensile strength were 21.2%, 19%, and 5.8% at the fly ash replacement ratios of 0, 20%, and 40%, respectively. As the w/b ratio grows to 0.4, the corresponding decreasing rates of splitting tensile strength were 12.6%, 19.1%, and 17.2% when the fly ash contents were 0, 20%, and 40%, respectively. From Figure 11, the splitting tensile destruction section is relatively smooth, and it is obvious that the RCAs in the yellow circles are split in two along the destruction section, which shows that RCAs bear the main tensile stress in splitting tensile failure.

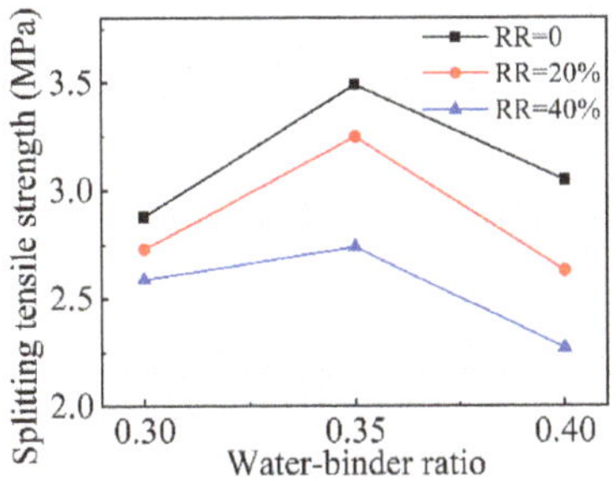

Figure 10. Variation of FRAC splitting tensile strength with w/b ratio.

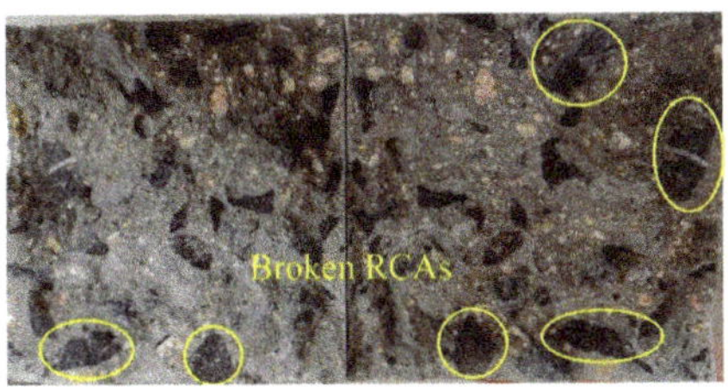

Figure 11. The typical destruction section of FRAC after splitting tensile destruction.

4.4.2. Effect of Fly Ash

Figure 12 sketches the variation trend of splitting tensile strength with the fly ash replacement ratio. It is manifest that the splitting tensile strength decreases almost linearly when the fly ash replacement ratio increases continuously. When the w/b ratio is 0.35, the splitting tensile strength value is the highest and the decrease amplitude is the largest, when the fly ash replacement ratios are 20% and 40%, compared with that without fly ash, and the splitting tensile strengths decrease by 6.9% and 21.5%, respectively. The splitting tensile strength decreases the most slowly as the w/b ratio is 0.3, compared with that without fly ash; the splitting tensile strengths decrease by 5.2% and 10.1% as the fly ash replacement ratios increase to 20% and 40%. As the w/b ratio increases to 0.4, compared to the splitting tensile strength of FRAC without fly ash, the decreased rates are 13.7% and 25.5% as the fly ash replacement ratios of FRAC are 20% and 40%, respectively. The study by Hashmi et al. [42] indicates that, as the w/b ratio is 0.45, the splitting tensile strengths of fly ash concrete with 25%, 40%, and 60% of fly ash are observed as 86%, 79%, and 72% of plain concrete at 28 days. In this paper, when the w/b ratio is 0.4, the splitting tensile strengths of fly ash concrete with 20% and 40% of fly ash are 86.3% and 74.5% of plain concrete, which is close to results of Hashmi. Kurda et al. [38] who collected and fitted tensile strengths of RAC incorporated with fly ash. Fitting results showed that the tensile strength decreased with the increasing incorporation contents of fly ash. For FRAC, the tensile strengths of FRAC with 20% and 40% fly ash are 90.7% and 84.0% of plain concrete. In this paper, when the w/b ratio is 0.35, the splitting tensile strengths of FRAC with 20% and 40% fly ash are 93.1% and 78.5% of plain FRAC, which is closest to the fitting result. Furthermore, the effect of fly ash on the splitting tensile strength of FRAC is affected by the w/b ratio. It can be seen that, when the w/b ratio is relatively low, using fly ash instead of cement has less of an effect on splitting tensile strength compared to FRAC with a high w/b ratio.

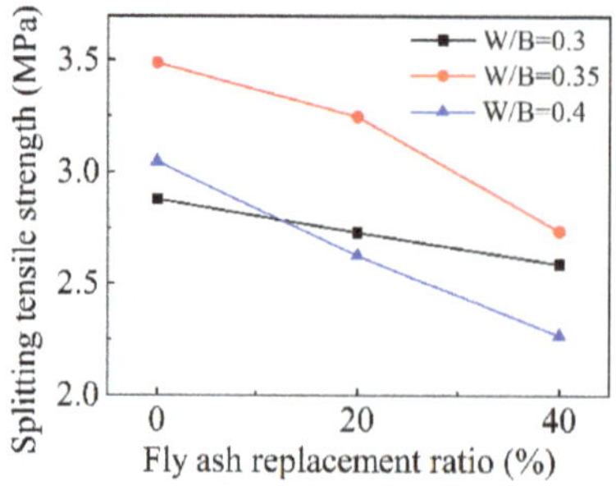

Figure 12. Variation of FRAC splitting tensile strength with fly ash replacement ratio.

4.4.3. The Relationship of Splitting Tensile Strength and Compressive Strength

The ratio of the splitting tensile strength to compressive strength of FRAC is listed in Table 9. The ratio ranges from 6.4% to 8.7%, which is similar to the result reported by Tangchirapat et al. [43], but lower than the ratio of NAC (approximately 10%). However, the ratio is influenced by the w/b ratio; with a w/b ratio of 0.4, the value reaches the maximum. The reason is that, when the w/b ratio is 0.4, the compressive strength of FRAC is much lower than other groups, while the value of the splitting tensile strength under different w/b ratios changes little, thus making the value of this ratio change more.

Table 9. Ratio of splitting tensile strength to compressive strength and relative difference.

NO	Specimen ID	f_{cu}	Experimental Values of $f_{t,sp}$	$f_{t,sp}/f_{cu}$	Calculated Values of $f_{t,sp}$	Relative Difference
1	C0.30F0	42.0	2.88	6.9%	3.16	10%
2	C0.30F20	42.4	2.73	6.4%	3.18	16%
3	C0.30F40	33.2	2.59	7.8%	2.70	4%
4	C0.35F0	46.7	3.49	7.5%	3.39	−3%
5	C0.35F20	43.1	3.25	7.5%	3.21	1%
6	C0.35F40	34.0	2.74	8.1%	2.75	0
7	C0.40F0	35	3.05	8.7%	2.80	−8%
8	C0.40F20	30.7	2.63	8.6%	2.57	−2%
9	C0.40F40	26.6	2.27	8.5%	2.33	3%

For the conversion formula for splitting tensile strength, the American Concrete Institute (ACI) [44] recommends the following:

$$f_{t,sp} = 0.59\sqrt{f_c'} \tag{3}$$

where $f_{t,sp}$ and f_c' are the splitting tensile strength and cylinder compressive strength, respectively, and the compressive strength of a standard cylinder is 1.05 times that of a cube.

The conversion formula for calculating the splitting tensile strength of ordinary concrete and high-strength concrete given by the Chinese standard is as follows [45]:

$$f_{t,sp} = 0.19 f_{cu}^{0.75} \tag{4}$$

Thus, the conversion formula between splitting tensile strength and compressive strength is assumed to be as following:

$$f_{t,sp} = A f_{cu}^{B} \tag{5}$$

The singular point in the data is removed, and the values of A and B are obtained by fitting. The value of the goodness of fit (R^2) is 0.839. Therefore, the model is valid.

The estimated value of the splitting tensile strength of FRAC can be calculated by the compressive strength by the following conversion formula:

$$f_{t,sp} = 0.265 f_{cu}^{0.663} \; (R^2 = 0.839) \tag{6}$$

The relative differences calculated by the following formula are listed in Table 9. From Table 9, it is illustrated that, when the w/b ratio is 0.35 and 0.4, and the relative difference is small, the calculated value of the fitting formula is in keeping with the experimental value.

$$\text{Releative difference} = \frac{\text{Calculated value} - \text{Experimental value}}{\text{Experimental value}} \tag{7}$$

4.5. Flexural Strength

4.5.1. Effect of w/b Ratio

The effect of the w/b ratio on the flexural strength of FRAC is shown in Figure 13. Without the admixture of fly ash, the flexural strength decreases as the w/b ratio increases. When the w/b ratio increases from 0.3 to 0.35, the flexural strength decreases slightly, and the decreasing rate is 1.8%. When the w/b ratio increases from 0.35 to 0.4, the flexural strength decreases obviously by 25%. When the fly ash replacement ratios are 20% and 40%, the flexural strengths increase first and then decrease as the w/b ratios increase. When the w/b ratio increases from 0.3 to 0.35, the flexural strengths of FRAC with 20% and 40% fly ash replacement ratios increase by 8.3% and 17.5%, respectively. In the same fly ash replacement ratio, compared to FRAC with a w/b ratio of 0.3, the flexural strengths of FRAC with a w/b ratio of 0.4 decrease significantly by 25% and 21.3%, respectively. The flexural strength of most concrete tended to decrease at low w/b ratios [46]. In this test, the relationship between flexural strength and w/b ratio is similar to that of compressive strength and also proves that the mechanical properties of FRAC at a w/b ratio of 0.35 are the most excellent.

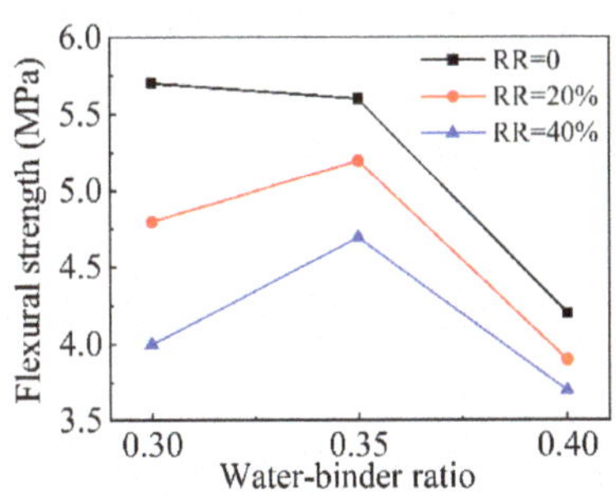

Figure 13. Variation of FRAC flexural strength with w/b ratio.

4.5.2. Effect of Fly Ash

Figure 14 sketches the effect of the fly ash replacement ratio on the flexural strength of FRAC. According to this figure, with the increase of the fly ash replacement ratio, flexural strength presents a linear downward trend, and the decline amplitude gradually slows down with the increase of the w/b ratio. When the w/b ratio is 0.3, the decrease is the most obvious, compared with that without fly ash, the flexural strength of FRAC with 20% and 40% fly ash replacement ratios decrease by 15.8% and 29.8%, respectively. When the w/b ratio is 0.35 and the fly ash replacement ratios are 20% and 40%, the flexural strengths decrease by 7.1% and 16.1%, respectively. As the w/b ratio increases to 0.4, the values of the decreasing rates of flexural strength of FRAC with 20% and 40% replacement further reduce to 7.1% and 11.9%, respectively. The use of fly ash exerts a negative influence on the flexural strength of FRAC, and the effect is more pronounced at a low w/b ratio. The study by Barbuta et al. [47] showed that when the w/b ratio of is under 0.48, the flexural strength reaches the highest value when the concrete is mixed with 10% fly ash, which is 20.3% higher than that of plain concrete. The flexural strengths of concrete with 20% and 40% fly ash are obtained as 84.1% and 57.1% of plain concrete. An investigation by Siddique et al. [48] showed that the flexural strength decreased by 39%, 48%, and 56% as the cement was replaced with 35%, 45%, and 55% of class F fly ash. In this experiment, although the flexural strength of FRAC decreases with the increase of fly ash, the decrease is far less than the results of the above two tests.

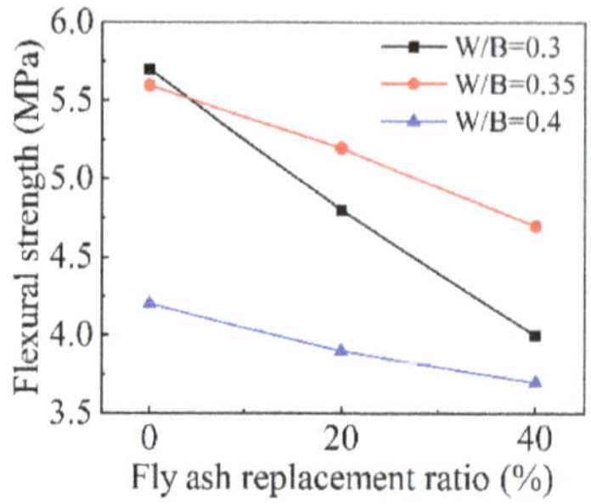

Figure 14. Variation of FRAC flexural strength with fly ash replacement ratio.

4.5.3. The Relationship of Flexural Strength and Compressive Strength

For the conversion formula between the flexural strength of ordinary concrete and the compressive strength of a standard cylinder, the American Concrete Institute (ACI) [49] recommends taking it as the following:

$$f_{tf} = 0.62\sqrt{f_c'} \tag{8}$$

Xiao and Tavakoli et al. [50,51] deem that the above standard is no longer applicable to the conversion formula between the flexural strength and compressive strength of RAC. The flexural strength and compressive strength values are linearly correlated. After a stepwise regression fitting, the flexural strength of FRAC can be expressed as the following:

$$f_{tf} = 0.25 f_{cu}^{0.80}\ (\mathrm{R}^2 = 0.832) \tag{9}$$

The R^2 value of this model is 0.832, which means that the compressive strength can explain 83.2% of the causes of variation in flexural strength. The relative difference between the calculated values obtained from this fitting formula and the experimental values is calculated by Equation (7) and included in Table 10. Most of the relative differences are within 10%, and the fitted formula is able to predict the value of flexural strength.

Table 10. Relative difference of calculated value and experimental value of flexural strength.

NO	Specimen ID	f_{cu}	Experimental Values of f_{tf}	Calculated Values of f_{tf}	Relative Difference
1	C0.30F0	42.0	5.7	4.97	−13%
2	C0.30F20	42.4	4.8	5.01	4%
3	C0.30F40	33.2	4.0	4.12	3%
4	C0.35F0	46.7	5.6	5.41	−3%
5	C0.35F20	43.1	5.2	5.08	−2%
6	C0.35F40	34.0	4.7	4.20	−11%
7	C0.40F0	35	4.2	4.30	2%
8	C0.40F20	30.7	3.9	3.87	−1%
9	C0.40F40	26.6	3.7	3.45	−7%

5. Conclusions

A novel mixture design method based on the workability of fully recycled aggregate concrete (FRAC) is proposed and verified by this experiment. The results show that the recycled mortar and FRAC produced by this method have a good strength. On this basis, the effects of the water-binder (w/b) ratio and class F fly ash on the mechanical properties recycled mortar and fully recycled aggregate concrete (FRAC) were studied. Based on the experimental results of this study, the following conclusions can be drawn:

1. The mixture design method of FRAC is mainly divided into three steps. Firstly, the optimal sand-cement ratio was determined by the expansion test of recycled mortar. Secondly, the sand ratio was calculated through the "Mortar abundant coefficient method". Finally, the amount of the FRAC material component can be directly obtained by the "Absolute volume method". This mixture design method can also be used for the preparation of recycled mortar
2. The recycled mortar produced by the new method has good fluidity and strength, and the strength can reach 85.7% to 92% of the standard mortar strength. In addition, the compressive strength ratio of FRAC to recycled mortar is mostly greater than 1, which shows that it is feasible to fabricate FRAC by this method.
3. The addition of class F fly ash can ameliorate the fluidity of recycled mortar, but it will reduce compressive strength; especially for mortar with a larger water-binder ratio, the effect to improve the fluidity is reduced, and the reduction of compressive strength is increased.
4. The mechanical properties of FRAC are best, as the water-binder ratio is 0.35. The ratio of compressive strength between recycled mortar and FRAC is from 0.84 to 1.22 for the same ratio, while the splitting tensile strength of FRAC is 6.4% to 8.7% of the compressive strength for the same ratio, which is less than that of NAC.
5. The addition of class F fly ash reduces the mechanical properties of FRAC, and the reduction increases almost linearly with the replacement ratio of fly ash. The main reason is that the calcium oxide content of fly ash is low, and the particle size is large; fly ash cannot be fully reacted at 28 days, resulting in a decrease in strength. Therefore, the strength evaluation time of FRAC with fly ash should be 90 days or longer. The effect of class F fly ash on the compressive strength and splitting tensile strength of FRAC is close to the previous experimental results, while the effect on the flexural strength of FRAC is lower than the previous experimental results. The large-scale application of class F fly ash in FRAC requires the further treatment of fly ash or the addition of other auxiliary materials.
6. The theory conversion formulas between the splitting tensile strength, flexural strength, and compressive strength of FRAC are deduced, respectively, and the results indicate that the compressive strength can explain more than 80% of the causes of variation in splitting tensile strength and flexural strength.

Author Contributions: L.Z.: methodology, writing-original draft. D.D.: investigation, formal analysis, writing, review, and editing. J.Z.: review and editing, supervision. G.Z.: review and editing. Z.W.: investigation, review and editing. All authors have read and agreed to the published version of the manuscript.

Funding: This research was funded by the National Natural Science Foundation of China (51808509) and the Program for Changjiang Scholars and Innovative Research Team in University of Minister of Education of China (IRT_16R67), the Henan science and technology research project (212102310402, 202102310281).

Institutional Review Board Statement: Not applicable.

Informed Consent Statement: Not applicable.

Data Availability Statement: All experimental data in this paper have been displayed in the Table of the paper.

Conflicts of Interest: The authors declare no conflict of interest.

References

1. Torres, A.; Brandt, J.; Lear, K.; Liu, J. A looming tragedy of the sand commons. *Science* **2017**, *357*, 970–971. [CrossRef] [PubMed]
2. Bendixen, M.; Best, J.; Hackney, C.; Iversen, L.L. Time is running out for sand. *Nature* **2019**, *571*, 29–31. [CrossRef]
3. Wang, L.; Wang, J.; Qian, X.; Chen, P.; Xu, Y.; Guo, J. An environmentally friendly method to improve the quality of recycled concrete aggregates. *Constr. Build. Mater.* **2017**, *144*, 432–441. [CrossRef]
4. Kim, J. Influence of quality of recycled aggregates on the mechanical properties of recycled aggregate concretes: An overview. *Constr. Build. Mater.* **2022**, *328*, 127071. [CrossRef]
5. Tang, Q.; Xiao, P.; Kou, C.; Lou, K.; Kang, A.; Wu, Z. Physical, chemical and interfacial properties of modified recycled concrete aggregates for asphalt mixtures: A review. *Constr. Build. Mater.* **2021**, *312*, 125357. [CrossRef]
6. Zhu, J.; Wu, S.; Zhong, J.; Wang, D. Investigation of asphalt mixture containing demolition waste obtained from earthquake-damaged buildings. *Constr. Build. Mater.* **2012**, *29*, 466–475. [CrossRef]
7. Pasandín, A.R.; Pérez, I. Mechanical properties of hot-mix asphalt made with recycled concrete aggregates coated with bitumen emulsion. *Constr. Build. Mater.* **2014**, *55*, 350–358. [CrossRef]
8. Grabiec, A.M.; Zawal, D.; Rasaq, W.A. The Effect of Curing Conditions on Selected Properties of Recycled Aggregate Concrete. *Appl. Sci.-Basel* **2020**, *10*, 4441. [CrossRef]
9. Al Ajmani, H.; Suleiman, F.; Abuzayed, I.; Tamimi, A. Evaluation of Concrete Strength Made with Recycled Aggregate. *Buildings* **2019**, *9*, 56. [CrossRef]
10. Nedeljković, M.; Visser, J.; Šavija, B.; Valcke, S.; Schlangen, E. Use of fine recycled concrete aggregates in concrete: A critical review. *J. Build. Eng.* **2021**, *38*, 102196. [CrossRef]
11. Gonçalves, P.; De Brito, J. Recycled aggregate concrete (RAC)—Comparative analysis of existing specifications. *Inst. Super. Técnico* **2010**, *62*, 339–346. [CrossRef]
12. Evangelista, L.; Guedes, M.; de Brito, J.; Ferro, A.; Pereira, M. Physical, chemical and mineralogical properties of fine recycled aggregates made from concrete waste. *Constr. Build. Mater.* **2015**, *86*, 178–188. [CrossRef]
13. Dang, J.; Zhao, J.; Hu, W.; Du, Z.; Gao, D. Properties of mortar with waste clay bricks as fine aggregate. *Constr. Build. Mater.* **2018**, *166*, 898–907. [CrossRef]
14. Dang, J.; Zhao, J. Influence of waste clay bricks as fine aggregate on the mechanical and microstructural properties of concrete. *Constr. Build. Mater.* **2019**, *228*, 116757. [CrossRef]
15. Park, K.-T.; Son, S.-H.; Han, C.-G. Effect of Recycled Aggregates Powder on the Properties of Zero Cement Mortar Using the Recycled Fine Aggregates and Fly-Ash. *J. Korea Inst. Build. Constr.* **2012**, *12*, 161–168. [CrossRef]
16. Montero, J.; Laserna, S. Influence of effective mixing water in recycled concrete. *Constr. Build. Mater.* **2017**, *132*, 343–352. [CrossRef]
17. Wagih, A.M.; El-Karmoty, H.Z.; Ebid, M.; Okba, S.H. Recycled construction and demolition concrete waste as aggregate for structural concrete. *HBRC J.* **2013**, *9*, 193–200. [CrossRef]
18. Wang, H.; Xu, J.; Liu, Y.; Sheng, L. Preparation of ceramsite from municipal sludge and its application in water treatment: A review. *J. Environ. Manag.* **2021**, *287*, 112374. [CrossRef]
19. Gupta, S.; Kua, H.W. Effect of water entrainment by pre-soaked biochar particles on strength and permeability of cement mortar. *Constr. Build. Mater.* **2018**, *159*, 107–125. [CrossRef]
20. Rodrigues, P.; Silvestre, J.D.; Flores-Colen, I.; Viegas, C.A.; de Brito, J.; Kurad, R.; Demertzi, M. Methodology for the Assessment of the Ecotoxicological Potential of Construction Materials. *Materials* **2017**, *10*, 649. [CrossRef]
21. Tripathi, R.D.; Vajpayee, P.; Singh, N.; Rai, U.N.; Kumar, A.; Ali, M.B.; Kumar, B.; Yunus, M. Efficacy of various amendments for amelioration of fly-ash toxicity: Growth performance and metal composition of Cassia siamea Lamk. *Chemosphere* **2004**, *54*, 1581–1588. [CrossRef] [PubMed]
22. Hemalatha, T.; Ramaswamy, A. A review on fly ash characteristics—Towards promoting high volume utilization in developing sustainable concrete. *J. Clean. Prod.* **2017**, *147*, 546–559. [CrossRef]
23. Şahmaran, M.; Li, V.C. Durability properties of micro-cracked ECC containing high volumes fly ash. *Cem. Concr. Res.* **2009**, *39*, 1033–1043. [CrossRef]
24. Lam, L.; Wong, Y.L.; Poon, C.S. Effect of Fly Ash and Silica Fume on Compressive and Fracture Behaviors of Concrete. *Cem. Concr. Res.* **1998**, *28*, 271–283. [CrossRef]
25. Poon, C.S.; Lam, L.; Wong, Y.L. A study on high strength concrete prepared with large volumes of low calcium fly ash. *Cem. Concr. Res.* **2000**, *30*, 447–455. [CrossRef]
26. American Society of Testing and Materials. *Standard Specification for Coal Fly Ash and Raw or Calcined Natural Pozzolan for Use in Concrete (ASTM C618-19)*; American Society of Testing and Materials: West Conshohocken, PA, USA, 2019.
27. *GB/T 1596-2017*; General Administration of Quality Supervision Inspection and Quarantine of the People's Republic of China; Standardization Administration of China. Fly Ash Used for Cement and Concrete (GB/T 1596-201). China Standards Press: Beijing, China, 2017.
28. Nayak, D.K.; Abhilash, P.P.; Singh, R.; Kumar, R.; Kumar, V. Fly Ash for sustainable construction: A review of fly ash concrete and its beneficial use case studies. *Clean. Mater.* **2022**, *6*, 100143. [CrossRef]

29. *JGJ 52-2006*; Ministry of Construction of the People's Republic of China. Standard for Technical Requirements and Test Method of Sand and Crushed Stone (or Gravel) for Ordinary Concrete (JGJ 52-2006). China Building Industry Press: Beijing, China, 2006.
30. *GB 50119-2013*; Ministry of Housing and Urban-Rural Development of China; General Administration of Quality Supervision Inspection and Quarantine of the People' s Republic of China. Code for concrete admixture application (GB 50119-2013). China Building Industry Press: Beijing, China, 2013.
31. *JGJ/T 70-2009*; Shaanxi Academy of Building Research. Standard for test method of basic properties of construction mortar (JGJ/T 70-2009). China Construction Industry Press: Beijing, China, 2009.
32. *GB/T 50081-2019*; Ministry of Housing and Urban-Rural Development of the People's Republic of China; State Administration for Market Regulation. Standard for Test Methods of Concrete Physical and Mechanical Properties (GB/T 50081-2019). China Construction Industry Press: Beijing, China, 2019.
33. Wang, G.; Kong, Y.; Sun, T.; Shui, Z. Effect of water-binder ratio and fly ash on the homogeneity of concrete. *Constr. Build. Mater.* **2013**, *38*, 1129–1134. [CrossRef]
34. Lee, C.Y.; Lee, H.K.; Lee, K.M. Strength and microstructural characteristics of chemically activated fly ash–cement systems. *Cem. Concr. Res.* **2003**, *33*, 425–431. [CrossRef]
35. Saha, A.K. Effect of class F fly ash on the durability properties of concrete. *Sustain. Environ. Res.* **2018**, *28*, 25–31. [CrossRef]
36. Corinaldesi, V.; Moriconi, G. Influence of mineral additions on the performance of 100% recycled aggregate concrete. *Constr. Build. Mater.* **2009**, *23*, 2869–2876. [CrossRef]
37. Saravanakumar, P.; Dhinakaran, G. Strength characteristics of high-volume fly ash-based recycled aggregate concrete(Article). *J. Mater. Civ. Eng.* **2013**, *25*, 1127–1133. [CrossRef]
38. Kurda, R.; de Brito, J.; Silvestre, J.D. Combined influence of recycled concrete aggregates and high contents of fly ash on concrete Properties. *Constr. Build. Mater.* **2017**, *157*, 554–572. [CrossRef]
39. Erdoğdu, K.; Türker, P. Effects of fly ash particle size on strength of portland cement fly ash mortars. *Cem. Concr. Res.* **1998**, *28*, 1217–1222. [CrossRef]
40. Chindaprasirt, P.; Jaturapitakkul, C.; Sinsiri, T. Effect of fly ash fineness on microstructure of blended cement paste. *Constr. Build. Mater.* **2007**, *21*, 1534–1541. [CrossRef]
41. Chindaprasirt, P.; Homwuttiwong, S.; Sirivivatnanon, V. Influence of fly ash fineness on strength, drying shrinkage and sulfate resistance of blended cement mortar. *Cem. Concr. Res.* **2004**, *34*, 1087–1092. [CrossRef]
42. Hashmi, A.F.; Shariq, M.; Baqi, A. Experimental and analytical investigation on the age-dependent tensile strength of low-calcium fly ash-based concrete. Department of Civil Engineering, Z.H. College of Engineering and Technology, Aligarh Muslim University, Aligarh, India. *Innov. Infrastruct. Solut.* **2021**, *6*, 72. [CrossRef]
43. Tangchirapat, W.; Rattanashotinunt, C.; Buranasing, R.; Jaturapitakkul, C. Influence of Fly Ash on Slump Loss and Strength of Concrete Fully Incorporating Recycled Concrete Aggregates. *J. Mater. Civ. Eng.* **2013**, *25*, 243–251. [CrossRef]
44. ACI Committee. *Building Code Requirements for Structural Concrete ACI 318 08 and Commentary*; American Concrete Institute: Farmington Hills, MI, USA, 2008; Volume 80.
45. *GB 50010-2015*; Ministry of Housing and Urban-Rural Development of the People's Republic of China. Code for Design of Concrete Structures (GB 50010-2015). China Construction Industry Press: Beijing, China, 2015.
46. Wille, K.; Naaman, A.E.; El-Tawil, S.; Parra-Montesinos, G.J. Ultra-high performance concrete and fiber reinforced concrete: Achieving strength and ductility without heat curing. *Mater. Struct.* **2012**, *45*, 309–324. [CrossRef]
47. Barbuta, M.; Bucur, R.; Serbanoiu, A.A.; Scutarasu, S.; Burlacu, A. Combined Effect of Fly Ash and Fibers on Properties of Cement Concrete. *Procedia Eng.* **2017**, *181*, 280–284. [CrossRef]
48. Siddique, R. Properties of concrete incorporating high volumes of class F fly ash and san fibers. *Cem. Concr. Res.* **2004**, *34*, 37–42. [CrossRef]
49. ACI Committee. *Building Code Requirements for Structural Concrete (ACI 318-05) and Commentary (ACI 318R-05)*; American Concrete Institute: Farmington Hills, MI, USA, 2004.
50. Xiao, J.; Li, J. Study on Relationships between Strength Indexes of Recycled Concrete. *J. Build. Mater.* **2005**, *8*, 197–201.
51. Tavakoli, M.; Soroushian, P. Strengths of recycled aggregate concrete made using field-demolished concrete as aggregate. *ACI Mater. J.* **1996**, *93*, 182–190.

materials

MDPI

Article

Evaluation of CBR of Graded Crushed Stone of Flexible Base Structural Layer Based on Discrete Element Model

Zhaoguang Hu [1], Heng Liu [2], Weiguang Zhang [3], Tianqing Hei [3,*], Xunhao Ding [3] and Zezhen Dong [3]

[1] China Road and Bridge Corporation, Dongcheng District, Beijing 100011, China
[2] School of Foreign Languages, Jiangsu Open University, Nanjing 210036, China
[3] School of Transportation, Southeast University, Nanjing 211189, China
* Correspondence: tqhei@seu.edu.cn; Tel.: +86-18851655528

Abstract: In order to study the mechanical properties of graded crushed stone, the discrete element method is used to simulate the CBR test of graded crushed stone. Aiming at the composition structure of graded crushed stone material, the PFC3D simulation software is used to construct the test model, and the process of constructing the virtual specimen model of the graded crushed stone discrete element model is discussed in detail. A servo mechanism is used to control the speed of the wall in the software, so as to control the virtual confining pressure imposed on graded crushed stone by the wall and simulate the real CBR test environment. The micro-parameter calibration of the virtual test is carried out by comparing the indoor and virtual CBR specimens of a single particle size specimen and three groups of graded crushed stone specimens. The comparison result shows that the stress–strain characteristics of the graded crushed rock obtained by the discrete element simulation during the uniaxial penetration process have a high degree of similarity, which can verify the accuracy of the model establishment. With the increase in the penetration depth, the penetration force of the aggregates of various particle sizes gradually increases, and the penetration force and the penetration depth are basically linear, and when the particle size is greater than 9.5 mm, the increase in particle size has little effect on the CBR test results. Under the certain conditions, the contact stiffness of graded crushed stone particles with particle sizes of 4.75 mm, 9.5 mm, 13.2 mm, 16 mm, and 19 mm should be 0.88×10^7 (N/m), 0.98×10^7 (N/m), 1.10×10^7 (N/m), 1.25×10^7 (N/m), and 2.05×10^7 (N/m), respectively. The recommended value of the contact stiffness of the small spherical particles increases with the increase in the particle size. This model can provide a basis for studying the micromechanical state of graded crushed stone and physical mechanics tests.

Keywords: graded crushed stone; gradation; the discrete element method; the CBR test

Citation: Hu, Z.; Liu, H.; Zhang, W.; Hei, T.; Ding, X.; Dong, Z. Evaluation of CBR of Graded Crushed Stone of Flexible Base Structural Layer Based on Discrete Element Model. *Materials* **2023**, *16*, 363. https://doi.org/10.3390/ma16010363

Academic Editor: Zhanping You

Received: 20 November 2022
Revised: 20 December 2022
Accepted: 21 December 2022
Published: 30 December 2022

1. Introduction

The application of graded crushed stone to the pavement structure base has many advantages, such as reducing reflection cracks and prolonging the service life of the pavement [1]. In order to obtain high-quality graded crushed stone, it is necessary to evaluate its performance with the help of the California bearing ratio CBR, resilience modulus, water permeability coefficient, and other indicators [2]. Chen [3] obtained from the actual test that the modified vibration molding method is in good agreement with the compaction degree of the actual project. Based on the indoor test of the modified vibration molding method, Chen proposed a series of performance design indexes for the design of graded crushed stone. Through a large number of repeated triaxial compression tests, Caroline [4] summarized some design rules of graded crushed stone grading for graded crushed stone design. Jiang [5] proposed a strength prediction equation for cement-graded crushed stone to calculate the strength development law of cement-graded crushed stone. Gustavo [6] presented a methodology for the experimental characterization of graded crushed stone, with a focus on mechanistic design. Xiao [7] investigated the effect of quality of graded crushed stone on conventional flexible pavement performance in Minnesota through a mechanistic–empirical pavement design approach. Zhang [8] proposed a random generation model to simulate the shear behavior of graded crushed stone.

Usually, researchers carry out virtual simulation experiments of various road materials based on particle discrete element software [9]. There are not only asphalt mixture simulations based on parametric shape and size and grading design [10], but also virtual numerical simulation tests of graded crushed stone for road bases.

Shen [11] evaluated the influence of skeleton stacking on the mechanical properties of the mixture by calculating the contact force chain and the average contact force. Chen [12] used the method of overlapping multiple spheres to characterize the irregular shape of the coarse aggregate, and quantitatively analyzed the effect of each particle size aggregate on the aggregate gap ratio, the number of contact points between aggregates, the contact force, and other parameters in the mixture after step-by-step filling.

Peng [13] established a two-dimensional meso model, and then studied the California bearing ratio test of graded crushed stone. Jiang [14] constructed a biaxial numerical test method for graded crushed stone to analyze the variation law of each index. Both Jiang and Zhang [9,15] independently established a numerical model for dynamic triaxial testing of graded crushed stone, which is used to simulate real tests, thereby reducing the number of real triaxial test models. Xiao [16] established a three-dimensional discrete element model considering the irregular shape of real particles to reveal the shear strength mechanism and evolution characteristics of skeleton hollow graded broken stone.

Cao [17] conducted a virtual triaxial test of graded crushed stone and analyzed the stress–strain relationship of the skeleton structure of graded crushed stone and the influence of gradation composition and skeleton porosity on the stress–strain curve.

Zhang [18] proposed a method to generate graded broken stone discrete element test pieces, which is used to simulate the laboratory test of the rebound modulus of graded broken stone.

Liu [19] obtained the basic law that the peak deviatoric stress of graded gravel increases with the increase in confining pressure through static triaxial virtual testing of graded gravel. The influence of grading on the elastic modulus was analyzed, and the nonlinear elastic modulus constitutive model equation was obtained by fitting.

Within the gradation range limited by the graded crushed stone base specification, the performance of the graded crushed stone is quite different [20]. This can be seen from the different CBR values of different grades of crushed stone. In order to obtain the grading of graded crushed stone with high CBR value, it is necessary to combine certain practical tests within the specified range, which will lead to a waste of more materials and time. Therefore, in order to solve these problems, it is necessary to design a numerical model to simulate a CBR test. At present, through the establishment of a virtual model, the research on the micro-effects of graded crushed stone became mature, but there is a lack of research on the macro-performance of graded crushed stone, such as CBR. This paper uses PFC3D to simulate the laboratory test of graded crushed stone, and builds a simulation model of the mechanical properties of graded crushed stone through the selection of logic language, the selection of contact model, the generation of graded particles, the realization of loading method, and the definition of material microscopic parameters. Through the comparison of a single-particle size specimen and three groups of graded crushed stone specimens with virtual CBR specimens, the micro-parameter calibration of the virtual test is carried out. The simulation analysis of this model can explain the relationship between the mechanical response state, mechanical parameters, and mechanical properties of graded crushed stone from a microscopic point of view, and provide a basis for subsequent physical and mechanical tests of graded crushed stone.

2. Basic Principles of Discrete Element Method

2.1. Force–Displacement Equation and Motion Relationship

The calculation of discrete element software is based on the most basic stress–strain relationship and the force–motion relationship with Newton's second law as the core. In each time step (small enough), the force state between particles changes, and physical quantities such as velocity, displacement, and strain are updated accordingly. These are all premised on the contact between the media and are calculated from the normal stiffness, tangential stiffness, contact overlap, and other parameters as follows:

$$F_i^n = k_i^n x_i^n n_i \tag{1}$$

$$F_i^s = -k_i^s \Delta x_i^s \tag{2}$$

$$\Delta x_i^s = V_i^s \Delta t \tag{3}$$

$$V_i^s = V_i - V_i^n \tag{4}$$

Among them, F_n^i and F_i^s are the normal and tangential components of the contact force at the contact point i, and the latter is calculated in increments; k_i^n and k_i^s are normal contact stiffness coefficients and tangential contact stiffness coefficients; x_i^n and Δx_i^s are the contact overlap and contact displacement increment at the contact point; n_i is the unit normal vector of the contact surface; V_i, V_i^n, vs. $_i$ are the contact point velocity and the normal and tangential components; Δt is the time step.

For a single particle, its motion state is calculated from the angular velocity and linear velocity, while for the entire structure or particle system, its velocity is calculated by Newton's second law, which is expressed as the following functional relationship:

$$w = f_1(a) \tag{5}$$

$$v = f_2(a) \tag{6}$$

$$V_i = f(w, v) \tag{7}$$

Among them, w and v are the angular velocity and linear velocity of the particle system; a is the acceleration of the particle system.

2.2. Contact Constitutive Model

Contacts in discrete elements can be divided into two types: "ball-ball" contact and "ball-wall" contact. The contact constitutive model is used to define the contact action. The contact constitutive model during simulation in PFC3D software (PFC3D-5.1, Itasca, IL US) mainly includes: linear contact stiffness model and slip model.

The linear contact stiffness model is a model proposed by Cundall in 1979 to characterize the interaction between aggregate particle units. The contact stiffness includes the normal stiffness coefficient k^n and the tangential stiffness coefficient k^s, and its value is determined by the stiffness coefficients of the entities A and B at both ends of the contact, as shown in Equations (8) and (9).

$$k^n = \frac{k_n^A k_n^B}{k_n^A + k_n^B} \tag{8}$$

$$k^s = \frac{k_s^A k_s^B}{k_s^A + k_s^B} \tag{9}$$

Among them, k_n^A and k_n^B are the normal contact stiffnesses of the two-end contact entities; and k_s^A and k_s^B are the tangential contact stiffnesses of the contact two-end entities.

The slip model is a contact action model for judging whether the entities at the two ends of the contact slip, and it is mainly calculated based on the normal contact force, the tangential contact force, and the friction coefficient. The formula for calculating the maximum tangential contact force that the contact point can bear is shown in Equation (10). When $|F_i^s| > F_{max}^s$, relative sliding occurs between the two contact entities.

$$F_{max}^s = \mu |F_i^n| \tag{10}$$

Among them, F_{max}^s is the maximum tangential contact force that the contact point can bear; μ is the friction coefficient; and F_i^n is the normal contact force.

3. Basic Principles of Discrete Element Method

3.1. Virtual Mold Setting

Use the "generate" command to generate aggregates with a specific gradation, and determine the number of generated particles and the number of attempts to ensure that particles do not overlap. In order to be similar to the form and size of the graded crushed stone specimen in the actual indoor test, the virtual mold adopts a cylindrical mold, which is composed of a cylindrical wall, an upper plane wall and a lower plane wall, as shown in Figure 1. Among them, the cylindrical wall is formed by splicing triangular plane walls. The more the number of triangular walls, the closer the cylindrical wall is to a smooth cylindrical surface, and the higher the calculation accuracy is. At the same time, the increase in the number will reduce the calculation efficiency. The number and precision of the triangular walls in this paper are set by the resolution parameter.

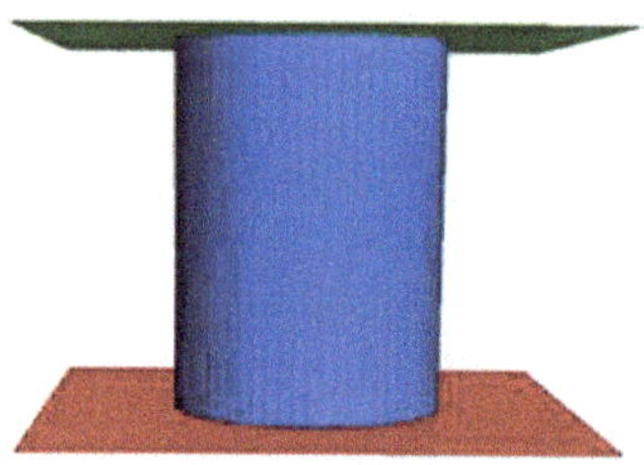

Figure 1. Schematic diagram of CBR test mold.

The diameter of the cylindrical specimen is 150 mm, which is consistent with the CBR test of the intermediate grade crushed stone in the "Geotechnical Test Method Standard GB/T 50123-2019" (MOHURD 2019) [21]. Compared with the indoor test, the random generation of aggregate particles in the virtual test requires sufficient space, so the height of the virtual mold is set to 1.6 times the height of the target specimen, and the height of the CBR test mold is set to 194 mm.

3.2. Specimen Setting of Graded Particles

The height, radius, initial void ratio, and particle size of the test piece are known, and on this basis, a digital test piece of graded crushed stone is generated. In order to better fit the actual specimen forming process, after the virtual mold is formed, the graded crushed stone specimen with the specified gradation is formed in three layers, and the thickness of each layer is equal, which is 1/3 of the final height of the specimen.

3.2.1. Calculation of The Number of Particles Corresponding to Gradation

In the actual indoor test, the gradation control is realized by calculating and weighing the specified mass of aggregates with different particle size ranges. In this paper, in the PFC software, the number of particles is also calculated according to the principle of equal mass. During the simulation, the mass relationship is converted into a volume relationship by assigning different densities to particles of different sizes. Since the graded crushed stone specimens are made of the same material, the aggregates have the same density.

According to the percentage passing of each sieve hole in the gradation, the total volume of the aggregate is allocated to each grade of aggregate according to the percentage retained, and then according to the aggregate density and the corresponding particle size to calculate the number of particles needed to generate each grade of aggregate. The formula for calculating the number of particles is shown in Equations (14)–(17).

$$V = V_0(1 - poros) \tag{11}$$

$$V_i = V\alpha_i \tag{12}$$

$$\alpha_i = P_{i+1} - P_i \tag{13}$$

$$N_i = \frac{V_i}{v_i} \tag{14}$$

$$v_i = \frac{4\pi R_{ave}^3}{3} \tag{15}$$

$$R_{ave} = \frac{d_{i+1} + d_i}{4} \tag{16}$$

In the formula, V_0, V, V_i, and v_i—the volume of the specimen, the total volume of the specimen particles (excluding voids), the total volume of the i-th grade particles, and the volume of a single particle of the i-th particle. α_i, p_{i+1}, and p_i are the volume fraction of the i-th grade particles to the specimen, respectively, and the sieve pass rate of the i + 1-th grade and the i-th grade; N_i is the number of i-th grade particles; poros is the porosity of the specimen; R_{ave}, d_{i+1}, and d_i are the average particle size of the i-th grade particles, respectively (to simplify the calculation, it is regarded as the average value of the maximum and minimum particle size of this grade of particles), in summary:

$$N_i = \frac{\pi \cdot 0.12 \cdot 0.15^2(1 - poros)\Delta P}{4} \cdot \left(\frac{4\pi\left(\frac{d_{i+1}+d_i}{4}\right)}{3}\right)^{-1} \tag{17}$$

3.2.2. Generation Process of Target Graded Crushed Stone Specimens

In order to obtain the uniform compaction effect of graded crushed stone, the model is compacted three times in the PFC software, and the number of aggregate particles generated each time is one third of the target particle number of each grade. The first (bottommost) layer of aggregate particles is formed first. The newly generated particles are in a loose state and have a large void ratio, occupying the space from the bottom of the mold to 1/2 the height of the mold. When generating particles, according to the actual situation when generating particles, specify the number of attempts through the tries command.

In this paper, vibration is used for particle compaction to ensure more reasonable distribution of aggregate particles of different sizes and closer contact [22,23]. The particles are given a gravity of 9.8 m/s^2. The frequency of 5 Hz was selected for vibration, and the vibration velocities of 5 cm/s and 1 cm/s were firstly applied in the z and x directions, and then the vibration velocities of 5 cm/s and 1 cm/s were applied in the z and y directions to realize the redistribution of the pellets. After the vibration is completed, a wall parallel to the mold base is generated at the highest point of the particles as a loading plate, and the loading plate is given a downward speed to form a load on the aggregate particles, simulating the actual graded crushed stone compaction process, until the aggregate particles are pressed to the specified height. After the compaction is completed, the loading plate is given a slight upward speed to move it up slowly until it is not in contact with the aggregate particles and is then removed to provide space for the second layer of compaction.

Repeat the above steps to generate the second layer and the third layer of aggregate particles. The generation range of the second layer of aggregate particles is from the top of the first layer of aggregate to 0.75 times the height of the mold, and the load plate presses it to 2/3 of the height of the test piece. The generation range of the third layer aggregate particles is from the top of the second layer aggregate to the height of the mold, and the load plate presses it to the height of the target specimen. After the above steps, the virtual test piece of the target cylinder is finally formed, and the specific molding process of the test piece is shown in Figure 2.

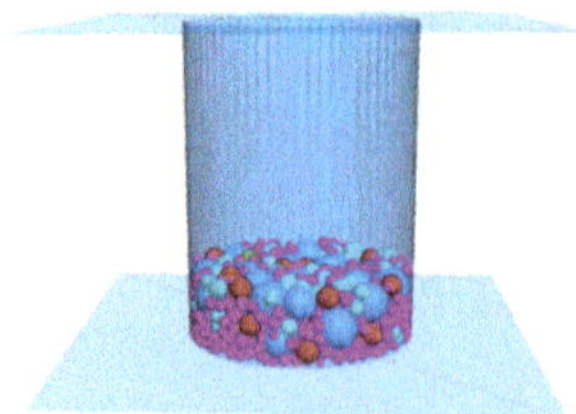

(**a**) After the first level is compacted with crushed stone particles.

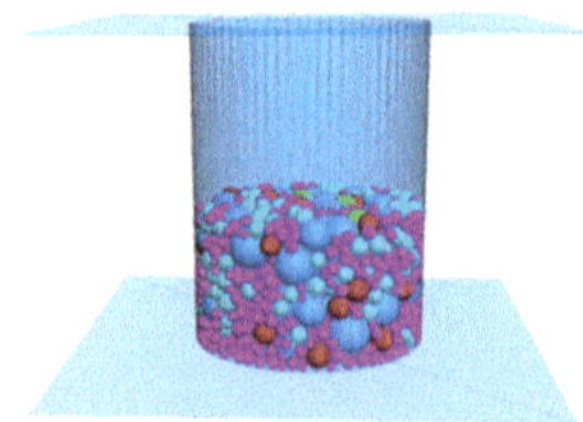

(**b**) Just generated second level with gravel particles.

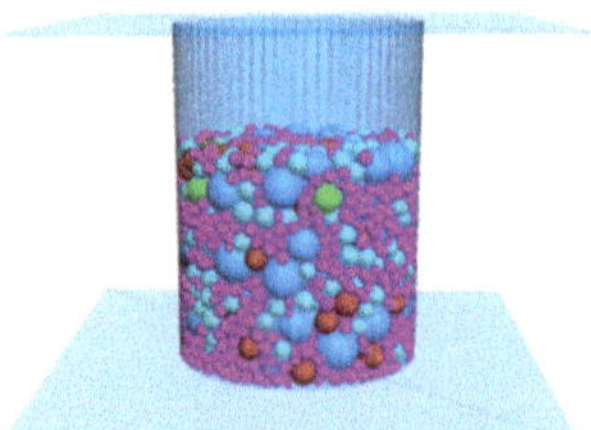

(**c**) The third level is equipped with crushed stone particles after compaction.

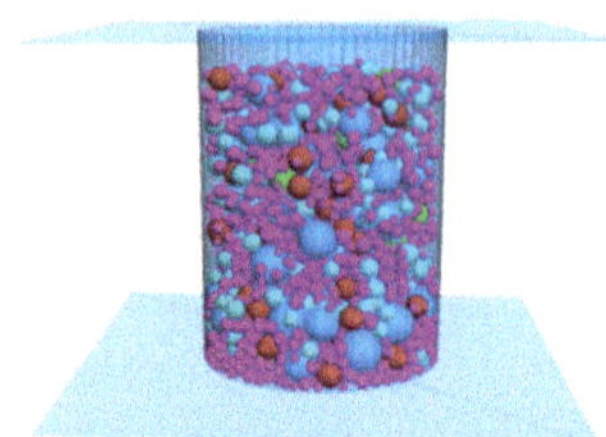

(**d**) Forming of graded crushed stone specimens.

Figure 2. Graded gravel specimen forming process.

3.3. Setting of Wall Speed and Confining Pressure

The speed of the wall is controlled by a servo mechanism to ensure that the wall exerts a specific pressure, which is used to simulate the compaction process of graded crushed stone [24]. The confining pressure is realized by two parameters: target confining pressure and allowable error. Servo control the wall speed and calculate the stress on the wall in real time and compare it with the required applied stress. If the two errors are less than the allowable error parameter, it indicates that the confining pressure was achieved. If the two errors exceed the allowable error parameter, the cycle will continue until the requirements are met.

The above comparison of wall velocity and stress error is carried out continuously during the virtual test. Assuming that there is a corresponding relationship between the wall velocity and the load applied to the wall, the wall velocity can be expressed as:

$$\dot{u}^{(wall)} = G\left(\sigma_{measured} - \sigma_{required}\right) = G\Delta\sigma \tag{18}$$

Among them, $\dot{u}^{(wall)}$ is the wall velocity; $\sigma_{measured}$ is the measured stress value; $\sigma_{\mathrm{required}}$ is the predetermined stress value; and $\Delta\sigma$ is the stress difference. G is the parameter of the servo mechanism, and the dimension is dim G = $L^2M^{-1}T$, which is used to control the speed of the wall.

The wall and the sphere are in contact with each other, and the wall moves in the direction of the sphere to receive the resistance force from the sphere, and the sphere receives the reaction force from the wall at the same time. In the time step Δt, the reaction force increment, of $\Delta F^{(wall)}$ of the wall acting on the sphere is:

$$\Delta F^{(wall)} = k_n^{(average)} N_c \dot{u}^{(wall)} \Delta t \tag{19}$$

Among them, $\Delta F^{(wall)}$ is the increment of the force on the wall; $k_n^{(average)}$ is the average stiffness of the contact between the wall and the sphere; N_c is the number of contacts between the wall and the sphere; $\dot{u}^{(wall)}$ is the wall velocity; and Δt is the time step.

According to the stress calculation formula, the calculation formula of the stress increment applied by the wall is:

$$\Delta\sigma^{(wall)} = \frac{\Delta F^{(wall)}}{A} = \frac{k_n^{average} N_c \dot{u}^{(wall)} \Delta t}{A} \tag{20}$$

Among them, $\Delta\sigma^{(wall)}$ is the stress increment applied by the wall; and A is the area of the wall where the load is applied.

In order to ensure that the expected load is applied to the wall, the value of $\Delta\sigma^{(wall)}$ should be strictly controlled. In this paper, the accuracy of the stress applied to the wall is controlled by taking the safety factor $\alpha(\alpha < 1)$, that is, the increment $\Delta\sigma^{(wall)}$ of the stress applied to the wall should be less than the difference $\Delta\sigma$ between the actual measured stress of the wall and the predetermined stress of the wall:

$$\left|\Delta\sigma^{(wall)}\right| = \alpha|\Delta\sigma| \tag{21}$$

Substitute Equations (18) and (20) into Equation (21) to get:

$$\alpha|\Delta\sigma| = \frac{k_n^{(average)} N_c G |\Delta\sigma| \Delta t}{A} \tag{22}$$

Eliminate $|\Delta\sigma|$ in Equation (22), we get:

$$G = \frac{\alpha A}{k_n^{(average)} N_c \Delta t} \tag{23}$$

3.4. Simulation of CBR Loading Conditions

During the CBR test, four pairs of semi-circular ring load blocks with a diameter of 150 mm and a center hole diameter of 52 mm should be placed above the specimen, and each pair weighs 1.25 kg, as shown in Figure 3a. Use MATLAB (2020a, MathWorks, Natick, MA, USA) software to generate circles with a diameter of 150 mm and a diameter of 52 mm on the x-y plane, divide the circle into 36 equal parts, and derive the coordinates of each bisected point. In the PFC software, the wall create command is used to input the coordinates of the above-mentioned equal points, and the generated small triangles finally form a circular wall, as shown in Figure 3b. After the annular load plate is generated, the above-mentioned servo method is used to simulate the gravity effect of the load block based on the wall stress.

(**a**) The actual effect generated in the PFC software. (**b**) The actual effect generated in the PFC software.

Figure 3. Schematic diagram of CBR test circle load block.

A small cylinder with a diameter of 5 cm and a height of 5 cm and a disc with a diameter of 5 cm were combined as the indenter, and the initial position was located above the center of the specimen, as shown in Figure 4. Considering the actual running speed and calculation time, the penetration speed of the virtual test is set to 2 mm/min and adjusted according to the subsequent loading conditions. After loading, with the monitored penetration depth as the control condition, when it reaches 8 mm, the loading stops.

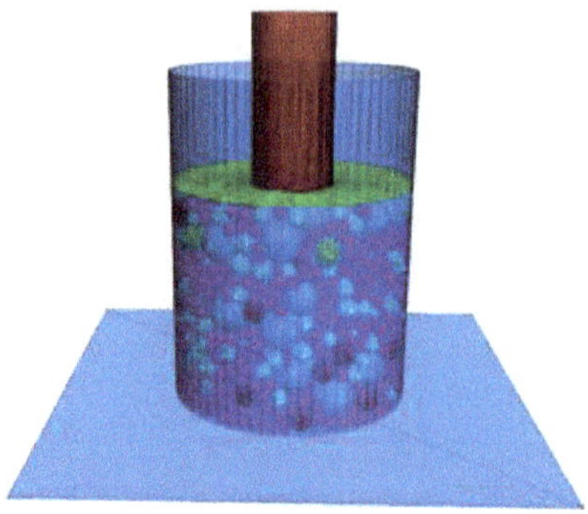

Figure 4. Virtual CBR test loading schematic.

4. Calibration of microscopic parameters

4.1. Indoor CBR Test

Carrying out indoor CBR experiments can provide data support for determining the micro-parameter correction of discrete element CBR in virtual experiments.

4.1.1. Material selection and specimen grading

Taking into account factors such as material and cost, limestone is selected as the research object. The following test pieces were formed in the laboratory test:

(1) Prepare coarse aggregate specimens with a single particle size, including 4.75 mm, 9.5 mm, 13.2 mm, 16 mm, and 19 mm;

(2) The middle value and upper and lower limits of grading type G-A-4 in the "Highway Pavement Base Construction Technical Specifications" (JTG/T F20-2015) [25] shall be selected, the fine aggregate part of the grading type shall be removed, and the proportion of the part above 4.75 mm shall be expanded. The gradation of the indoor test specimens is shown in Table 1.

Table 1. Specimen grading for indoor test.

Gradation Types	Mass Percentage (%) Passing Through the Following Square Mesh Sieve (mm)					
	26.5	**19**	**16**	**13.2**	**9.5**	**4.75**
Gradation one	100	82	70	60	40	0
Gradation two	100	75	63	52	34	0
Gradation three	100	70	57	44	27	0

4.1.2. Forming Steps of The Test Piece

In this study, the static compaction method was used to form graded crushed stone specimens with a size of 150 mm in diameter and 120 mm in height. Before static pressing, put the crushed stone into the test tube in three steps, and strike 10 times with a small hammer each time to make the aggregate initially compact. After putting all the crushed stone into the test tube, it was placed under a press to form under static pressure, with a load of 15 KN and a static pressure of 10 s, as shown in Figure 5a.

(**a**) Hydrostatic forming test piece. (**b**) CBR test.

Figure 5. Indoor CBR test procedure.

4.1.3. Indoor CBR Test

The formed specimen was placed on the digital CBR tester, and four pairs of semi-circular load blocks were placed on the surface of the specimen. Adjust the height of the lifting platform on the bottom of the test piece and the position of the sensor so that the sensor is just placed on the edge of the test tube at the beginning of the test. Set the loading speed of the loading indenter to 1 mm/s, set the loading stop condition as the penetration depth reaching 8 mm, and carry out the CBR test (see Figure 5b). Three indoor tests are repeated for each single particle size, and the final results are taken as the average of the three tests. After the test, the instrument will automatically output the relationship between the loading indenter pressure and the penetration depth (with the penetration depth of 0.5 mm as the interval) to draw the stress–displacement curve. The indoor CBR test results are shown in Figure 6.

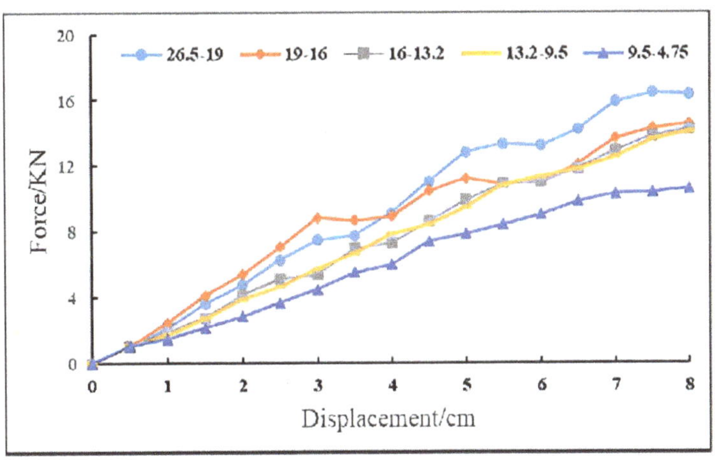

Figure 6. Indoor CBR test results.

Observing the test curve, it can be seen that as the penetration depth (displacement) increases, the penetration force of each grade of particle size aggregate increases gradually, and the penetration force and penetration depth are basically linear. This is because as the penetration progresses, the gaps between the graded crushed stone particles gradually decrease, and the entrapment effect is strengthened. To obtain the same displacement, the force required by the load head must increase [26]. Among the coarse aggregates with a single particle size in the fifth grade, except for the 4.75 mm particle size aggregate, which has a slightly smaller penetration force, the penetration force of other particle sizes is not much different. The penetration force of the 19 mm particle size aggregate is slightly larger than that of other particle sizes when the penetration depth is greater than 4 mm. Therefore, it is not recommended to blindly pursue large-sized aggregate skeletons in order to increase the load bearing capacity when selecting the gradation of crushed stone.

4.2. Calibration Results of Microscopic Parameters

In order to make the virtual specimen of discrete element model simulate the actual test situation, the contact stiffness, friction coefficient and other microscopic parameters are calibrated by comparing the penetration force penetration depth curves of virtual CBR test and indoor CBR test. The number of virtual tests is not certain. The error between the virtual test results and the indoor test results is controlled within 5% through multiple iterations, so as to calibrate the contact stiffness, friction coefficient, and other microscopic parameters of the discrete element entities (walls and balls). Combined with studies [27,28], it is found that among the microscopic parameters of normal phase stiffness, tangential stiffness, and friction coefficient between the sphere and the wall, the normal contact stiffness of the sphere has the greatest influence on the penetration force. Therefore, in this paper, the normal stiffness and tangential stiffness of the wall are set to 1×10^{10} (N/m), the friction coefficient is 0.7, and the spherical friction coefficient is 0.5. By changing the normal and tangential stiffness of the sphere (general normal stiffness = tangential stiffness), the virtual test results closest to the laboratory test results are obtained. The selection of parameters is shown in Table 2.

Table 2. Preliminary table of micro-parameter values.

	Sphere	Wall
Normal stiffness (N/m)	$2 \times 10^5/2 \times 10^6/2 \times 10^7/2 \times 10^8$	1×10^{10}
Tangential stiffness (N/m)	$2 \times 10^5/2 \times 10^6/2 \times 10^7/2 \times 10^8$	1×10^{10}
Friction coefficient	0.5	0.7

The virtual uniaxial penetration tests with 4.75 mm, 9.5 mm, 13.2 mm, 16 mm, and 19 mm grain sizes were carried out according to the steps of virtual CBR test model construction using different spherical method/tangential stiffness in the table, and the comparison results with the indoor tests are shown in Figure 7.

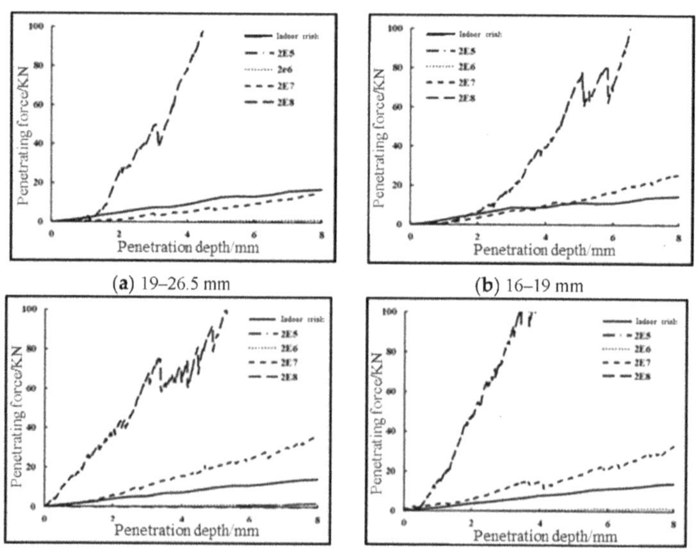

(**a**) 19–26.5 mm (**b**) 16–19 mm (**c**) 13.2–16 mm (**d**) 9.5–13.2 mm

Figure 7. *Cont.*

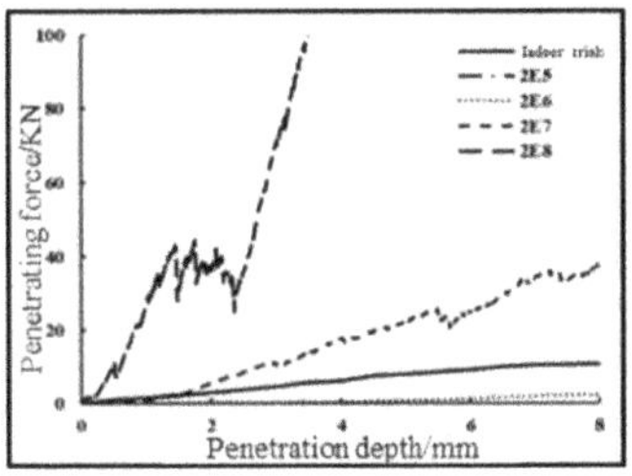

(**e**) 4.75–9.5 mm

Figure 7. Comparison of indoor and virtual CBR test results for single particle size specimens with different microscopic parameters.

It can be seen that the magnitude of the penetration force obtained from the virtual test is positively correlated with the ball stiffness, and when the ball stiffness is 2×10^8 (N/m), the penetration force obtained from the simulation is much larger than the actual value, which is 15–25 times that of the actual value, and is undesirable. When the sphere stiffness is 2×10^5 (N/m) and 2×10^6 (N/m), the simulated penetration force is much smaller than the actual value, which is about 1/1000 and 1/100 of the actual value, which is not desirable.

When the sphere stiffness is 2×10^7 (N/m), the simulated penetration force is slightly smaller than that of the indoor test for 19 mm particle size graded aggregates; for 16 mm particle size graded aggregates, the simulated penetration force is slightly larger than that of the indoor test; for 13.2 mm, 9.5 mm, and 4.75 mm particle size graded aggregates, the simulated penetration force is larger than that of the indoor test. The simulated penetration force for 13.2 mm, 9.5 mm, and 4.75 mm graded aggregates is greater than the indoor test results.

Therefore, based on the results of the above analysis, the microscopic parameters were further optimized. For the graded crushed stone with the particle size of 19mm and 16mm, the values larger than and smaller than 2×10^7 shall be selected for further verification, and 0.05×10^7 shall be taken as the change step of the stiffness of the small ball each time. For the other three groups of particle size test pieces, take 2×10^6 and 2×10^7 as the sphere stiffness value interval, and continuously use the dichotomy method to approximate the final value of the sphere stiffness. Compare the simulation results with the indoor test results, when the difference between the two is less than 5% of the indoor test results, the parameter optimization is stopped, and the small ball stiffness is considered as the final value of small ball stiffness for the corresponding particle size at this time.

The optimization results of the microscopic parameters for different particle sizes of graded aggregates are shown in Table 3, and the comparison between the virtual CBR test results of graded aggregates using the microscopic parameters in the table and the indoor CBR test results is shown in Figure 8, and the errors are within the permissible range. It can be seen that the results of the virtual test and the indoor test are roughly consistent. There are two reasons for the deviation. Firstly, that during the indoor test, the data are read every 0.5 mm and are connected into a curve due to the equipment limitation, while in the virtual test, the results are recorded every step, which can be regarded as an almost continuous curve. Secondly, the position of each particle with the same grading is different, which leads to the lateral slip of gravel particles at different positions during the penetration process, resulting in up and down fluctuations at different positions of the curve.

Table 3. Micro-parameter values.

	Sphere					Wall
Aggregate Size (mm)	**4.75**	**9.5**	**13.2**	**16**	**19**	**All**
Normal stiffness (N/m)	0.88×10^7	0.98×10^7	1.10×10^7	1.25×10^7	2.05×10^7	1×10^{10}
Tangential stiffness (N/m)	0.88×10^7	0.98×10^7	1.10×10^7	1.25×10^7	2.05×10^7	1×10^{10}
Friction coefficient			0.5			0.7

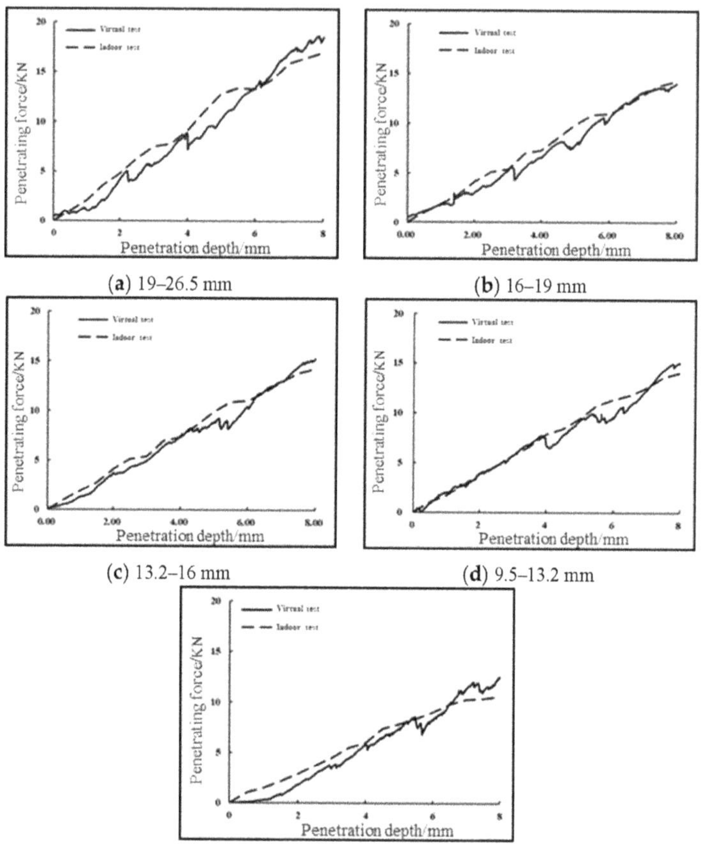

(**a**) 19–26.5 mm (**b**) 16–19 mm (**c**) 13.2–16 mm (**d**) 9.5–13.2 mm (**e**) 4.75–9.5 mm

Figure 8. Final comparison of indoor and virtual CBR test results for single particle size specimens.

In order to further verify the correctness of the calibrated microscopic parameters, three groups of gradations of the upper, middle, and lower limits of G-A-4 are obtained in "Technical Rules for Construction of Highway Pavement Base" (JTG/T F20-2015) (MOT 2015). The fine aggregate part is then removed. Next, the coarse aggregate part is enlarged in equal proportion, and the virtual CBR test is carried out using the microscopic parameters shown in the Table 3, and the results are compared with the indoor test results. For the penetration force–penetration depth curve of the virtual CBR test and the indoor CBR test, the error of the penetration forces of gradation one, gradation two, and gradation three under different penetration depths shall not exceed 5%, 15%, and 10%, respectively. The results are shown in Figure 9.

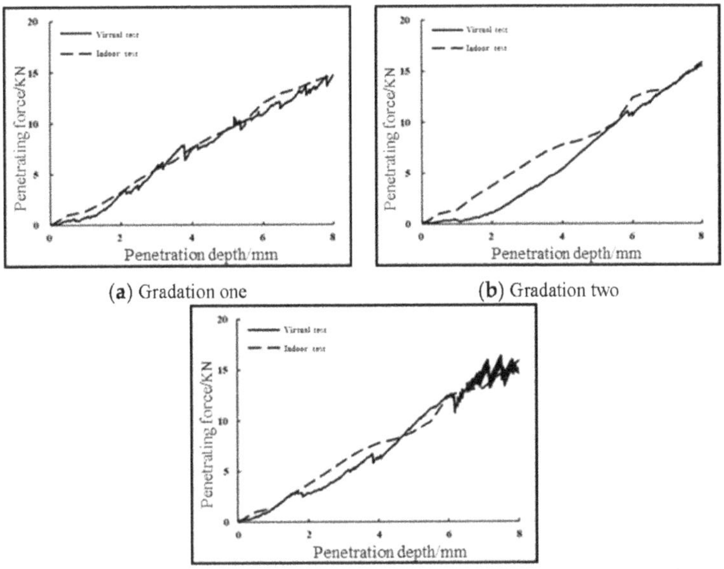

(**a**) Gradation one (**b**) Gradation two

(**c**) Gradation three

Figure 9. Comparison of indoor and virtual CBR test results for the G-A-4 three-group grading.

According to the comparison between the virtual test and the indoor test, it can be seen that the numerical test results of CBR test can be well matched with the indoor test results. The correctness of the calibrated micro-parameters in Table 3 is further verified, so the stress–strain characteristics during the uniaxial penetration of graded crushed stone can be predicted relatively quickly through indoor simulation tests.

5. Conclusions

In this paper, the virtual model of graded crushed stone is improved from the interaction of micro-graded crushed stone particles to the performance of macro-graded crushed stone, and the performance prediction model of two-dimensional graded crushed stone CBR is extended to the performance prediction model of three-dimensional graded crushed stone CBR, which provides a basis for the physical and mechanical tests of graded crushed stone in the future.

The discrete element software PFC3D is used to simulate the main calculation principle, specimen forming method, and loading method of the test of graded crushed stone, and combined with the indoor CBR test, the microscopic parameters of the virtual test are calibrated. The main research conclusions are as follows:

(1) According to the non-bonding characteristics of graded crushed stone, when simulating in PFC3D software, the linear stiffness model and the slip model are used. When the specimen is simulated and formed, enough space is reserved for the compaction process, and the height of the specimen mold wall is 1.6 times the target height of the specimen. The aggregate particles are put in three times. After each time, gravity should be applied first, then vibration should be applied, and then a pressure plate should be formed for compaction. After compaction, it is recommended to slowly remove the pressure plate to prevent large disturbance of the particles.

(2) The pressure cannot be directly applied in the PFC3D software, so the servo mechanism is used to apply a constant load through the relationship between the servo parameter G, speed, and stress. The loading of the CBR test is a strain-controlled mode, which is accomplished by imparting a constant speed to the indenter.

(3) As the penetration depth (displacement) increases, the penetration force of each grade of particle size aggregate gradually increases, and the penetration force and penetration depth are basically linear. According to the relationship between the penetration force of coarse aggregates with a single particle size of five grades, when the particle size is larger than 9.5 mm, the increase in particle size has little effect on the CBR test results. Therefore, it is not necessary to blindly pursue a large particle size in order to obtain a larger bearing capacity.

(4) Combined with the indoor CBR test, the micro-parameter calibration method of the discrete element simulation test of graded crushed stone is proposed. Among the microscopic parameters, the contact stiffness of the spherical particles has the greatest influence on the simulation results. The calibration found that when the normal and tangential contact stiffness of the wall is 1e10, the friction coefficient of the wall is 0.7, and the friction coefficient of the sphere is 0.5, the contact stiffness of graded crushed stone particles with particle sizes of 4.75 mm, 9.5 mm, 13.2 mm, 16 mm, and 19 mm should be 0.88×10^7, 0.98×10^7, 1.10×10^7, 1.25×10^7, and 2.05×10^7, respectively.

Author Contributions: Conceptualization, Z.H. and H.L.; Methodology, T.H.; Validation, Z.D and W.Z; Investigation, X.D.; Resources, X.D.; Data Curation, Z.H.; Writing—Original Draft Preparation, T.H.; Writing—Review & Editing, Z.D.; Visualization, W.Z.; Supervision, Z.H.; Project Administration, H.L.; Funding Acquisition, H.L. All authors have read and agreed to the published version of the manuscript.

Funding: This research is supported by the National Key Research and Development Project, under Grant No. 2020YFA0714302, the project is also supported by the National Natural Science Foundation of China with grant number of 52278443.

Institutional Review Board Statement: Not applicable.

Informed Consent Statement: Not applicable.

Data Availability Statement: Some or all data, models, or code that support the findings of this study are available from the corresponding author upon reasonable request, including specific dimensional stiffness.

Conflicts of Interest: The authors declare no conflict of interest.

References

1. Wang, L.; Feng, D.C. Methods to improve the use performance of graded gravel base. *China J. Highw. Transp.* **2016**, *4*, 40–45.
2. Jiang, Y.L.; Li, D.; Ma, Q.W.; Cao, H.H.; Ren, J.L. Experimental study on the factors influencing the mechanical properties of graded aggregates. *J. Transp. Sci. Eng.* **2010**, *26*, 6–13.
3. Chen, L.B.; Yan, E.H.; Xu, J.; Ma, T.; Zeng, J.C.; Gong, Y. Research on Design Indicators for Graded Crushed Stone Mixture Based on Vibration Molding Method. *Adv. Mater. Sci. Eng.* **2020**, *2020*, 5179563. [CrossRef]
4. Lima, C.; Motta, L. Study of Permanent Deformation and Granulometric Distribution of Graded Crushed Stone Pavement Material. *Procedia Eng.* **2016**, *143*, 854–861. [CrossRef]
5. Falla, G.C.; Leischner, S.; Blasl, A.; Erlingsson, S. Characterization of unbound granular materials within a mechanistic design framework for low volume roads. *Transp. Geotech.* **2017**, *13*, 2–12. [CrossRef]
6. Xiao, Y.; Tutumluer, E.; Siekmeier, J. Mechanistic-Empirical Evaluation of Aggregate Base and Granular Subbase Quality Affecting Flexible Pavement Performance in Minnesota. *Transp. Res. Rec.* **2011**, *2227*, 97–106. [CrossRef]
7. Jiang, Y.; Wang, H.; Yuan, K.; Li, M.; Yang, M.; Yi, Y.; Fan, J.; Tian, T. Research on Mechanical Properties and Influencing Factors of Cement-Graded Crushed Stone Using Vertical Vibration Compaction. *Materials* **2022**, *15*, 2132. [CrossRef]
8. Zhang, J.; Li, J.; Yao, Y.; Zheng, J.; Gu, F. Geometric anisotropy modeling and shear behavior evaluation of graded crushed rocks. *Constr. Build. Mater.* **2018**, *183*, 346–355. [CrossRef]
9. Zhang, D.; Huang, X.M.; Tian, F. Experimental study on the factors influencing the mechanical properties of graded aggregates. *J. Highw. Transp. Res. Dev.* **2014**, *31*, 39–42+49.
10. Li, J.; Zhang, J.; Qian, G.; Zheng, J.; Zhang, Y. Three-Dimensional Simulation of Aggregate and Asphalt Mixture Using Parameterized Shape and Size Gradation. *J. Mater. Eng. C* **2019**, *31*, 04019004. [CrossRef]
11. Shen, S.; Yu, H. Characterize packing of aggregate particles for paving materials: Particle size impact. *Constr. Build. Mater.* **2011**, *25*, 1362–1368. [CrossRef]
12. Chen, J.; Huang, X.M. Evaluation of skeletal structure of aggregates using discrete element method. *J. Southeast Univ. Nat. Sci. Ed.* **2012**, *42*, 761–765.

13. Peng, A.P.; Li, L.; Zhang, Z. Study on the meso-mechanical behavior of California bearing ratio of graded gravel with discrete element method. *J. Railw. Sci. Eng.* **2019**, *16*, 2467–2474. [CrossRef]
14. Jaing, Y.J.; Li, S.W.; Chen, Z.J.; Fang, J. Biaxial Numerical Simulation on Graded Broken Stone and Its Influence Factors. *J. Tongji Univ. Nat. Sci. Ed.* **2015**, *43*, 1045–1050.
15. Jiang, Y.-J.; Ni, C.-Y.; Zhang, Y.; Yi, Y. Numerical investigation of the plastic deformation behaviour of graded crushed stone. *PLoS ONE* **2021**, *16*, e0258113. [CrossRef] [PubMed]
16. Xiao, Y.J.; Wang, X.M.; Yu, Q.D.; Hua, W.J.; Chen, X.B. Discrete Element Modeling of Shear Strength Evolution Characteristics of Unbound Permeable Aggregate Base Materials. *J. China Highw. Eng.* **2022**, 1–23. Available online: http://kns.cnki.net/kcms/detail/61.1313.U.20220414.1050.002.html (accessed on 15 April 2022).
17. Cao, M.M.; Tan, Q.Q.; Liu, J.L.; Huang, W. Methods to improve the use performance of graded gravel base. *East Road* **2019**, *1*, 109–114.
18. Zhang, H.L.; Chen, G.B.; Zhang, Z.X. Simulations on the Triaxial Repeated Load Tests of Graded Crushed Stones. In *International Workshop on Energy and Environment in the Development of Sustainable Asphalt Pavements*; Highway School, Chang'an University: Xi'an, China; China Railway Engineering Consultants Group Co., Ltd.: Zhengzhou, China, 2010.
19. Liu, J.L. Analysis of Structural Behavior of Inverted Base Asphalt Pavement. Master's Thesis, Southwest Jiaotong University, Chengdu, China, 2016.
20. Yuan, J.; Shao, M.H.; Huang, X.M. Grade selection of graded aggregates. *Highway* **2005**, *4*, 40–45.
21. *GB/T 50123-2019*; Standard for Soil Method. Ministry of Housing and Urban-Rural Development of the People's Republic of China: Beijing, China; Standards Press of China: Beijing, China, 2019.
22. Wang, L.; Xie, X.G.; Li, C.J. Comparative vibration and compaction molding tests on the performance of graded aggregates. *China J. Highw. Transp.* **2007**, *6*, 19–24.
23. Zhou, W.F.; Li, Y.W.; Zhang, X.L.; Ma, B. Road performance and design criteria of graded aggregates based on vibration forming. *J. Chongqing Jiaotong Univ. Nat. Sci.* **2009**, *28*, 559–564.
24. Liu, H.T.; Cheng, X.H. Discrete element analysis of the size effect of coarse-grained soils. *Rock Soil Mech.* **2009**, *30*, 287–292.
25. *JTG/T F20-2015[S]*; Technical Guidelines for Construction of Highway Roadbases. Ministry of Transport of the People's Republic of China: Beijing, China; China Communications Press: Beijing, China, 2015.
26. Xu, G.H.; Gao, H.; Wang, Z.R. Analysis of continuous dynamic monitoring of vibratory compaction process of graded aggregates. *Chin. J. Geotech. Eng.* **2005**, *11*, 35–37.
27. Jiang, Y.J.; Li, S.C.; Wang, T.L. Numerical simulation method for dynamic triaxial test of graded aggregates. *J. Southeast Univ. Nat. Sci. Ed.* **2013**, *43*, 604–609.
28. Li, S.C. Numerical Simulation of Granular Flow for the Shakedown Behavior of Graded Aggregates. Master's thesis, Chang'an University, Xi'an, China, 2013.

materials

MDPI

Article

Effects of Thermal Treatment on the Mechanical Properties of Bamboo Fiber Bundles

Jie Cui [1,2], Daixin Fu [2,3], Lin Mi [2,3], Lang Li [2,3,*], Yongjie Liu [2,3], Chong Wang [2], Chao He [2], Hong Zhang [2], Yao Chen [2] and Qingyuan Wang [1,2,3,*]

1 Institute of New Energy and Low-Carbon Technology, Sichuan University, Chengdu 610065, China
2 Failure Mechanics and Engineering Disaster Prevention, Key Laboratory of Sichuan Province, Sichuan University, Chengdu 610065, China
3 Key Laboratory of Deep Earth Science and Engineering, Ministry of Education, Sichuan University, Chengdu 610065, China
* Correspondence: lilang@scu.edu.cn (L.L.); wangqy@scu.edu.cn (Q.W.)

Abstract: Bamboo is known as a typical kind of functional gradient natural composite. In this paper, fiber bundles were extracted manually from various parts of the stem in the radial direction, namely the outer, middle, and inner parts. After heat treatment, the mechanical properties of the fiber bundles were studied, including the tensile strength, elastic modulus, and fracture modes. The micromechanical properties of the fiber cell walls were also analyzed. The results showed that the mean tensile strength of the bamboo fiber bundles decreased from 423.29 to 191.61 MPa and the modulus of elasticity increased from 21.29 GPa to 27.43 GPa with the increase in temperature. The elastic modulus and hardness of the fiber cell walls showed a positive correlation with temperature, with the modulus of elasticity and the hardness increasing from 15.96 to 18.70 GPa and 0.36 to 0.47 GPa, respectively. From the outside to the inside of the bamboo stems, the tensile strength and elastic modulus showed a slight decrease. The fracture behavior of the fiber bundles near the outside approximates ductile fracture, while that of the bundles near to the inside tend to be a brittle fracture. The fracture surfaces of the bamboo bundles and the single fibers became smoother after heat treatment. The results show that bamboo fiber bundles distributed near the outside are most suitable for industrial development under heat treatment at 180 °C. Therefore, this study can provide a reasonable scientific basis for the selective utilization, functional optimization, and bionic utilization of bamboo materials, which has very important theoretical and practical significance.

Keywords: moso bamboo; thermal treatment; fiber bundles; facture behavior; tensile properties

Citation: Cui, J.; Fu, D.; Mi, L.; Li, L.; Liu, Y.; Wang, C.; He, C.; Zhang, H.; Chen, Y.; Wang, Q. Effects of Thermal Treatment on the Mechanical Properties of Bamboo Fiber Bundles. *Materials* **2023**, *16*, 1239. https://doi.org/10.3390/ma16031239

Academic Editor: Jānis Andersons

Received: 11 January 2023
Revised: 27 January 2023
Accepted: 29 January 2023
Published: 31 January 2023

1. Introduction

In the past several decades, fiber-reinforced composites have been widely used in various construction fields, transportation fields, furniture items, and many other fields [1]. Glass and carbon fibers are popular due to their good mechanical and thermal properties [2]. However, these fibers require complex preparation processes and high costs, which can cause serious environmental problems especially when the products are scrapped [3]. Therefore, natural fibers are considered to be a more sustainable alternative for structural composites [4]. As a kind of natural plant fiber, bamboo fiber is an ideal fiber reinforcement material due to its easy availability and low cost, as well as its high strength and aspect ratio [5]. Because of its great significance in environmental protection and resource conservation, bamboo fiber has gradually become a current research hotspot [6].

From a biomaterial perspective, the gradient composite structure of bamboo is a masterpiece of nature from the micro- to macro-scale, as shown in Figure 1. On the cell scale, each sublayer of the cell wall can be regarded as a nanofiber-reinforced composite, with cellulose microfibrils acting as the reinforcement and flexible hemicellulose and lignin acting as the matrix; on the tissue scale, bamboo fibers and parenchymal cells with multiple wall layers are formed by laminar composites between cell wall sublayers [7]; and on the macro-scale, bamboo fibers are bonded to form fiber sheaths by middle lamellar and together with vessels and other conducting tissues to form vascular bundles, which are distributed in flexible parenchyma tissues in forming two-phase composites, as shown in Figure 1b During the reinforcing phase, vascular bundles determine almost all of the mechanical properties, and their distribution density increases continuously from the inside to the outside in the radial direction, which makes bamboo material have the characteristic of being a mechanical gradient functional material [8].

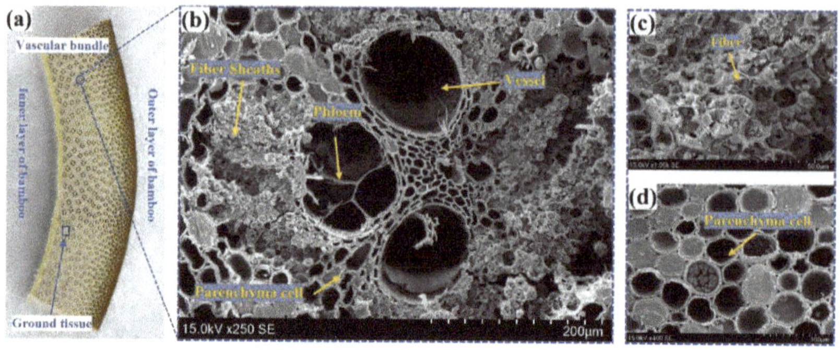

Figure 1. The macroscopic and microscopic structure of moso bamboo: (**a**) radial section of bamboo; (**b**) microstructure from the cross-section; (**c**) fiber sheaths observed from the cross-section; (**d**) parenchymal cells observed from the cross section.

For the development of natural bamboo fiber composites, currently, it is crucial to study the macroscopic and microscopic mechanical properties of bamboo fibers as well as their chemical properties. Currently, bamboo fibers are extracted by chemical and steam explosion methods [9], as shown in Table 1. Chen et al. [10] tested the tensile strength and elastic modulus of single bamboo fibers extracted by a chemical method and the results were 1.77 GPa and 26.85 GPa, respectively, while the tensile strength and elastic modulus of the fiber bundles were 0.61 GPa and 23.56 GPa, respectively. The tensile strength and elastic modulus of the bamboo fibers extracted by the steam explosion method were 0.62 to 0.86 GPa and 35.45 GPa, respectively [11]. However, the tissue structure and chemical composition of bamboo fibers are inevitably affected by either chemical or steam explosion treatments [2], which affects their mechanical properties. Shao et al. [12] prepared tensile specimens of bamboo vascular bundles without physical and chemical treatments and measured their tensile strength and elastic modulus to be 0.48 GPa and 33.9 GPa, respectively, but the paper and glue deformed slightly during the stretching process which affected the overall experimental results. Meanwhile, Li et al. [13] investigated the effect of radial distribution on vascular bundles extracted by alkali treatment, and the results showed that the longitudinal tensile strength and elastic modulus of vascular bundles increased linearly from the inside to the outside of the bamboo stem.

In practical applications, heat treatment is generally used to improve dimensional stability, mold resistance, and durability [14]. The heat treatment of bamboo can be divided into steam heat treatment, oil heat treatment, and air or inert gas treatment [15]. In the current study, the main focus is on the effects of heat treatment on chemical composition and less so on the macroscopic and microscopic mechanical properties of bamboo. Azadeh et al. [16] tested the mechanical properties of bamboo in six temperature ranges from 100 °C to 225 °C and found no significant changes in mechanical properties below 150 °C, while dimensional stability and mildew resistance were improved. The relative crystallinity of cellulose increases as well as the mechanical properties after steam heat treatment at 180 °C. For example, the hardness of the cell wall increased from 0.69 GPa to 0.84 GPa [17] and the elastic modulus increased from 17.3 GPa to 21.2 GPa [18].

Table 1. Extraction methods and mechanical properties of bamboo fibers and fiber bundles.

Materials	Extraction Method	Tensile Strength (GPa)	Elastic Modulus (GPa)	References
Bamboo fibers	Chemical method	1.77	26.85	[10]
	Steam explosion method	0.62–0.86	35.45	[11]
Bamboo fiber bundles	Chemical method	0.61	23.56	[10]
	Manual extraction method	0.482	33.9	[12]

In summary, bamboo fibers have a wide range of applications and good prospects for development. In addition, heat treatment technology can improve the mechanical properties of bamboo fiber cell walls while effectively improving dimensional stability. According to previous studies, the traditional fiber bundles tensile test is performed with paper as the clamping section, which will produce relative slip. Moreover, data on the mechanical properties of bamboo fiber bundles after heat treatment are very scarce due to increased experimental difficulty as the bamboo fiber bundles become brittle after heat treatment. In addition, the effect of heat treatment on the mechanical properties of fiber bundles in terms of radial distribution needs to be explored further.

In this study, natural bamboo fiber bundles were extracted manually, and the conventional tensile test method was optimized. The macroscopic and microscopic mechanical properties, chemical composition, fracture mode, and the change law of the desired properties in the radial direction of bamboo fiber bundles after heat treatment were investigated by nanoindentation, FTIR, SEM, and other characterization methods. This study improves the accuracy of the fiber tensile tests and helps to select the appropriate bamboo fiber part and heat treatment temperature in the production of bamboo fibers, which has some referential significance for optimizing the production of bamboo-fiber-reinforced composites and the future development of advanced composites.

2. Materials and Methods

2.1. Specimen Preparation

3-year-old moso bamboo from Sichuan Province China was adopted as the raw material. Bamboo stems at a height of about 2.9 m to 3.2 m above the ground were selected, as shown in Figure 2a. After harvesting, the stem was dried naturally for 90 days under laboratory conditions (25–33 °C and 60–70% relative humidity), and the final moisture content was measured to be 8%. Then, the stem was split into 10 mm-wide strips and divided into three equal parts in the radial direction (Figure 2b). According to the location of the parts, they were named as the outer layer, middle layer, and inner layer. To separate the fiber bundles from the stem, the strips were squeezed repeatedly with a roller and then soaked in distilled water for one week; then, the fiber bundles in the bamboo strips were manually peeled out.

To measure the hardness and elastic modulus of the fibers in different parts, the test area of the nano-indentation specimen was 10 × 5 mm (10 mm was the radial direction), and the thickness of the bamboo block was 3 mm.

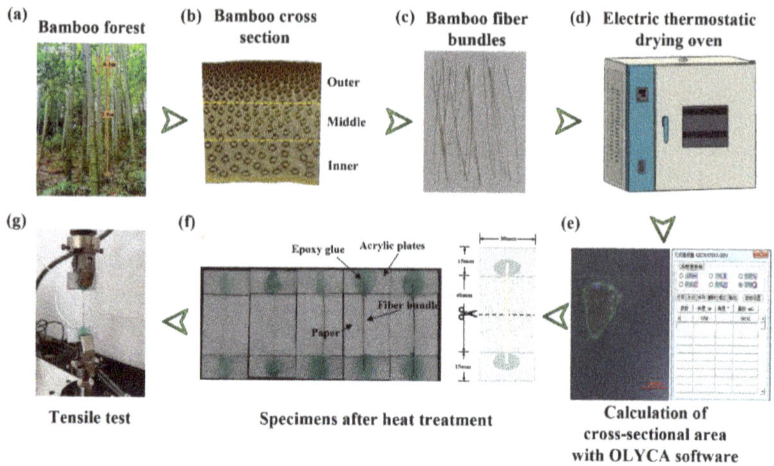

Figure 2. Processing steps and experimental process.

2.2. Heat Treatment

The chemical composition of bamboo remains unchanged when the temperature is lower than 160 °C and the highest application temperature of heat treatment at present is 220 °C [15]. Therefore, 160 °C, 180 °C, 200 °C, and 220 °C were chosen as the heat treatment temperatures. Before heat treatment, the above specimens were heated to 65 °C in the electric thermostatic drying oven for 24 h. Then, the dry specimens were subjected to the target treatment temperature for one hour. After heat treatment was completed, the temperature was naturally cooled down to the ambient temperature in the oven.

2.3. Tensile Test

Fiber bundles with a length of 60 mm and uniform thickness were selected, and the cross-section of fiber bundles was photographed with an optical microscope (Olympus GX53, Olympus Corporation, Tokyo, Japan). Then, the cross-sectional area of each fiber bundle was analyzed with OLYCA m3 software, as shown in Figure 2e. The two ends of the fiber bundle were glued to 30 mm × 15 mm × 1 mm acrylic plates with epoxy resin glue (3M DP6310, Minnesota Mining and Machinery Manufacturing Company, Saint Paul, USA) and the bonding depth was 10 mm. Therefore, the standard distance section of the tensile specimen was 40 mm. The other two acrylic plates were glued to both ends with acrylic plate special glue, and the fiber bundles were sandwiched in the two acrylic plates to minimize slippage during the stretching process. A piece of paper in the size of 70 mm × 30 mm was glued between the two acrylic plates to protect the fiber bundle from being broken and then cut before loading, as shown in Figure 2f. The test equipment was a fatigue tester (MMT-250, Shimadzu, Kyoto, Japan) from the brand Shimadzu with a loading capacity of 250 N. The loading speed of the specimens was 0.12 mm/min. Six specimens were tested for each group of specimens. According to the statistics, this method could make 96.7% of the fiber bundles break in the effective area, which greatly improved the experimental success rate.

2.4. Nanoindentation Test

A nanoindentation test (KLA iNano, KLA Corporation, Chandler, AZ, USA) performed by a Berlovich indenter was used to study the hardness and elastic modulus of the bamboo fibers. Before testing, the surface of the specimen was finely polished. In all of the tests, the maximum load was 500 μN, and the rate of loading and unloading was 50 nm/s. The load was held for 6 s at the maximum load, which was used to eliminate the effect of creep. Forces and displacements were recorded simultaneously to obtain load–depth curves. Parameters such as elastic modulus and hardness were calculated according to the Olive–Pharr method by the following equation [19]:

$$H = \frac{P}{A} \quad (1)$$

$$A = 24.56h_c^2 \quad (2)$$

$$h_c = h - \varepsilon\frac{P}{S} \quad (3)$$

$$E_r = \frac{\sqrt{\pi}}{2\beta} \times \frac{S}{\sqrt{A}} \quad (4)$$

$$\frac{1}{E_r} = \frac{1-v^2}{E} + \frac{1-v_i^2}{E_i} \quad (5)$$

where P is the peak indentation load; A is the projected area of the indentation at the maximum indentation depth; h_c is the contact depth derived from the synthetic load–displacement curve; h is the maximum penetration depth; ε is the geometric constant of the indenter. For conical indenters, the empirical value of ε is 0.75; S = dP/dh (stiffness) is the slope of the upper half of the unloading phase of the load–displacement curve; E_r is the fold modulus; commercial nanoindenters are usually taken as β = 1.034 for Berkovich indenters; the elastic modulus E_i and Poisson's ratio v_i of diamond indenters are 1440 GPa and 0.07, respectively [20]. To calculate E (the elastic modulus of the sample), it is necessary to know the Poisson's ratio v of the sample in the direction of the test. In general, the elastic modulus is not sensitive to the value of the Poisson's ratio, and the cell walls of moso bamboo are much softer than diamonds, which leads to it taking no account of the effect of the Poisson's ratio v of the sample [21]. Each probe indentation point was on a single fiber, and 30 valid data were selected for each of the outer, middle, and inner areas.

2.5. SEM Test

A scanning electron microscope (Hitachi SU3500, Tokyo, Japan) was used to examine the fracture morphology of the fiber bundle specimens under an accelerating voltage of 15 kV. The samples were gold coated and fixed on the carrier table with conductive adhesive.

2.6. FTIR Test

Fiber bundles after heat treatment at different temperatures were collected and an FTIR analysis was conducted to analyze the chemical composition change. Firstly, the bamboo was grinded to powder with a grinding bowl; then, the powder was dealt with by a tablet machine to obtain tablet samples for FTIR analysis. The analysis was conducted with FTIR from the brand Invenio at Sichuan University. The scanning speed was set to 16 times per minute and the resolution was 4 cm^{-1}. The spectra within the wavenumber of 400~4000 cm^{-1} were recorded.

3. Results and Discussions

3.1. Tensile Strength and Elastic Modulus

The stress–strain curves, the tensile strength and elastic modulus of the bamboo fiber bundles at different heat treatment temperatures are shown in Figure 3 and Table 2. The curves at each temperature showed nearly perfect straightness characteristics with no slip or yielding bending segments. Therefore, the fracture of moso bamboo fiber bundles was a brittle fracture. The mean tensile strength of the unheated specimens was 423.29 MPa and the strength decreased gradually with the increase in temperature, and finally, it decreased to 191.61 MPa at 220 °C. However, the elastic modulus increased with the increasing treatment temperatures, which gradually increased from 21.29 GPa for the untreated specimen to 27.43 GPa for the treated specimen at 220 °C. In addition, the fastest increase in the elastic modulus was observed after 200 °C. In contrast, the two mechanical property differences of moso bamboo in the radial direction (Table 2) showed an increasing trend of both tensile strength and elastic modulus from the inside to the outside.

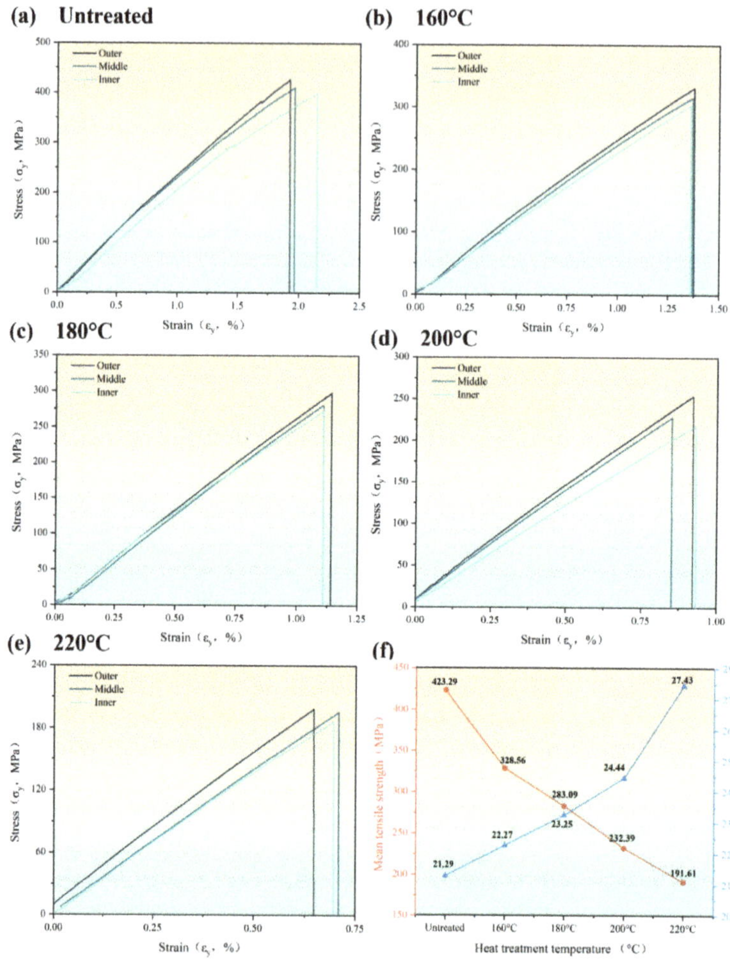

Figure 3. Tensile behavior of the untreated specimen: (**a**) the stress–strain curve of the untreated specimen; (**b**) the stress–strain curve of the specimen at 160 °C; (**c**) the stress–strain curve of the specimen at 180 °C; (**d**) the stress–strain curve of the specimen at 200 °C; (**e**) the stress–strain curve of the specimen at 220 °C; (**f**) the mean tensile strength and elastic modulus of bamboo bundles.

The above tensile experiments were performed for the macroscopic mechanical properties of bamboo fiber bundles. To further analyze the differences and changes in the internal mechanical properties of bamboo materials, the elastic modulus and hardness of individual fiber cell walls in different parts before and after heat treatment were measured by nanoindentation tests, and the results are shown in Figure 4.The elastic modulus and hardness of bamboo fibers were positively correlated with the temperature, and the elastic modulus and hardness of the fiber cell walls increased by 17.17% and 30.56% from the initial 15.96 GPa and 0.36 GPa values at 220 °C, respectively. In addition, the growth rate of both mechanical properties after 200 °C and the changes in the radial direction were also consistent with the results of tensile experiments. The increase in the elastic modulus of microscopic fiber cell walls from the inside to the outside of the bamboo diameter is essentially responsible for the increase in the elastic modulus of macroscopic fiber bundles. Moreover, the intercellular layer, which has weak adhesive strength, did not affect the fiber bundles composed of multiple fiber cells due to temperature changes, and the individual fibers still showed a consistent pattern of change with the fiber bundles.

Table 2. Tensile strength and elastic modulus at different heat temperatures.

Temperature (°C)	Area in Radial Section	Mean Tensile Strength of Fiber Bundles (MPa)	Mean Elastic Modulus of Fiber Bundles (GPa)	Mean Elastic Modulus of Fiber Cell Wall (GPa)	Mean Hardness of Fiber Cell Wall (GPa)
	Outer	442.36	22.75	16.58	0.37
Untreated	Middle	421.93	20.87	15.82	0.35
	Inner	400.46	20.26	15.48	0.36
	Outer	322.49	23.70	17.09	0.40
160 °C	Middle	314.41	22.26	16.69	0.37
	Inner	310.69	21.01	16.26	0.36
	Outer	295.94	24.05	17.48	0.43
180 °C	Middle	280.58	23.88	16.88	0.41
	Inner	272.24	22.44	16.52	0.38
	Outer	246.38	25.33	18.02	0.44
200 °C	Middle	228.53	24.95	17.39	0.42
	Inner	217.90	23.31	17.21	0.37
	Outer	193.17	28.00	19.14	0.48
220 °C	Middle	190.43	27.36	18.67	0.47
	Inner	185.68	26.41	18.27	0.46

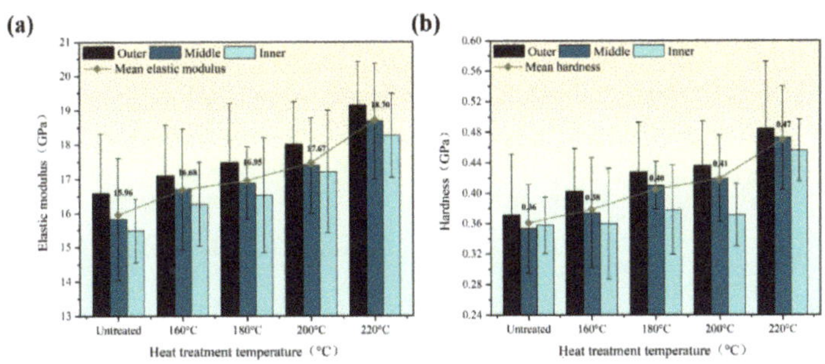

Figure 4. The micromechanical properties of the fiber cell walls in different areas in the radial direction: (**a**) elastic modulus and (**b**) hardness.

According to previous studies, the tensile strength of untreated moso bamboo fiber bundles ranges from 290–610 MPa and the elastic modulus ranges from 15.85–28.5 GPa [12,22,23], so the test results of this experiment were relatively accurate. The reasons for the discrepancies were not only due to basic factors such as the age, height, and origin of the bamboo, but also the differences in extraction methods and testing methods. Traditional chemical and physical extraction methods lead to changes in fiber cell structure, and the specimen used by this paper as the gripping end inevitably led to slippage during the tensile process. The mean elastic modulus and hardness of untreated bamboo cell walls ranges from 10.4–17.3 GPa and 0.22–0.49 GPa, respectively [17,18,24], and the results of this experiment were also within reasonable limits. The elastic modulus values obtained from the nanoindentation were smaller than those measured in the tensile experiments. On the one hand, the nanoindentation test only measured the elastic modulus of the cell wall of the fiber part; on the other hand, the surface of bamboo is relatively fragile, and the cell wall would be damaged inevitably in the process of sample preparation, resulting in a small measurement value. It has also been shown that the widely used Oliver–Pharr analysis method greatly underestimates the longitudinal elastic modulus of anisotropic plant cell walls [25], but this discrepancy is not a major concern of this work.

In addition, the mechanical properties and change law of bamboo fiber bundles after heat treatment were studied for the first time, while ensuring the accuracy of the experimental results and increasing experimental precision. This study provides favorable reference data for the later development of bamboo-fiber-reinforced fiber composites and the application of heat treatment technology.

3.2. FTIR Analysis

In previous studies, it has been shown that cellulose is the determining factor affecting the longitudinal mechanical properties of fibers [26], while lignin contributes to the elastic modulus [27]. The study of the changes in chemical functional groups and the compositional structure of bamboo fiber bundles during different heat treatments was of great importance for the analysis of their mechanical properties.

Hemicellulose was easier to decompose than any other chemical component due to its amorphous structure [18], and the non-conjugated C=O stretching vibration of hemicellulose xylan was represented at a peak of 1737 cm^{-1} [28], as shown in Figure 5. The intensity of the characteristic peak decreased with increasing temperature, indicating the degradation of hemicellulose. Cellulose has characteristic peaks at 3340 cm^{-1} (O-H stretching vibration) and 900 cm^{-1} (C-H bending vibration). The intensity of the peaks at 3340 cm^{-1} and 900 cm^{-1} decreased slightly but not significantly with increasing temperature, showing that the degradation degree of cellulose was small, which was due to the good stability of cellulose [29]. The main reason for the degradation of cellulose was that a large number of hydroxyl groups in the amorphous region of bamboo cellulose underwent an oxidation reaction under the combined action of high-temperature heat treatment and oxygen-containing conditions, which leads to the polymerization of the free hydroxyl groups and the formation of aldehyde, ketone, or carboxyl groups [15]. Due to the relatively stable benzene ring structure of lignin [30], the degradation of cellulose and hemicellulose after heat treatment and lignin condensation reactions thus led to an increase in the relative lignin content [31]. The intensity of the absorption bands of lignin at 1512 cm^{-1} (C-H aromatic skeleton vibration [32]) and 1242 cm^{-1} (C-O stretching vibration [33]) remained essentially unchanged until 200 °C and only started to decrease at 220 °C.

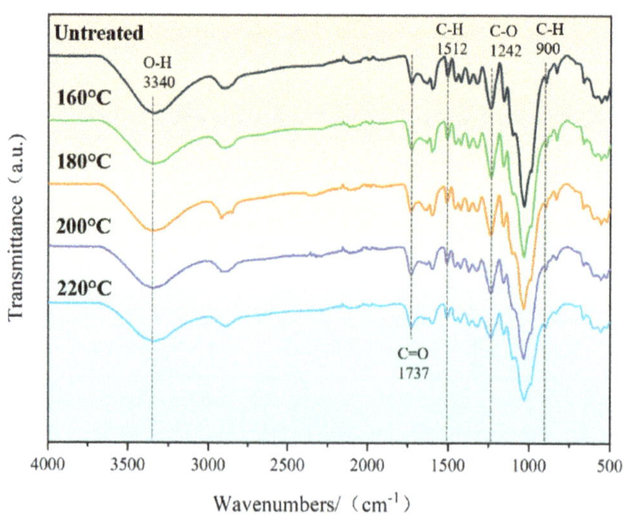

Figure 5. FTIR analysis of untreated and heat-treated specimens.

As indicated by the FTIR analysis, both cellulose and hemicellulose underwent some degree of degradation with the increase in heat treatment temperature, which led to a significant decrease in the tensile strength of the fiber bundles. The increase in the relative lignin content was chemically responsible for the increase in the elastic modulus of the fiber bundles. Due to the decomposition of hemicellulose during heat treatment, the elastic modulus of the cell wall increased, which also resulted in a reduction of hydrogen bonding on the fiber surface, thus enhancing the micromechanical properties of the bamboo cell wall [18].

3.3. *Analysis on Fracture Surface*

Tensile damage of bamboo fiber bundles begins with relative sliding between cell walls. On a molecular scale, the abundance of hydroxyl groups in the cellulose molecular chains [34] leads to relative sliding between the cell walls of bamboo fibers involving abundant hydrogen bond formation, breakage, and recombination. As the load continues to increase and the stress reaches a level sufficient to break the covalent bonds, the internal covalent bonds between C-C and C-O receive disruption and larger molecules become smaller ones. After that, the stress continues to increase, with the cellulose crystalline region disrupted and the main skeleton disintegrating at the molecular level, which eventually leads to cell wall destruction and single fiber breakage. To better study the damage modes of different specimens, SEM was used to record the fracture of each specimen.

From the fracture surfaces of the specimens (Figure 6), before 200 °C, the fracture of the specimens near the outside was serrated, accompanied by a large amount of fiber extraction, and the fracture behavior could be regarded as a ductile fracture, while the fracture of the specimens near the inside was relatively neat and resembled a brittle fracture. After 200 °C, the fracture of the fiber bundles in different areas showed a flatter shape, and only a small number of fibers were extracted from the fracture near the outside of the tensile specimen. For the same part of the bamboo, the higher the heat treatment temperature, the smoother the fracture surface of bamboo bundles.

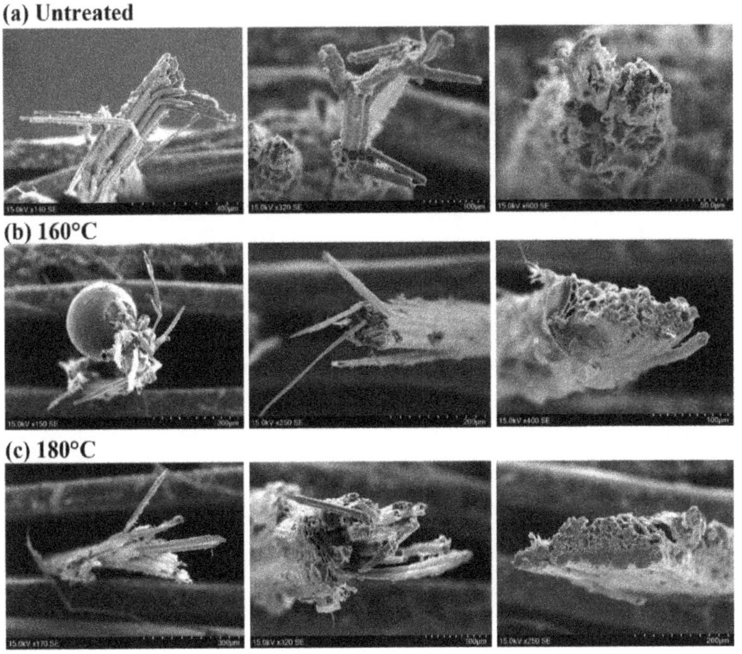

Figure 6. *Cont.*

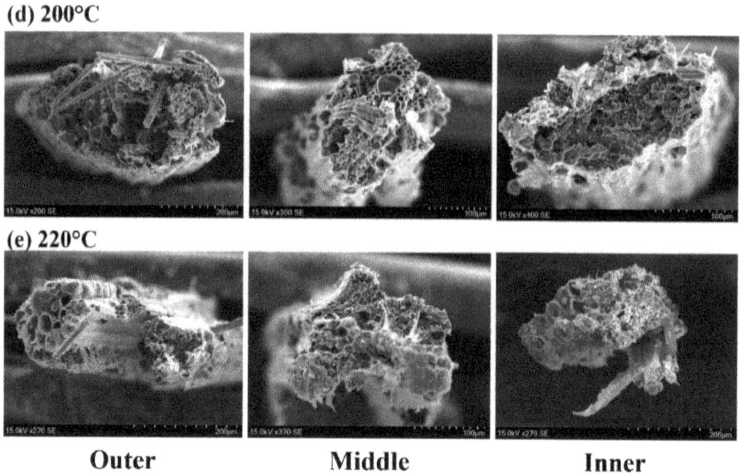

Figure 6. Overview of fracture surfaces in different areas in the radial direction with different temperatures.

The fracture modes of moso bamboo single fibers can be classified into two modes: multilevel delamination fracture (Figure 7a—outer, Figure 7b—outer and middle) and near-flat fracture (Figure 7c–e). Multilevel delamination fracture is due to the different deformation capacities of thick and thin layers in the fiber cell wall containing different microfibril angles (MFAs) and lignin content [35,36]. As the tensile force increased, the thick and thin layer interfaces continuously separated, tore, and even delaminated, resulting in the formation of a multilevel delamination fracture mode. This fracture pattern generally occurred in the outer and middle extracted fiber bundles before 180 °C. With the increase in the heat treatment temperature and the degradation of internal chemical composition and small molecules, the bonding of the cell wall layers gradually loosened along with the appearance of the pores. Both the extracted fibers (Figure 7c) and the flat fracture surface (Figure 7d,e) showed the mode of near-flat fracture.

Temperature causes some degree of damage to the fibers, usually in the form of increased gaps between cell wall layers and the loss of bulk tissue, which become more pronounced as the temperature increases. In addition, it is most noticeable at 220 °C and not significant before 220 °C, which is also consistent with previous studies [33]. Furthermore, it is known that the changes in the cell structure caused by heat treatment are the physical cause of the decrease in tensile strength. Interestingly, under the same heat treatment temperature, the fracture morphology of single fibers at different sites also differed slightly, with the fracture near the inner side being rougher than the outer side, indicating a greater influence by temperature. This phenomenon becomes more pronounced at higher temperatures, especially at 220 °C (Figure 8e). Differences in fiber fracture of different parts of bamboo after high temperature heat treatment explain the phenomenon that the tensile strength of the outer fiber bundles is higher than that of the inner ones at the same temperature. In addition, the schematic fracture mode of the bamboo bundles and fibers is shown in Figure 8.

The temperature could cause some degree of damage to the fibers, usually in the form of an increased number of pores between the cell wall layers and the shedding of bulk tissue, which became more pronounced as the temperature increased. Therefore, the changes in cell structure that were caused were the physical cause of the decrease in tensile strength after heat treatment. It is interesting that the fracture morphology of single fibers in different areas also differed slightly under the same heat treatment temperature, with the fracture surface near the inside of the bamboo stem being rougher than the outside, showing that the temperature had a greater influence on the inner fibers of the bamboo stem. This phenomenon becomes more pronounced at higher temperatures, especially at 220 °C (Figure 7e). The structural differences in the different areas with temperature explained the phenomenon that the tensile strength of the outer fiber bundles is greater than that of the inner ones at the same temperature.

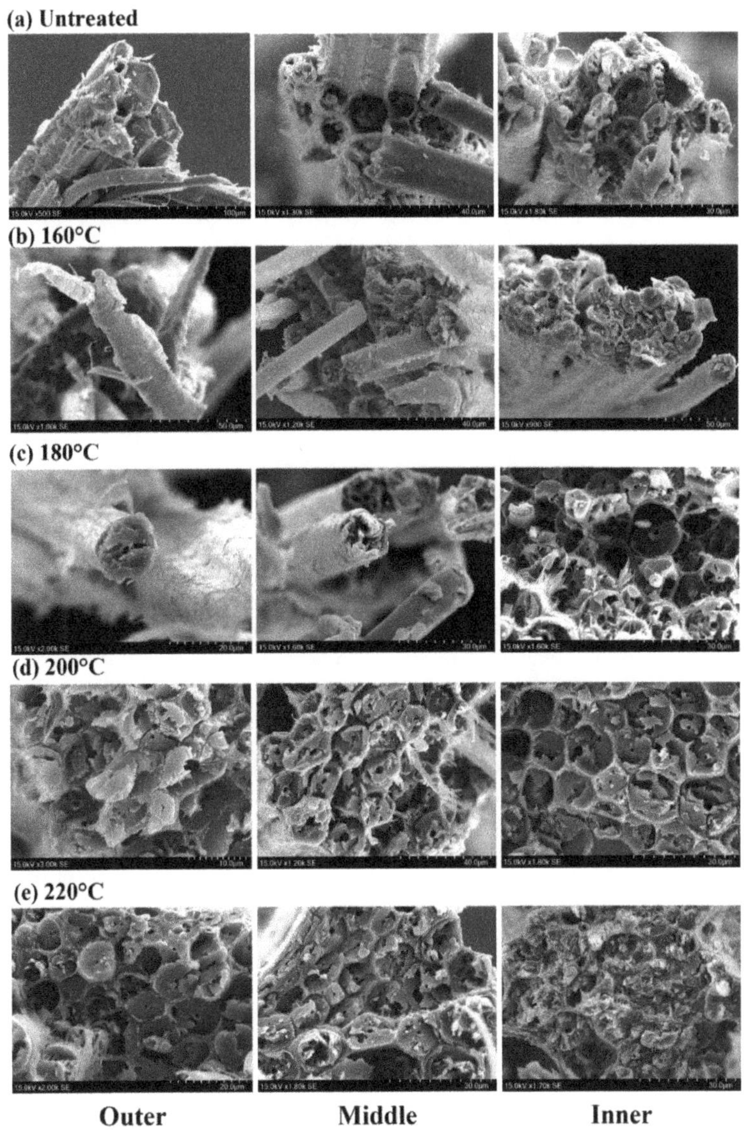

Figure 7. Overview of fracture surfaces of fibers in different areas in the radial direction with different temperatures.

It is worth noting that the stress–strain curve of the fiber bundles showed the characteristics of brittle fracture when the heat treatment temperature was below 200 °C, while the outer and partially inner fiber bundle fractures exhibited the mode of ductile fracture. Tensile damage of bamboo fibers bundles begins with relative sliding between cell walls.

As can be seen in Figure 3a–d, the stress–strain curve of the fiber bundles contains very small stepwise jitters. When the curve jitters, some of the fibers have relative slippage and internal damage is formed. At this time, the overall specimen was not damaged, and brittle fracture occurred to the fibers when the ultimate strength was reached. The fracture mode was characterized by a ductile fracture due to the slippage between the fiber cells during the stretching process.

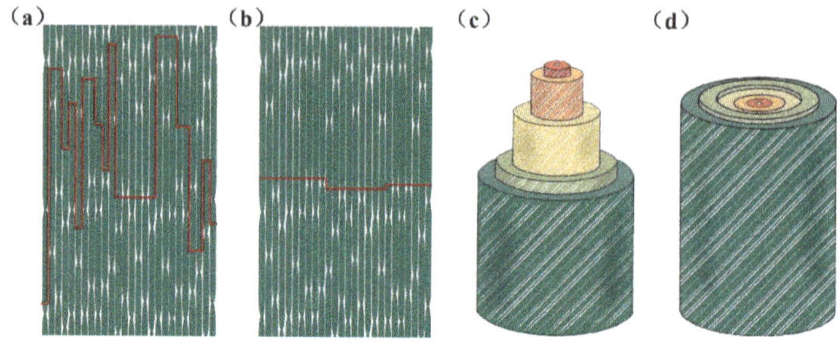

Figure 8. Schematic fracture mode of bamboo bundles and fibers: (**a**) ductile fracture mode of bamboo bundles; (**b**) brittle fracture mode of bamboo bundles; (**c**) multilevel delamination fracture; (**d**) near-flat fracture.

More pores and a loss of bulk tissue appeared in the fiber fracture with the increase in the heat treatment temperature, indicating a decrease of fiber strength. These pores were squeezed by the two wall layers and played a buffering role throughout the tensile process. The highest number of pores between the cell walls was found at 220 °C. During the tensile process, the slip between the fibers had not yet occurred when the pores were not reduced completely, while the fibers were not able to withstand the load and were destroyed. The fracture was relatively flat because there was no slip, which results in a brittle fracture at high temperatures. In addition, the characteristics of the stress–strain curve correspond to the microstructure.

From the previous analysis, higher heat treatment temperatures lead to the degradation of cellulose and hemicellulose thus decreasing the tensile properties of bamboo fibers and increasing the elastic modulus and hardness. At the same time, the fiber bundles distributed on the outer side have better mechanical properties. In practice, the heat treatment temperature can be increased if a higher elastic modulus and stiffness are required. On the contrary, the heat treatment temperature can be reduced if a higher tensile strength is required. In general, a heat treatment temperature of 180 °C is a good choice.

4. Conclusions

This study aimed to investigate tensile properties, elastic modulus, and tissue structure as well as the variation law in the radial distribution of artificially extracted natural fiber bundles by heat treatment techniques. The following conclusions can be drawn from the results:

FTIR shows that an increase in temperature leads to degradation of chemical composition, with a decrease in the relative content of cellulose and hemicellulose and an increase in the relative content of lignin. The tensile strength of the bamboo fiber bundles gradually decreased with the increasing temperature from 432.29 MPa at the beginning to 191.61 MPa at 220 °C, which was mainly due to the decrease in the relative content of cellulose and hemicellulose. The elastic modulus of bamboo fiber bundles showed an opposite trend, increasing from 21.29 GPa to 27.43 GPa with the increasing temperature, which was mainly due to the decrease in relative lignin content. In the radial direction, the elastic modulus and tensile strength of the fiber bundles near the outside of the bamboo were greater than those near the inside. In addition, the elastic modulus of the cell wall showed the same law as the elastic modulus of the macroscopic fiber bundles.

The fracture behavior of the bamboo fiber bundles near the outside resembled ductile fracture, while the fracture behavior of the bamboo fiber bundles near the inside approximated brittle fracture. The fracture mode of the fibers also changed from multilevel delamination fracture to near-flat fracture when the temperature increased, and the differences in the fiber cell wall structure in the radial direction became more obvious. The change in intrinsic structural organization was the main reason for the decrease in tensile strength. Remarkably, the fracture of fiber bundles is a brittle fracture, while it produces a fracture in ductile fracture mode at low temperatures because of cell wall slippage. Therefore, bamboo fiber bundles distributed near the outside are most suitable for industrial development under heat treatment at 180 °C.

The bond strength between the cell walls and the significant increase in elastic modulus at 200 °C–220 °C still need to be further explored in future research work. Quantitative analysis of chemical composition and physical modeling of cell wall fracture will be the focus of future research work. This study is of great importance for the selection of heat treatment temperatures, the utilization of bamboo materials, and the development of green and sustainable infrastructure construction materials.

Author Contributions: Conceptualization, Q.W. and L.L.; methodology, J.C.; software, J.C. and L.M.; validation, Y.L. and Y.C.; formal analysis, H.Z., C.H. and C.W.; investigation, J.C.; resources, L.M. and D.F.; data curation, J.C.; writing—original draft preparation, J.C.; writing—review and editing, J.C., L.L., Y.L., C.W. and Q.W.; visualization, J.C.; supervision, L.L.; project administration, L.L.; funding acquisition, L.L. and Q.W. All authors have read and agreed to the published version of the manuscript.

Funding: This work was supported by the National Natural Science Foundation of China (No. 12002226, No. 11832007) and Applied Basic Research Programs of Sichuan Province (No. 2022NSFSC1977, No. 2022JDJQ0011 and No. 2022NSFSC0337).

Institutional Review Board Statement: Not applicable.

Informed Consent Statement: Not applicable.

Data Availability Statement: Not applicable.

Acknowledgments: The kind of assistance of Yani Xie and Yi He from the Analytical and Testing Center of Sichuan University is greatly appreciated.

Conflicts of Interest: The authors declare no conflict of interest.

References

1. Han, G.C.; Kumar, S. Making Strong Fibers. *Science* **2008**, *319*, 908–909.
2. Osorio, L.; Trujillo, E.; Van Vuure, A.W.; Verpoest, I. Morphological aspects and mechanical properties of single bamboo fibers and flexural characterization of bamboo/epoxy composites. *J. Reinf. Plast. Compos.* **2011**, *30*, 396–408. [CrossRef]
3. Umair, S. Environmental Impacts of Fiber Composite Materials: Study on Life Cycle Assessment of Materials used for Ship Superstructure. *Annan Nat.* **2006**.
4. Li, Z.; Chen, C.; Xie, H.; Yao, Y.; Zhang, X.; Brozena, A.; Li, J.; Ding, Y.; Zhao, X.; Hong, M.; et al. Sustainable high-strength macrofibres extracted from natural bamboo. *Nat. Sustain.* **2021**, *5*, 235–244. [CrossRef]
5. Faruk, O.; Bledzki, A.K.; Fink, H.-P.; Sain, M. Biocomposites reinforced with natural fibers: 2000–2010. *Prog. Polym. Sci.* **2012**, *37*, 1552–1596.
6. Huang, X.; Netravali, A. Biodegradable green composites made using bamboo micro/nano-fibrils and chemically modified soy protein resin. *Compos. Sci. Technol.* **2009**, *69*, 1009–1015. [CrossRef]
7. Ray, A.K.; Mondal, S.; Das, S.K.; Ramachandrarao, P. Bamboo—A functionally graded composite-correlation between microstructure and mechanical strength. *J. Mater. Sci.* **2005**, *40*, 5249–5253. [CrossRef]
8. Akinbade, Y.; Harries, K.A.; Flower, C.V.; Nettleship, I.; Papadopoulos, C.; Platt, S. Through-culm wall mechanical behaviour of bamboo. *Constr. Build. Mater.* **2019**, *216*, 485–495. [CrossRef]
9. Popat, T.V.; Patil, A.Y. A review on bamboo fiber composites. *IRE J.* **2017**, *1*, 54–72.
10. Chen, H.; Cheng, H.; Wang, G.; Yu, Z.; Shi, S.Q. Tensile properties of bamboo in different sizes. *J. Wood Sci.* **2015**, *61*, 552–561. [CrossRef]
11. Biswas, S.; Ahsan, Q.; Cenna, A.; Hasan, M.; Hassan, A. Physical and mechanical properties of jute, bamboo and coir natural fiber. *Fibers Polym.* **2013**, *14*, 1762–1767. [CrossRef]

12. Shao, Z.-P.; Fang, C.-H.; Huang, S.-X.; Tian, G.-L. Tensile properties of Moso bamboo (*Phyllostachys pubescens*) and its components with respect to its fiber-reinforced composite structure. *Wood Sci. Technol.* **2009**, *44*, 655–666. [CrossRef]
13. Li, H.; Shen, S. The mechanical properties of bamboo and vascular bundles. *J. Mater. Res.* **2011**, *26*, 2749–2756. [CrossRef]
14. Zhang, Y.; Yu, Y.; Lu, Y.; Yu, W.; Wang, S. Effects of heat treatment on surface physicochemical properties and sorption behavior of bamboo (*Phyllostachys edulis*). *Constr. Build. Mater.* **2021**, *282*, 122683. [CrossRef]
15. Li, Z.-Z.; Luan, Y.; Hu, J.-B.; Fang, C.-H.; Liu, L.-T.; Ma, Y.-F.; Liu, Y.; Fei, B.-H. Bamboo heat treatments and their effects on bamboo properties. *Constr. Build. Mater.* **2022**, *331*, 127320. [CrossRef]
16. Azadeh, A.; Ghavami, K.; García, J.J. The influence of heat on mechanical properties of *Dendrocalamus giganteus* bamboo. *J. Build. Eng.* **2021**, *43*, 102613. [CrossRef]
17. Yuan, T.; Huang, Y.; Zhang, T.; Wang, X.; Li, Y. Change in Micro-Morphology and Micro-Mechanical Properties of Thermally Modified Moso Bamboo. *Polymers* **2022**, *14*, 646. [CrossRef]
18. Yuan, T.; Wang, X.; Lou, Z.; Zhang, T.; Han, X.; Wang, Z.; Hao, X.; Li, Y. Comparison of the fabrication process and macro and micro properties of two types of crack-free, flatten bamboo board. *Constr. Build. Mater.* **2022**, *317*, 125949. [CrossRef]
19. Oliver, W.C.; Pharr, G.M. An improved technique for determining hardness and elastic modulus using load and displacement sensing indentation experiments. *J. Mater. Res.* **2011**, *7*, 1564–1583. [CrossRef]
20. Donnelly, E.; Baker, S.P.; Boskey, A.L.; van der Meulen, M.C. Effects of surface roughness and maximum load on the mechanical properties of cancellous bone measured by nanoindentation. *J. Biomed. Mater. Res. A* **2006**, *77*, 426–435. [CrossRef]
21. Huang, Y.; Fei, B.; Wei, P.; Zhao, C. Mechanical properties of bamboo fiber cell walls during the culm development by nanoindentation. *Ind. Crops Prod.* **2016**, *92*, 102–108. [CrossRef]
22. Lods, L.; Richmond, T.; Dandurand, J.; Dantras, E.; Lacabanne, C.; Durand, J.-M.; Sherwood, E.; Hochstetter, G.; Ponteins, P. Thermal stability and mechanical behavior of technical bamboo fibers/bio-based polyamide composites. *J. Therm. Anal. Calorim.* **2021**, *147*, 1097–1106. [CrossRef]
23. Wang, F.; Shao, Z. Study on the variation law of bamboo fibers' tensile properties and the organization structure on the radial direction of bamboo stem. *Ind. Crops Prod.* **2020**, *152*, 112521. [CrossRef]
24. Zou, L.; Jin, H.; Lu, W.-Y.; Li, X. Nanoscale structural and mechanical characterization of the cell wall of bamboo fibers. *Mater. Sci. Eng. C* **2009**, *29*, 1375–1379. [CrossRef]
25. Yu, Y.; Tian, G.; Wang, H.; Fei, B.; Wang, G. Mechanical characterization of single bamboo fibers with nanoindentation and microtensile technique. *Holzforschung* **2011**, *65*, 113–119. [CrossRef]
26. Bergander, A.; Salmén, L. Variations in Transverse Fibre Wall Properties: Relations Between Elastic Properties and Structure. *Holzforschung* **2005**, *54*, 654–660. [CrossRef]
27. Zhang, S. *Chemical Components Effect on Mechanical Properties of Wood Cell Wall*; Chinese Academy of Forestry: Beijing, China, 2011.
28. Piao, X.; Xie, M.; Duan, X.; Jin, C.; Wang, Z. Novel high-performance bamboo modification through nature rosin-based benzoxazine. *Constr. Build. Mater.* **2022**, *319*, 126123. [CrossRef]
29. Lee, C.-H.; Yang, T.-H.; Cheng, Y.-W.; Lee, C.-J. Effects of thermal modification on the surface and chemical properties of moso bamboo. *Constr. Build. Mater.* **2018**, *178*, 59–71. [CrossRef]
30. Wang, X.; Luo, X.; Ren, H.; Zhong, Y. Bending failure mechanism of bamboo scrimber. *Constr. Build. Mater.* **2022**, *326*, 126892. [CrossRef]
31. Okon, K.E.; Lin, F.; Chen, Y.; Huang, B. Effect of silicone oil heat treatment on the chemical composition, cellulose crystalline structure and contact angle of Chinese parasol wood. *Carbohydr Polym* **2017**, *164*, 179–185. [CrossRef]
32. Yang, T.-C.; Lee, T.-Y. Effects of density and heat treatment on the physico-mechanical properties of unidirectional round bamboo stick boards (UBSBs) made of Makino bamboo (*Phyllostachys makinoi*). *Constr. Build. Mater.* **2018**, *187*, 406–413. [CrossRef]
33. Feng, Q.; Huang, Y.; Ye, C.; Fei, B.; Yang, S. Impact of hygrothermal treatment on the physical properties and chemical composition of Moso bamboo (*Phyllostachys edulis*). *Holzforschung* **2021**, *75*, 614–625. [CrossRef]
34. Song, J.; Chen, C.; Zhu, S.; Zhu, M.; Dai, J.; Ray, U.; Li, Y.; Kuang, Y.; Li, Y.; Quispe, N.; et al. Processing bulk natural wood into a high-performance structural material. *Nature* **2018**, *554*, 224–228. [CrossRef]
35. Ahvenainen, P.; Dixon, P.G.; Kallonen, A.; Suhonen, H.; Gibson, L.J.; Svedstrom, K. Spatially-localized bench-top X-ray scattering reveals tissue-specific microfibril orientation in Moso bamboo. *Plant Methods* **2017**, *13*, 5. [CrossRef] [PubMed]
36. Adler, D.C.; Buehler, M.J. Mesoscale mechanics of wood cell walls under axial strain. *Soft Matter* **2013**, *9*, 7138–7144. [CrossRef]

materials

MDPI

Article

Compressive Strength Prediction of Rice Husk Ash Concrete Using a Hybrid Artificial Neural Network Model

Chuanqi Li [1,*], Xiancheng Mei [2], Daniel Dias [1], Zhen Cui [2] and Jian Zhou [3,*]

[1] Laboratory 3SR, CNRS UMR 5521, Grenoble Alpes University, 38000 Grenoble, France; daniel.dias@univ-grenoble-alpes.fr

[2] Institute of Rock and Soil Mechanics, Chinese Academy of Sciences, Wuhan 430071, China; xcmei@whrsm.ac.cn (X.M.); zcui@whrsm.ac.cn (Z.C.)

[3] School of Resources and Safety Engineering, Central South University, Changsha 410083, China

* Correspondence: chuanqi.li@univ-grenoble-alpes.fr (C.L.); j.zhou@csu.edu.cn (J.Z.)

Abstract: The combination of rice husk ash and common concrete both reduces carbon dioxide emission and solves the problem of agricultural waste disposal. However, the measurement of the compressive strength of rice husk ash concrete has become a new challenge. This paper proposes a novel hybrid artificial neural network model, optimized using a reptile search algorithm with circle mapping, to predict the compressive strength of RHA concrete. A total of 192 concrete data with 6 input parameters (age, cement, rice husk ash, super plasticizer, aggregate, and water) were utilized to train proposed model and compare its predictive performance with that of five other models. Four statistical indices were adopted to evaluate the predictive performance of all the developed models. The performance evaluation indicates that the proposed hybrid artificial neural network model achieved the most satisfactory prediction accuracy regarding R^2 (0.9709), VAF (97.0911%), RMSE (3.4489), and MAE (2.6451). The proposed model also had better predictive accuracy than that of previously developed models on the same data. The sensitivity results show that age is the most important parameter for predicting the compressive strength of RHA concrete.

Keywords: rice husk ash; concrete; compressive strength; reptile search algorithm with circle mapping; artificial neural network

Citation: Li, C.; Mei, X.; Dias, D.; Cui, Z.; Zhou, J. Compressive Strength Prediction of Rice Husk Ash Concrete Using a Hybrid Artificial Neural Network Model. *Materials* **2023**, *16*, 3135. https://doi.org/10.3390/ma16083135

Academic Editor: René de Borst

Received: 29 March 2023
Revised: 12 April 2023
Accepted: 14 April 2023
Published: 16 April 2023

1. Introduction

Concrete is globally still one of the most highly demanded materials in construction and other industries [1]. By 2018, the production of concrete exceeded 10 billion cubic meters [2]. As a main component, the production of cement rose to 4 billion tons in 2020 [3]. Although cement provides the necessary strength for concrete, carbon dioxide (CO_2) produced in the forging process is a heavy burden (approximately 7%) on the atmosphere. Considering the harm of CO_2 to the environment and human beings, energy conservation and emission reduction have become normal goals in concrete application. Searching for ovel materials to replace parts of cement, namely, supplementary cementitious materials (SCMs), is one of the most effective ways to solve this problem.

Most available SCM options are derived from byproducts associated with industrial and agricultural processes, such as palm-oil fuel [4,5], olive-oil [6,7], and fly [8,9] ash, silica fume [10,11], seed shells [12], dispersed coconut fibers [13], and other types of powder [14–20]. Among these novel SCMs, the combination of rick husk ash (RHA) and conventional concrete has received much attention [21–23]. First, RHA is one of the main byproducts of agricultural production. Conventional stacking could pollute the air and groundwater [24], but adding it to concrete is a reasonable and innovative way to recycle. Second, the pozzolanic nature of RHA helps in improving the durability and strength of concrete [25]. However, the addition of RHA has an important effect on concrete performance [26], especially compressive strength, which directly affects the durability and stability of structures in construction and other industries. Madandoust et al. [27] used RHA to replace 20% of cement to study the strength of concrete. Their results showed that the short-term compressive strength of RHA concrete was reduced, but the long-term compressive strength was increased. Ahsan and Hossain [28] compared cement performance at different RHA replacement rates (10% and 20%). They found that replacing 10% of cement with RHA was optimal because the interfacial transition zone was more effectively densified with the silica content of RHA. However, Noaman et al. [29] reported that that replacing cement with 15% RHA could maximize concrete's compressive strength. Furthermore, determining the mixing ratio of other components in concrete production with cement and RHA is complicated; thus, it is both necessary and challenging to determine the strength of concrete.

The most accurate strength measurement method of concrete is the compressive test in the laboratory. However, the production and maintenance of concrete samples is complicated and time-consuming, and wastes workers and material resources [30]. For example, a group of experiments require two to three professionals to complete. In order to improve calculation efficiency and site limitation, a method based on an empirical formula was developed to estimate compressive strength that was especially praised by field workers. Islam et al. [31] developed an empirical formular by using the least-squares approach to calculate the compressive strength of RHA concrete. Their results showed that the formular achieved good predictive performance with a correlation coefficient (R) of 0.816. Liu et al. [32] utilized six empirical equations to estimate the compressive strength of concrete containing RHA with different replacement values. However, the limitation of the empirical formula is that it cannot accurately express the complex nonlinear relationship between the considered parameters and compressive strength [33].

In recent years, artificial-intelligence methods with machine learning (ML) as a mainstream technologies have been widely used to solve the problem of concrete strength prediction [34–40]. Azimi-Pour et al. [41] utilized four types of support vector machine (SVM) models to predict the compressive strength of fly ash concrete. The performance results indicated that the radial basis function (RBF)-based SVM model had the highest accuracy for a coefficient of determination (R^2) equal to 0.9932. Zhang et al. [42] improved the random forest (RF) model to predict the compressive strength of lightweight concrete (LWC). The extreme learning machine (ELM) model was applied for the compressive strength prediction of lightweight foamed concrete [43]. Compared with these models, an artificial neural network (ANN) model with a simple structure, and good capabilities for processing high-dimensional data and complex parameter relationships is more favored in predicting the concrete strength of RHA [44–47]. Getahun et al. [48] developed an ANN model to forecast the 28-day compressive strength of a composite concrete mixture with RHA and reclaimed asphalt pavement (RAP). The prediction results illustrated that the ANN model could accurately fit the relationship between the considered components and the strength, as evidenced by excellent performance indices: R was 0.9811 and the root mean square error (RMSE) was 0.648. To optimize the selection scheme of the ANN model on weight and bias values, and further improve model performance, many scholars modified this model using numerous optimization algorithms for predicting concrete strength, e.g., grey wolf optimization (GWO) [49,50], particle swarm optimization (PSO) [51,52], the genetic algorithm (GA) [53], the whale optimization algorithm (WOA) [54], and simulated annealing (SA) with PSO [55]. For the strength prediction of RHA concrete, Andalib et al. [56] utilized the bat algorithm (BA), PSO, and teaching–learning-based optimization (TLBO) algorithm to optimize the ANN model for predicting compressive strength. The performance results showed that all optimized ANN models achieved satisfactory prediction accuracy, especially the BA–ANN model (RMSE = 5.898); Hamidian et al. [33] proposed four hybrid ANN models to estimate the compressive strength of RHA concrete. On the basis of the results of the performance analysis of all models, the PSO-with-two differential-mutations (PSOTD)-based ANN model achieved superior performance than that of other models, indicated by the higher R^2 values (0.9697). There are still many newly developed and excellent optimization-algorithm-based populations that have not been applied to the strength prediction of RHA concrete. Population initialization also needs attention to maximize the predictive potential of ANN models.

Therefore, this paper utilizes circle mapping (CM) to improve the optimization performance of the reptile search algorithm (RSA). A novel hybrid ANN model optimized with CMRSA is proposed to estimate the compressive strength of RHA concrete. The predictive accuracy of four ML models and an empirical model was compared. These ML models consisted of optimized and common models: seagull optimization algorithm (SOA)-based SVM (SOA–SVM) and RF (SOA–RF) models, an ANN model, and an ELM model. Four statistical indices, regression analysis, error comparison, and the Taylor diagram were adopted to evaluate the predictive performance of all models in order to determine the optimal model. Lastly, sensitivity analysis was performed to select the most important parameter for predicting the compressive strength of RHA concrete.

2. Data and Methods

2.1. Rice Husk Ash Concrete

RHA concrete cannot be produced without the use of other materials. For example, cement is used to provide sufficient strength for concrete, water is key to controlling concrete compactness in the mixing process, and the aggregate maintains concrete volume stability. To assess the compressive strength of RHA concrete, Iftikhar et al. [57] combined cement (kg/m^3), RHA (kg/m^3), a superplasticizer (kg/m^3), an aggregate (kg/m^3), and water (kg/m^3) to produce a series of concrete samples. Freshly poured concrete needs to be cured, and its strength must be measured after a certain time. Therefore, age (days) is also an important variable in predicting concrete strength. In this paper, 192 compressive-strength data from Iftikhar et al. [57] were utilized to evaluate RHA concrete. The statistical information of these variables and the compressive strength of the target concrete samples is listed in Table 1.

Table 1. Statistical information on each variable for predicting RHA concrete strength.

Variables	Statistical Indices				
	Min	Max	Mean	Median	St. D
Cement	249.0	783.0	409.02	400.00	105.47
RHA	0.0	171.0	62.33	57.00	41.55
Superplasticizer	0.0	11.3	3.34	1.85	3.52
Aggregate	1040.0	1970.0	1621.51	1725.00	267.77
Water	120.0	238.0	193.54	203.00	31.93
Age	1.0	90.0	34.57	28.00	33.52
Compressive strength	16.0	104.1	48.14	45.95	17.54

Note: Min, minimal values; Max, maximal values; St. D, standard deviation.

For establishing the prediction model, all variables except compressive strength were taken as the input parameters. The interdependence of the input parameters must be evaluated to simplify the model and maintain prediction accuracy. The correlation coefficient is widely used to describe dependence [58–60]. If the correlation coefficient between any two input parameters exceeds 0.8, parameter deletion should be considered. Table 2 shows the calculation results of correlation coefficient values between input parameters. The maximal correlation coefficient value was 0.549, induced by water and the aggregate. Therefore, all input parameters could be considered for generating a prediction model for estimating the compressive strength of RHA concrete.

Table 2. Correlation coefficient values between all considered parameters.

Variables	Cement	RHA	Superplasticizer	Aggregate	Water	Age	Compressive Strength
Cement	1	−0.219	0.253	−0.238	0.083	−0.106	0.370
RHA		1	−0.021	−0.139	0.136	−0.033	−0.023
Superplasticizer			1	−0.205	0.268	−0.000	0.301
Aggregate				1	−0.549	−0.063	0.147
Water					1	0.011	−0.244
Age						1	0.495
Compressive strength							1

2.2. Reptile Search Algorithm

RSA is a novel metaheuristic optimization-algorithm-based algorithm proposed by Abualigah et al. [61]. This algorithm was inspired by the hunting behavior of crocodiles to solve the optimization problem. As the apex predators in amphibious environments, the behavior of crocodiles has long attracted the attention of scientists. Crocodiles are highly mobile, and can thereby quickly chase and attack prey, especially at night. The crocodile's excellent night vision and body shape with little resistance benefit this feature [61]. Second, crocodiles are highly intelligent animals, which endows them with high recognition and high perception capabilities. For instance, crocodiles wait where prey is frequent, such as near a river. Crocodile hunting is also group behavior, and teams with a clear division of labor enable individuals to obtain enough food. Therefore, the first step in performing a hunting campaign is to initialize the population in the search space as follows:

$$C_{ij} = rand \cdot (U_b - L_b) + L_b \tag{1}$$

where C_{ij} represents the j-th position of the i-th crocodile; U_b and L_b represent the upper and lower bounds of the search space, respectively; *rand* is a random number. The setting of *rand* indicates that the individual position is randomly determined to find the prey. However, population diversity and the possible search area are limited by this random initialization-method-based mechanism [62]. To solve this problem, various types of chaos mapping were combined to establish the different distributions of individuals in the search space [63,64]. In this paper, circle mapping, with the advantages of stability and coverage rate, was utilized to optimize the population initialization of RSA.

$$C_{ij} = C_{ij} + H - \left(\frac{G}{2\pi}\right) sin\left(2\pi C_{ij}\right) mod(1) \tag{2}$$

where H and G represent the externally applied frequency and strength of nonlinearity, respectively.

After determining the initial positions of individuals, the exploration command was executed to find and encircle prey in the search space (see Figure 1a). In this phase, two strategies could be selected by the crocodiles to search the entire area as much as possible. The mathematical expressions of these strategies are as follows:

$$C_{ij}^{t+1} = \begin{cases} Best^t - \partial_{ij}^t \cdot \alpha - F_{ij}^t \cdot rand, & t \le \frac{T}{4} \\ Best^t \cdot C_{ij} \cdot \eta^t \cdot rand, & t \le \frac{T}{2} \text{ and } t > \frac{T}{4} \end{cases} \tag{3}$$

where C_{ij}^{t+1} represents the j-th position of the i-th crocodile at the t + 1 iteration; T is the maximal iteration value; $Best^t$ indicates the best position at the current (t) iteration, ∂_{ij}^t represents an internal parameter, namely, the hunting operator for the j-th position of the i-th crocodile at the current iteration; α represents a related parameter to exploration accuracy, which was equal to 0.1 in this paper; F_{ij}^t and η^t are the reduce function and evolutionary sense, respectively. The former is used to narrow the search in a limited space, and the latter is a probability ratio.

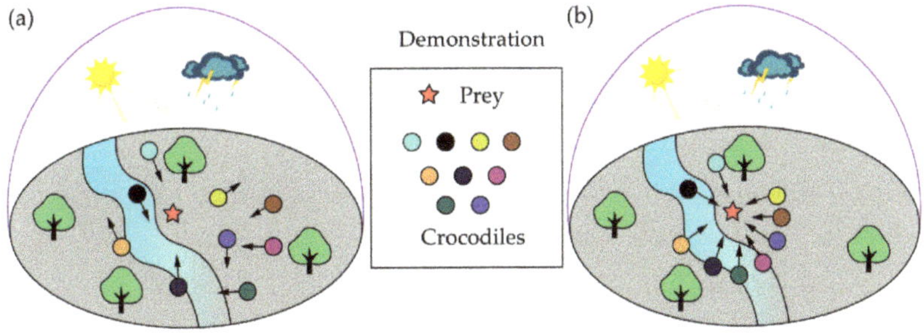

Figure 1. Illustration of hunting behavior for RSA proposed by Abualigah et al. [61]: (**a**) exploration; (**b**) exploitation.

Once the prey is encircled by crocodiles, the hunting (i.e., exploitation, as shown in Figure 1b) can be performed, which uses two strategies, coordination and cooperation, to determine the optimal crocodile position. Two strategies in this phase are mathematically expressed as follows:

$$C_{ij}^{t+1} = \begin{cases} Best^t \cdot P_{ij}^t \cdot rand, & t \leq \frac{3T}{4} \text{ and } t > \frac{T}{2} \\ Best^t - \partial_{ij}^t \cdot \mu - F_{ij}^t \cdot rand, & t \leq T \text{ and } t > \frac{3T}{4} \end{cases} \tag{4}$$

where P_{ij}^t represents the percentage difference between the best and current positions, and μ is a small value in RSA. In general, the aim of the combination of coordination and cooperation is to avoid falling into local optima.

3. Development of the Novel CMRSA–ANN Model

In this paper, the ANN model was generated to accurately predict the strength of RHA concrete. However, the design of an ANN structure has an important effect on predictive performance. In particular, the determination of weights and biases among the input, hidden, and output layers is difficult and challenging [65]. The improved RSA using Circle mapping (CM) was utilized to find the optimal weights and biases for the ANN model. To that end, a novel prediction model, CMRSA–ANN model, was proposed to predict the compressive strength of RHA concrete. Before running the model, a total of 192 data were randomly divided into training and test sets at a 4 to 1 ratio, i.e., 154 data were utilized to train the model, and 38 data to verify the model performance. All data can be found in the Supplementary materials. Since the units of all used parameters were different, the necessary normalization could avoid this impact on performance development. Thus, all parameters were normalized in the range from −1 to 1. For the optimization-algorithm-based population, population size is the most important internal parameter that needs to be determined during iterations [66–68]. To find the global optimal solution, six population sizes (25, 50, 75, 100, 125, and 150) were adopted to conduct the optimization process for the ANN model. We set up 300 iterations to ensure that the optimal solution could be found and remain stable. In general, the fitness value was utilized to represent the solution calculated with the optimization algorithm. In this paper, RMSE was used to generate a fitness function for evaluating optimization performance. The flowchart of developing CMRSA–ANN models to predict the compressive strength of RHA concrete is shown in Figure 2.

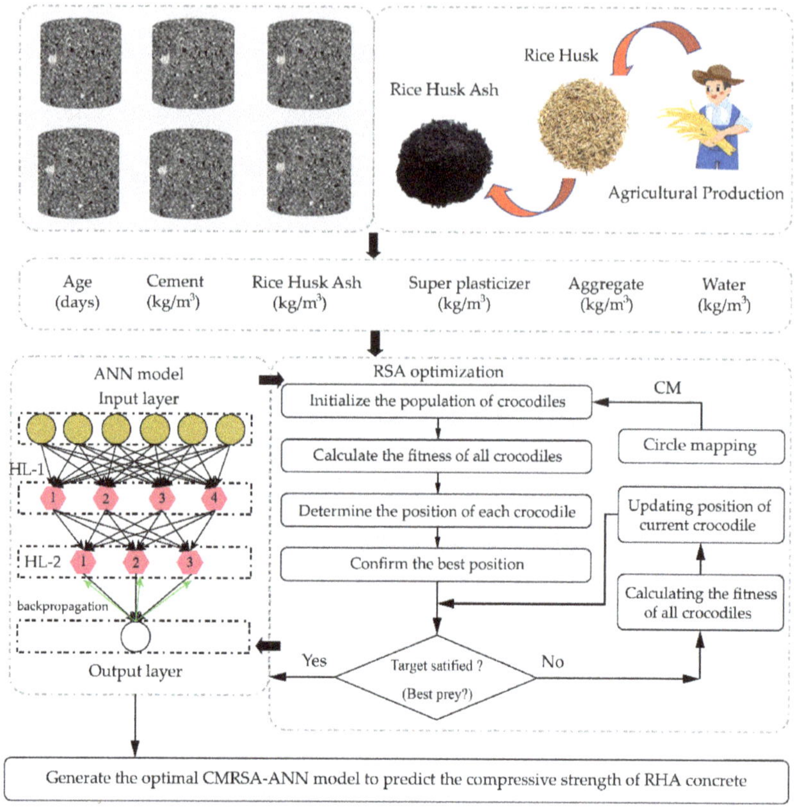

Figure 2. Flowchart of generating CMRSA–ANN model to predict the compressive strength of RHA concrete.

Other models (SOA–SVM, SOA–RF, ANN, ELM, and an empirical formular) were also developed to predict concrete strength, and their prediction results were compared with those of the CMRSA–ANN model. To select the best prediction model, four statistical indices were considered to evaluate the predictive performance of each model: R^2, RMSE, variance accounted for (VAF), and mean absolute error (MAE). The definition of these indices can be found in the literature [69,70], and their formulars are expressed as follows:

$$R^2 = 1 - \frac{\left[\sum\limits_{t=1}^{T} (C_t - c_t)\right]^2}{\left[\sum\limits_{t=1}^{T} (C_t - \overline{C})\right]^2} \tag{5}$$

$$\text{VAF} = \left[1 - \frac{\text{var}(C_t - c_t)}{\text{var}(C_t)}\right] \times 100 \tag{6}$$

$$\text{RMSE} = \sqrt{\frac{1}{T}\sum_{t=1}^{T}(C_t - c_t)^2} \tag{7}$$

$$\text{MAE} = \frac{1}{T}\sum_{t=1}^{T}|C_t - c_t| \tag{8}$$

where T is the maximal number of samples; C_t and c_t are the values of the t-th measured and predicted, respectively; $\overline{C}$ is the average of the measured values.

4. Prediction Model Development

Before applying the ideal proposed model to predict the compressive strength of RHA concrete, all models were developed using the same training set (80% of the database). The detailed development process of each model is shown in this section.

4.1. ANN Model

For a common ANN model, the basic structure is composed of input, hidden, and output layers. Compared with the number of hidden layers, one input layer and one output layer are fixed collocations in single-target regression tasks. Two hidden layers are often utilized to solve similar prediction problems [71–73]. Furthermore, the number of neurons in each hidden layer greatly impacts ANN model performance. Hence, a series of tests were carried to select the suitable ANN structure and the corresponding neurons for predicting the compressive strength of RHA concrete. In this paper, the hidden layers were 1 or 2, the range of neuron numbers was from 2 to 12, the activation function was set to sigmoid, and the backpropagation algorithm was utilized to improve the prediction accuracy. As a result, 10 tests with different ANN models were established, and their performance was represented by using R^2 and RMSE, as shown in Table 3. The ANN model with two hidden layers (four neurons in the first hidden layer and three neurons in the second hidden layer) had the best performance, with a higher R^2 (0.8772) and lower RMSE (5.8632) than those of other models.

Table 3. Performance of the ANN model with different hidden layers and neuron numbers.

Tests	Structure		Performance	
	HL-1	**HL-2**	**R^2**	**RMSE**
1	2	/	0.8322	6.8525
2	4	/	0.7839	7.7690
3	6	/	0.8100	7.2921
4	8	/	0.8225	7.0476
5	10	/	0.8554	6.3611
6	4	3	0.8772	5.8632
7	4	6	0.8312	6.8726
8	6	8	0.8025	7.4350
9	8	10	0.8143	7.2101
10	10	12	0.8338	6.8193

Note: HL-1, first hidden layer; HL-2, second hidden layer.

4.2. CMRSA–ANN Model

Although the best structure was determined in the ANN model development, it is difficult to choose weights and biases between layers to minimize prediction error. Therefore, the CMRSA optimization algorithm was utilized to optimize the initial ANN model with two hidden layers (four neurons in the first hidden layer and three neurons in the second hidden layer); the framework is shown in Figure 2. Six hybrid CMRSA–ANN models with different population sizes were run for 300 iterations. The iteration curve of each model is shown in Figure 3a. Figure 3b shows that the CMRSA–ANN model with a population size of 75 had the lowest fitness value among all hybrid ANN models. As a result, this model was used to predict the compressive strength of RHA concrete in this paper.

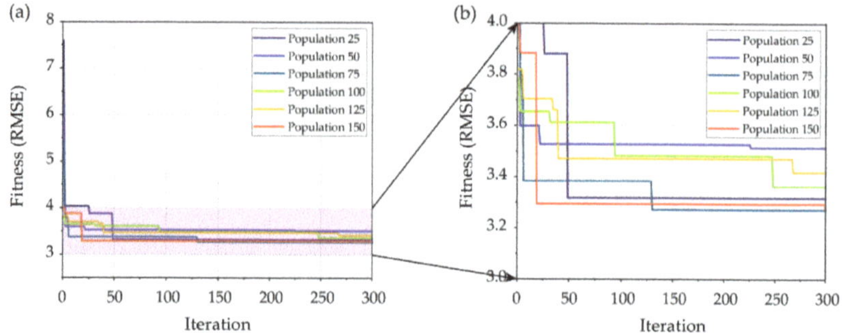

Figure 3. CMRSA–ANN model development: (**a**) iteration curves; (**b**) local comparison.

4.3. SOA–SVM Model

The development of the SOA–SVM model is similar to that of the CMRSA–ANN model. For the SVM model, two main hyperparameters, the regularization parameter (C) and kernel coefficient (γ) of the used kernel function, are key players to improving the model performance [74,75]. In this paper, the popular radial basis function (RBF) was considered as the kernel function of the SVM model. To determine the optimal hyperparameter combination of the SVM model, the range of these parameters was 0 to 100. Thew population sizes and iteration number of SOA were set to be the same as those of the CMRSA. The development results of the SOA–SVM models are shown in Figure 4. The best SOA–SVM model had a population of 75 in the training phase and had a lower fitness value of RMSE than that of other models.

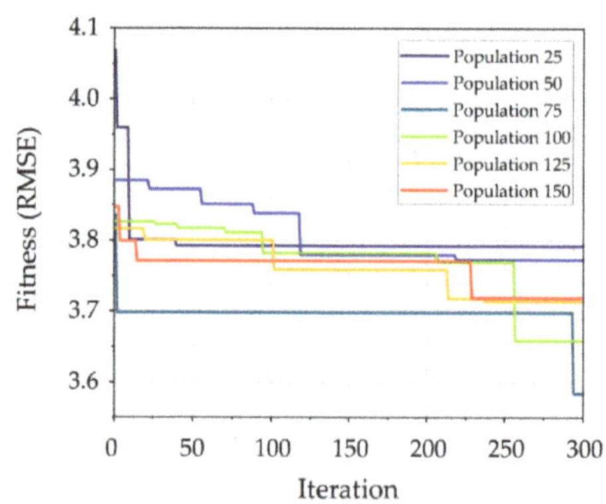

Figure 4. SOA–SVM model development.

4.4. SOA–RF Model

Ensemble models such as the RF model could achieve good performance in solving classification and regression problems; a detailed introduction of the RF model can be found in the literature [58,76]. The unique tree structure and bootstrap sampling allow for the RF performance to be determined by all trees and resist overfitting [74]. In the development of SOA–RF models, the main purpose is to find the best hyperparameter combination of the RF model, i.e., the number of tress (*Nt*) and the random features (*Maxdepth*). In this paper, the tree-number range was from 1 to 100, and the random-feature range was from 1 to 10. Figure 5 shows the optimization results of thew SOA–RF models based on different population sizes after 300 iterations. The OA–RF model containing a population of 75 achieved the most satisfactory performance, as shown by having the lowest fitness value of RMSE. Therefore, this SOA–RF model was considered to predict the compressive strength of RHA concrete.

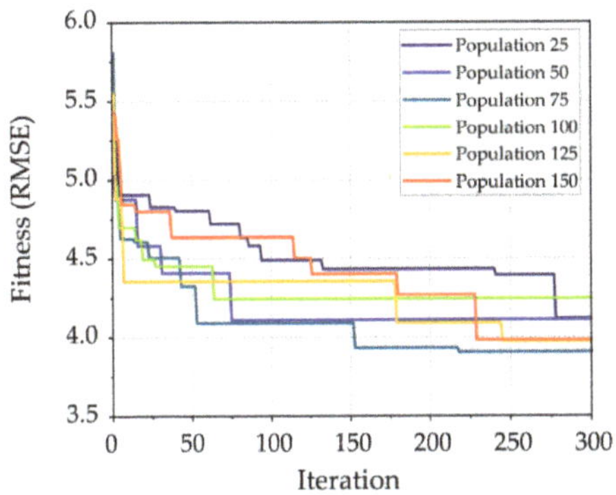

Figure 5. SOA–RF model development.

4.5. ELM Model

The ELM model is a special neuron network with a single hidden layer for solving regression problems. The predictive performance of the ELM model is only controlled by the selection of neuron numbers in the hidden layer. To that end, 10 ELM models with various neuron numbers in the hidden layer were generated to predict concrete strength. Table 4 lists the predictive performance of each ELM model in the training phase. ELM models with large neuron numbers achieved better performance than that of models with smaller neuron numbers. However, the best ELM model was in the 9th test, when the neurons were 100. The performance indices of this model were more reliable than those of other models, i.e., R^2 is equal to 0.8932 and RMSE is equal to 5.4682.

Table 4. Performance of an ELM model with different neuron numbers.

Tests	Neuron Numbers	Performance	
		R^2	RMSE
1	20	0.5268	11.5078
2	30	0.6460	9.9534
3	40	0.7327	8.6492
4	50	0.7595	8.2046
5	60	0.7851	7.7555
6	70	0.7997	7.4873
7	80	0.8589	6.2835
8	90	0.8373	6.7479
9	100	0.8932	5.4682
10	110	0.8788	5.8235

4.6. Empirical Model

The empirical model is an effective method that uses the relevant parameters to quickly achieve the target calculations. In this paper, six input parameters were considered into the empirical formular using multivariate linear regression, as expressed in Equation (9). The training performance of the developed empirical model is shown in Figure 6.

$$Y = -0.47317 + 0.297X1 + 0.0779X2 - 0.0732X3 - 0.145X4 + 1.524X5 + 0.0154X6 \quad (9)$$

where Y represents the compressive strength. X1–X6 represent the age, cement, RHA, water, superplasticizer, and aggregate, respectively.

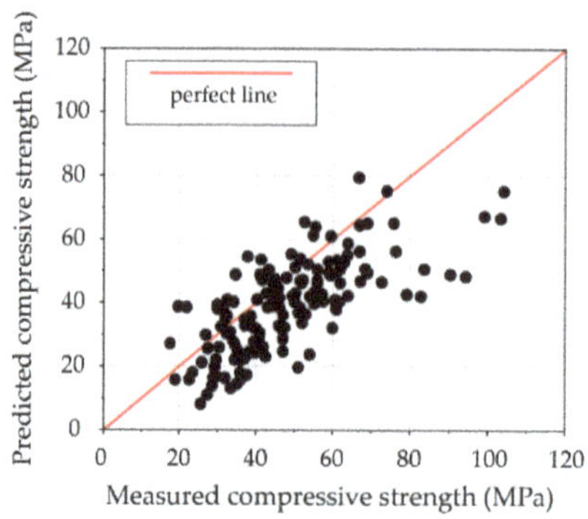

Figure 6. Performance of the empirical model using the training set.

5. Results and Discussion

After training the proposed models, the predictive performance of each model was properly evaluated. Figure 7 shows the prediction curves of all models for estimating the compressive strength of RHA concrete in the training phase. The difference between the prediction curve of the empirical model and the training curve was the greatest among all models. The similarity between the prediction curves of three hybrid models and training was relatively higher, especially in the CMRSA–ANN model.

However, all trained models needed to be further tested to ensure the retention of the excellent predictive ability. Table 5 illustrates the evaluation results of each model using four performance indices in both the training and the testing phases. The performance results from using the training set show that the CMRSA–ANN model was the best prediction model, as it had the highest values of R^2 and VAF (0.9679 and 96.7884%), and the lowest values of RMSE and MAE (2.9991 and 2.3169). Following this model, two other hybrid models (SOA–SVM and SOA–RF) also had superior predictive accuracy than that of the unoptimized ML (ANN and ELM) and empirical models. On the other hand, the proposed CMRSA–ANN model still achieved better predictive performance than that of other models, indicated by the higher values of R^2 and VAF (0.9709 and 97.0911%), and the lower values of RMSE and MAE (3.4489 and 2.6451). Although the performance of the SOA–SVM and SOA–RF models in the testing phase was worse than that using the training set, they still achieved higher predictive accuracy than that of the unoptimized ML models. The ANN model achieved better performance than that of the ELM model in the testing phase, proving that the prediction accuracy of the ELM model is unstable for solving regression problems.

Table 5. Performance evaluation of prediction models using training and test sets.

Model	Performance (Training Set)				Model	Performance (Test Set)			
	R^2	VAF %	RMSE	MAE		R^2	VAF %	RMSE	MAE
ANN	0.8772	87.7619	5.8632	4.1423	ANN	0.8572	86.0686	7.6353	5.2808
CMRSA–ANN	0.9679	96.7884	2.9991	2.3169	CMRSA–ANN	0.9709	97.0911	3.4489	2.6451
SOA–SVM	0.9595	96.0957	3.3651	1.2528	SOA–SVM	0.9494	95.0044	4.5436	3.0904
SOA–RF	0.9224	92.2384	4.6610	3.2359	SOA–RF	0.8941	89.5048	6.5743	4.8037
ELM	0.8932	89.3163	5.4682	4.0644	ELM	0.7020	70.6826	11.0294	8.5905
Empirical	0.2023	50.0783	14.9418	12.0202	Empirical	0.3716	57.3263	16.0169	13.1709

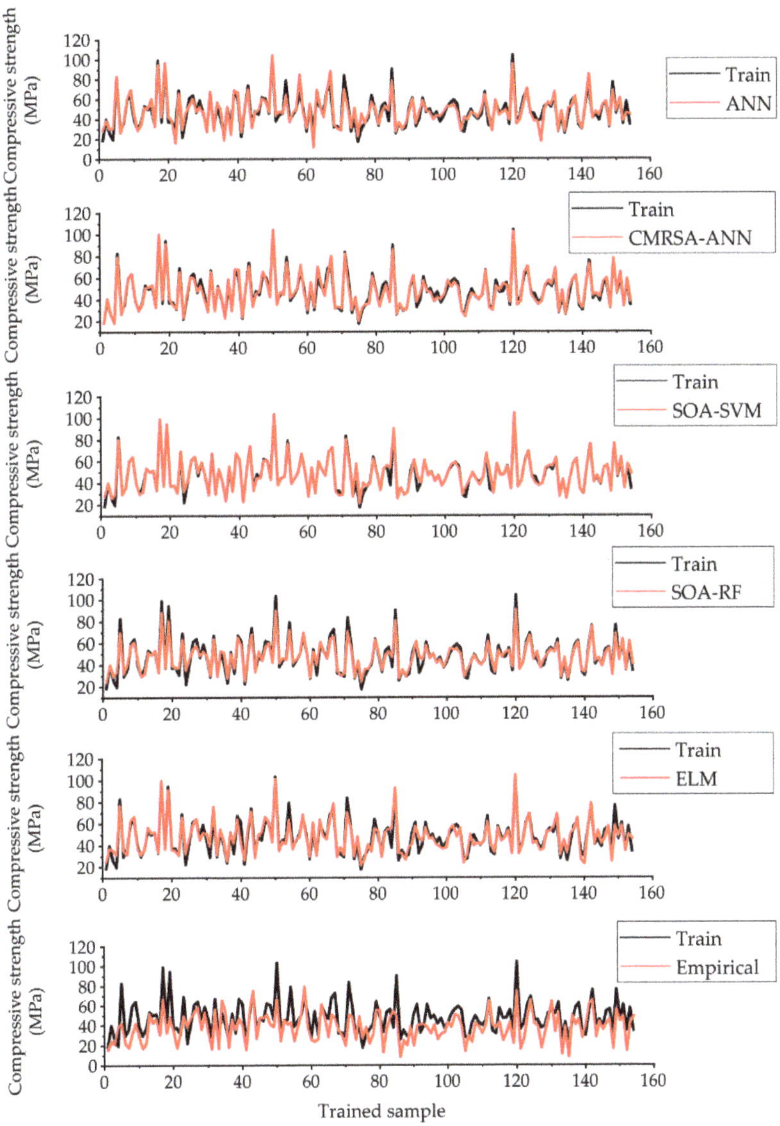

Figure 7. Prediction curves of the developed models using the training set.

Regression relationships can also be used to evaluate and compare the predictive performance of models. Figure 8 shows the regression results of each model using the test set. In each regression plot, one perfect and two limited lines were used to evaluate the regression relationship between the values from the prediction model and the measured values. For instance, the data point determined by the best model with a prediction accuracy of 100% could lie on the perfect line. Observations based on this criterion show that the CMRSA–ANN model achieved better predictive performance than that of other models, indicated by the greater number of data points close to the perfect line and within the limited lines. Furthermore, the performance of the SOA–SVM and SOA–RF models was better than that of the ANN, ELM, and empirical models, but they could not perform accurate predictions for small values of compressive strength (less than 30 MPa).

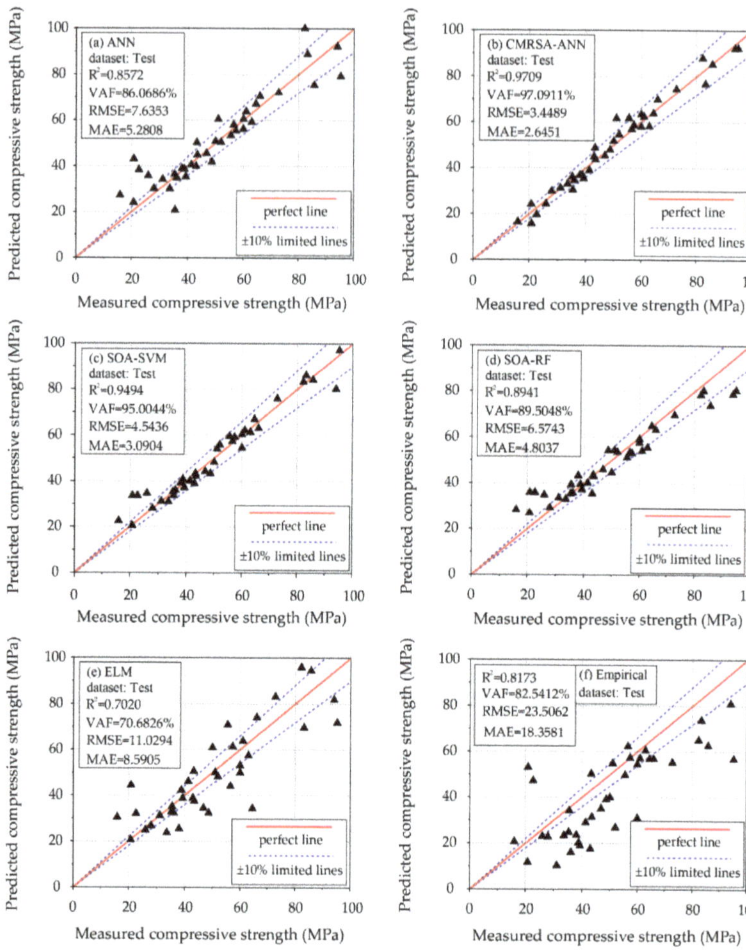

Figure 8. Regression results of all developed models using the test set.

For the regression problem, the error between the predicted and measured values was one of the most concerning performance indices. Although perfect predictions rarely exist, one of the purposes of training and testing is to shrink the prediction error of each target as much as possible. The measured and predicted values of the compressive strength of RHA concrete are listed in Table 6. Figure 9 illustrates the error distribution of each prediction model in the testing phase. The error by the CMRSA–ANN model was mainly concentrated within 10 MPa and accounted for the highest proportion within 5 MPa. The error distribution of the SOA–SVM model was similar to that of the CMRSA–ANN model in a small range where the error was less than 10 MPa, while there were some larger errors between 10 and 15 MPa. The error distribution from the empirical model was undoubtedly the most unsatisfactory, as it both accounted for the lowest proportion of small errors and had many excessive errors.

Table 6. Prediction results of the compressive strength of RHA concrete using the developed models.

No.	Measured	Predicted					
		ANN	CMRSA–ANN	SOA–SVM	SOA–RF	ELM	Empirical
1	82.20	100.09	87.90	83.20	78.36	95.91	64.96
2	72.80	72.04	74.38	75.93	69.61	83.15	55.26
3	43.50	44.85	43.59	42.07	42.81	37.48	31.36
4	48.70	41.82	47.91	43.25	54.26	32.36	39.11
5	16.00	27.32	16.50	22.53	28.28	30.40	20.62
6	85.70	75.39	85.31	84.08	73.81	94.46	62.58
7	43.00	39.93	44.88	38.92	35.43	38.80	17.57
8	33.60	30.16	33.05	31.06	33.00	23.74	23.33
9	94.00	92.21	92.18	80.18	78.79	81.86	81.00
10	31.10	34.15	31.28	31.15	33.57	31.17	10.31
11	57.30	55.35	58.61	59.18	52.92	61.25	57.31
12	41.30	40.49	38.96	39.98	40.03	46.12	28.98
13	20.80	24.08	24.19	20.48	26.86	20.58	11.78
14	22.70	38.28	19.66	33.55	35.84	32.02	47.24
15	38.80	38.68	36.91	40.64	39.48	42.21	20.21
16	60.00	60.42	63.35	54.54	59.20	49.98	54.46
17	55.50	53.28	61.66	59.50	51.22	70.86	49.78
18	61.00	63.75	62.30	62.09	54.07	63.77	57.04
19	63.00	59.20	58.12	61.35	55.48	57.46	60.53
20	66.00	70.44	69.78	63.07	63.24	74.16	56.84
21	52.00	50.39	54.58	55.85	53.52	48.25	26.83
22	43.30	50.25	48.77	43.25	43.49	50.56	50.28
23	26.00	35.75	24.35	34.61	34.82	24.83	22.97
24	64.50	67.10	63.77	66.99	64.85	34.38	56.76
25	35.30	36.41	36.21	35.36	35.36	32.85	24.82
26	83.20	88.88	76.67	86.11	80.21	69.73	73.68
27	50.00	50.77	51.73	48.15	44.58	60.87	39.90
28	56.50	57.93	56.92	57.31	53.43	44.06	62.27
29	35.50	20.85	30.46	33.68	39.28	35.10	34.13
30	36.10	34.92	34.59	36.03	35.78	32.35	16.05
31	20.90	42.96	15.75	33.59	35.93	44.32	53.03
32	51.00	60.39	61.63	54.00	54.33	50.00	54.81
33	95.20	79.24	92.32	97.05	80.43	71.92	56.78
34	28.00	30.20	29.90	27.95	29.14	26.68	22.74
35	60.00	56.15	57.96	60.36	57.60	53.10	30.99
36	46.80	45.49	45.32	44.13	45.88	34.48	35.04
37	39.30	35.34	35.41	37.21	37.24	38.70	18.93
38	38.00	38.96	36.98	39.21	43.23	25.48	23.56

Taylor diagrams are used in visually comparing the predictive performance of multiple models. In a Taylor diagram, a model with high prediction accuracy is close to the position of the target value. The position of each model is determined with three indices, i.e., St. D., RMSE, and R. Therefore, the model performance can be evaluated and compared with multiple indices. Figure 10 displays the evaluation results of all developed models in the Taylor diagram. The CMRSA–ANN model was the closest to the position of the test set. Following this model, models sorted by distance are SOA–SVM, SOA–RF, ANN, ELM, and empirical. These results indicate that the CMRSA optimization algorithm is successful in improving the predictive performance of ANN models. The optimized SVM and RF models had better predictive accuracy than that of the unoptimized ANN and ELM models. Therefore, it is feasible to use the hybrid optimization model to predict the compressive strength of RHA concrete. CMRSA–ANN was selected as the optimal model in this paper.

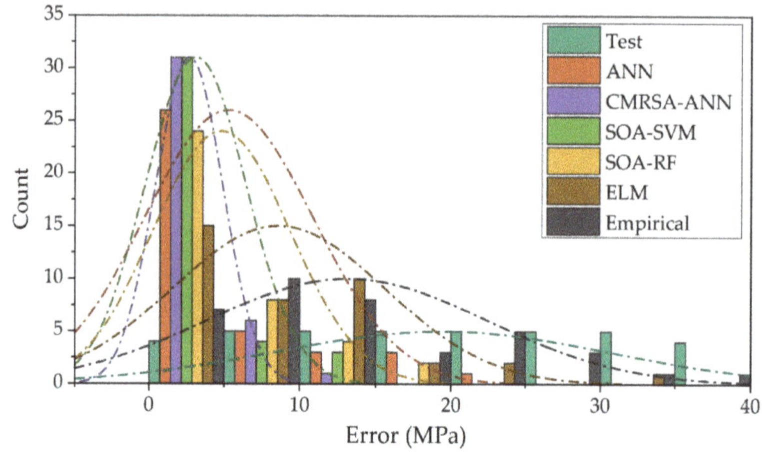

Figure 9. Error distribution of each prediction model in the testing phase.

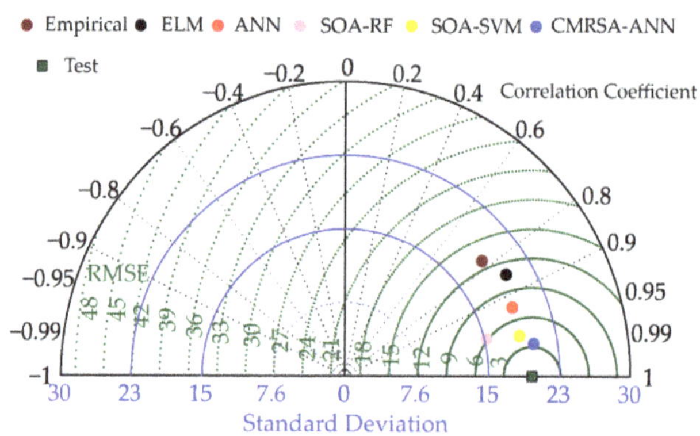

Figure 10. Performance comparison between prediction models using a Taylor diagram.

Nevertheless, the importance or sensitivity of each input parameter to the prediction of compressive strength is unknown, which is detrimental to further improving concrete properties. Therefore, sensitivity analysis was conducted to evaluate the impact of each input parameter on the output. In this paper, calculation method PAWN, proposed by Pianosi and Wagener [77,78], was adopted to calculate the importance score of the input parameters. Figure 11 illustrates the sensitivity results of the compressive strength prediction of the CMRSA–ANN model. Age was the most important parameter, with the highest score (0.351), for predicting the compressive strength of RHA concrete. After age, parameters ranked by influence are cement (0.300), the superplasticizer (0.292), water (0.279), RHA (0.227), and the aggregate (0.225). This result is consistent with that obtained by Iftikhar et al. [57].

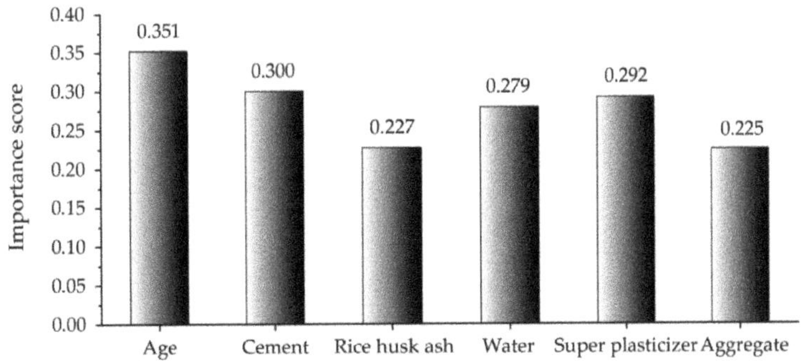

Figure 11. Importance score of each parameter based on the CMRSA-ANN model.

In order to verify the effectiveness and superiority of the prediction model, the predictive performance of the other models developed using the same database was compared with that of the CMRSA–ANN model proposed in this paper, and the results are shown in Table 7. The proposed model had superior predictive performance than that of the published models, indicated by the higher R^2 value. These results also indicate that the CMRSA–ANN model could better explain the relationship between the input parameters and the compressive strength of RHA concrete.

Table 7. Performance comparison of previous works and the proposed model.

Reference	Model	Performance (R^2)
Iftikhar et al. [57]	GEP	0.9670
	RFR	0.9130
Amin et al. [79]	DT	0.8900
	BgR	0.9200
	ADB	0.9100
This paper	CMRSA–ANN	0.9709

Note: GEP, gene expression programming; RFR, random forest regression; DT, decision trees; BgR, bagging regressors; ADB, AdaBoost regressors.

6. Conclusions

The combination of RHA and concrete not only solves the problem of carbon dioxide emissions from cement production and reduces the pressure of waste accumulation, but could also be widely used as a green building material. To evaluate the performance of RHA concrete, we proposed a novel hybrid CMRSA–ANN model to predict the compressive strength of RHA concrete. We utilized 192 concrete data to train the model and test its performance. Furthermore, four ML models and an empirical model were developed, and their prediction results were compared with those of the proposed model. The main conclusions of this paper are as follows:

(1) The proposed hybrid CMRSA–ANN model achieved the best prediction accuracy for R^2 (0.9679 and 0.9709), VAF (96.7884% and 97.0911%), RMSE (2.9991 and 3.4489), and MAE (2.3169 and 2.6451) among all models in the both the training and the testing phases. The performance comparison between the proposed and optimized ANN models also indicated that the CMRSA could effectively improve the prediction ability of the ANN model.

(2) The empirical model could not better explain the relationship between the input parameters and the compressive strength of RHA concrete. Therefore, the empirical model was not suitable as a conventional means to evaluate concrete performance.

(3) The hybrid SOA–SVM and SOA–RF models achieved better performance than that of the unoptimized ANN and ELM models, indicated by a higher R^2 (0.9491 and 0.8941) and VAF (95.0044% and 89.5048%), and lower RMSE (4.5436 and 6.5743) and MAE (3.0904

and 4.8037) in the testing phase. It is effective and necessary to use an optimization (such as population-based) algorithm to improve the performance of ML models.

(4) Age was the most important input parameter for predicting the compressive strength of RHA concrete. However, other input parameters with similar importance scores should also be given high priority.

The purpose of this paper was to propose a new method for predicting RHA concrete strength, and the mining of the potential relationship among the data themselves through hybrid algorithm combination and optimization. However, the limitation of this paper is that the amount of data used for training and testing the models was always insufficient. An increase in effective data could help in improving the ability of the model to learn the potential relationship between input and output parameters, and the diversification of the test data could better verify the model performance. Therefore, adding more experimental data is an effective way to further improve the prediction accuracy of the model. Combinations of other optimization algorithms and different ML models in the performance prediction of RHA concrete are also worth comparing.

Supplementary Materials: The following supporting information can be downloaded at: https://www.mdpi.com/article/10.3390/ma16083135/s1.

Author Contributions: Conceptualization: C.L. and X.M.; methodology: C.L. and J.Z.; investigation: Z.C., D.D. and J.Z.; writing—original draft preparation: C.L. and X.M.; writing—review and editing: C.L., X.M., Z.C., D.D. and J.Z.; visualization: C.L. and D.D.; funding acquisition: C.L. All authors have read and agreed to the published version of the manuscript.

Funding: The first author was funded by the China Scholarship Council (grant no. 202106370038).

Institutional Review Board Statement: Not applicable.

Informed Consent Statement: Not applicable.

Data Availability Statement: The data used in this study are from published research: Iftikhar et al. [57].

Acknowledgments: The authors want to thank all the members who give us lots of help and cooperation.

Conflicts of Interest: The authors declare no conflict of interest.

References

1. Mei, X.; Cui, Z.; Sheng, Q.; Zhou, J.; Li, C. Application of the Improved POA-RF Model in Predicting the Strength and Energy Absorption Property of a Novel Aseismic Rubber-Concrete Material. *Materials* **2023**, *16*, 1286. [CrossRef] [PubMed]
2. Thomas, B.S. Green concrete partially comprised of rice husk ash as a supplementary cementitious material–A comprehensive review. *Renew. Sustain. Energy Rev.* **2018**, *82*, 3913–3923. [CrossRef]
3. Sheheryar, M.; Rehan, R.; Nehdi, M.L. Estimating CO_2 emission savings from ultrahigh performance concrete: A system dynamics approach. *Materials* **2021**, *14*, 995. [CrossRef] [PubMed]
4. Hamada, H.M.; Thomas, B.S.; Yahaya, F.M.; Muthusamy, K.; Yang, J.; Abdalla, J.A.; Hawileh, R.A. Sustainable use of palm oil fuel ash as a supplementary cementitious material: A comprehensive review. *J. Build. Eng.* **2021**, *40*, 102286. [CrossRef]
5. Amran, M.; Murali, G.; Fediuk, R.; Vatin, N.; Vasilev, Y.; Abdelgader, H. Palm oil fuel ash-based eco-efficient concrete: A critical review of the short-term properties. *Materials* **2021**, *14*, 332. [CrossRef]
6. Tayeh, B.A.; Hadzima-Nyarko, M.; Zeyad, A.M.; Al-Harazin, S.Z. Properties and durability of concrete with olive waste ash as a partial cement replacement. *Adv. Concr. Constr.* **2021**, *11*, 59–71.
7. Hakeem, I.Y.; Agwa, I.S.; Tayeh, B.A.; Abd-Elrahman, M.H. Effect of using a combination of rice husk and olive waste ashes on high-strength concrete properties. *Case Stud. Constr. Mater.* **2022**, *17*, e01486. [CrossRef]
8. Herath, C.; Gunasekara, C.; Law, D.W.; Setunge, S. Performance of high volume fly ash concrete incorporating additives: A systematic literature review. *Constr. Build. Mater.* **2020**, *258*, 120606. [CrossRef]
9. Teixeira, E.R.; Camões, A.; Branco, F.G. Synergetic effect of biomass fly ash on improvement of high-volume coal fly ash concrete properties. *Constr. Build. Mater.* **2022**, *314*, 125680. [CrossRef]
10. Mehta, A.; Ashish, D.K. Silica fume and waste glass in cement concrete production: A review. *J. Build. Eng.* **2020**, *29*, 100888. [CrossRef]
11. Khan, M.; Ali, M. Improvement in concrete behavior with fly ash, silica-fume and coconut fibres. *Constr. Build. Mater.* **2019**, *203*, 174–187. [CrossRef]

12. Beskopylny, A.N.; Shcherban, E.M.; Stel'makh, S.A.; Meskhi, B.; Shilov, A.A.; Varavka, V.; Özkılıç, Y.O.; Aksoylu, C.; Karalar, M. Composition Component Influence on Concrete Properties with the Additive of Rubber Tree Seed Shells. *Appl. Sci.* **2022**, *12*, 11744. [CrossRef]
13. Shcherban, E.M.; Stel'makh, S.A.; Beskopylny, A.N.; Mailyan, L.R.; Meskhi, B.; Shilov, A.A.; Chernil'nik, A.; Özkılıç, Y.O.; Aksoylu, C. Normal-Weight Concrete with Improved Stress–Strain Characteristics Reinforced with Dispersed Coconut Fibers. *Appl. Sci.* **2022**, *12*, 11734. [CrossRef]
14. Zeybek, Ö.; Özkılıç, Y.O.; Karalar, M.; Çelik, A.İ.; Qaidi, S.; Ahmad, J.; Burduhos-Nergis, D.D.; Burduhos-Nergis, D.P. Influence of replacing cement with waste glass on mechanical properties of concrete. *Materials* **2022**, *15*, 7513. [CrossRef] [PubMed]
15. Karalar, M.; Bilir, T.; Çavuşlu, M.; Özkiliç, Y.O.; Sabri Sabri, M.M. Use of recycled coal bottom ash in reinforced concrete beams as replacement for aggregate. *Front. Mater.* **2022**, *9*, 1064604. [CrossRef]
16. Qaidi, S.; Najm, H.M.; Abed, S.M.; Özkılıç, Y.O.; Al Dughaishi, H.; Alosta, M.; Sabri, M.M.S.; Alkhatib, F.; Milad, A. Concrete containing waste glass as an environmentally friendly aggregate: A review on fresh and mechanical characteristics. *Materials* **2022**, *15*, 6222. [CrossRef] [PubMed]
17. Çelik, A.İ.; Özkılıç, Y.O.; Zeybek, Ö.; Karalar, M.; Qaidi, S.; Ahmad, J.; Burduhos-Nergis, D.D.; Bejinariu, C. Mechanical Behavior of Crushed Waste Glass as Replacement of Aggregates. *Materials* **2022**, *15*, 8093. [CrossRef] [PubMed]
18. Karalar, M.; Özkılıç, Y.O.; Aksoylu, C.; Sabri MM, S.; Beskopylny, A.N.; Stel'makh, S.A.; Shcherban, E.M. Flexural behavior of reinforced concrete beams using waste marble powder towards application of sustainable concrete. *Front. Mater.* **2022**, *9*, 1068791. [CrossRef]
19. Çelik, A.L.İ.; Özkılıç, Y. Geopolymer concrete with high strength, workability and setting time using recycled steel wires and basalt powder. *Steel Compos. Struct.* **2023**, *46*, 689–707.
20. Acar, M.C.; Çelik, A.İ.; Kayabaşı, R.; Şener, A.; Özdöner, N.; Özkılıç, Y. Production of perlite-based-aerated geopolymer using hydrogen peroxide as eco-friendly material for energy-efficient buildings. *J. Mater. Res. Technol.* **2023**, *24*, 81–99. [CrossRef]
21. Santhosh, K.G.; Subhani, S.M.; Bahurudeen, A. Recycling of palm oil fuel ash and rice husk ash in the cleaner production of concrete-A review. *J. Clean. Prod.* **2022**, *354*, 131736. [CrossRef]
22. Amin, M.N.; Ahmad, W.; Khan, K.; Sayed, M.M. Mapping research knowledge on rice husk ash application in concrete: A scientometric review. *Materials* **2022**, *15*, 3431. [CrossRef] [PubMed]
23. Li, Q.; Song, Z. Prediction of compressive strength of rice husk ash concrete based on stacking ensemble learning model. *J. Clean. Prod.* **2023**, *382*, 135279. [CrossRef]
24. Ihedioha, J.N.; Ukoha, P.O.; Ekere, N.R. Ecological and human health risk assessment of heavy metal contamination in soil of a municipal solid waste dump in Uyo, Nigeria. *Environ. Geochem. Health* **2017**, *39*, 497–515. [CrossRef] [PubMed]
25. He, Z.H.; Yang, Y.; Yuan, Q.; Shi, J.Y.; Liu, B.J.; Liang, C.F.; Du, S.G. Recycling hazardous water treatment sludge in cement-based construction materials: Mechanical properties, drying shrinkage, and nano-scale characteristics. *J. Clean. Prod.* **2021**, *290*, 125832. [CrossRef]
26. Paris, J.M.; Roessler, J.G.; Ferraro, C.C.; DeFord, H.D.; Townsend, T.G. A review of waste products utilized as supplements to Portland cement in concrete. *J. Clean. Prod.* **2016**, *121*, 1–18. [CrossRef]
27. Madandoust, R.; Ranjbar, M.M.; Moghadam, H.A.; Mousavi, S.Y. Mechanical properties and durability assessment of rice husk ash concrete. *Biosyst. Eng.* **2011**, *110*, 144–152. [CrossRef]
28. Ahsan, M.B.; Hossain, Z. Supplemental use of rice husk ash (RHA) as a cementitious material in concrete industry. *Constr. Build. Mater.* **2018**, *178*, 1–9. [CrossRef]
29. Noaman, M.A.; Karim, M.R.; Islam, M.N. Comparative study of pozzolanic and filler effect of rice husk ash on the mechanical properties and microstructure of brick aggregate concrete. *Heliyon* **2019**, *5*, e01926. [CrossRef]
30. Mei, X.; Li, C.; Sheng, Q.; Cui, Z.; Zhou, J.; Dias, D. Development of a hybrid artificial intelligence model to predict the uniaxial compressive strength of a new aseismic layer made of rubber-sand concrete. *Mech. Adv. Mater. Struct.* **2022**, 1–18. [CrossRef]
31. Islam, M.N.; Mohd Zain, M.F.; Jamil, M. Prediction of strength and slump of rice husk ash incorporated high-performance concrete. *J. Civ. Eng. Manag.* **2012**, *18*, 310–317. [CrossRef]
32. Liu, C.; Zhang, W.; Liu, H.; Lin, X.; Zhang, R. A compressive strength prediction model based on the hydration reaction of cement paste by rice husk ash. *Constr. Build. Mater.* **2022**, *340*, 127841. [CrossRef]
33. Hamidian, P.; Alidoust, P.; Golafshani, E.M.; Niavol, K.P.; Behnood, A. Introduction of a novel evolutionary neural network for evaluating the compressive strength of concretes: A case of Rice Husk Ash concrete. *J. Build. Eng.* **2022**, *61*, 105293. [CrossRef]
34. Ozcan, G.; Kocak, Y.; Gulbandilar, E. Estimation of compressive strength of BFS and WTRP blended cement mortars with machine learning models. *Comput. Concr* **2017**, *19*, 275–282. [CrossRef]
35. Deshpande, N.; Londhe, S.; Kulkarni, S. Modeling compressive strength of recycled aggregate concrete by Artificial Neural Network, Model Tree and Non-linear Regression. *Int. J. Sustain. Built Environ.* **2014**, *3*, 187–198. [CrossRef]
36. Saha, P.; Debnath, P.; Thomas, P. Prediction of fresh and hardened properties of self-compacting concrete using support vector regression approach. *Neural Comput. Appl.* **2020**, *32*, 7995–8010. [CrossRef]
37. Dao, D.V.; Ly, H.B.; Vu HL, T.; Le, T.T.; Pham, B.T. Investigation and optimization of the C-ANN structure in predicting the compressive strength of foamed concrete. *Materials* **2020**, *13*, 1072. [CrossRef]
38. Bai, C.; Nguyen, H.; Asteris, P.G.; Nguyen-Thoi, T.; Zhou, J. A refreshing view of soft computing models for predicting the deflection of reinforced concrete beams. *Appl. Soft Comput.* **2020**, *97*, 106831. [CrossRef]

39. Yaman, M.A.; Abd Elaty, M.; Taman, M. Predicting the ingredients of self compacting concrete using artificial neural network. *Alex. Eng. J.* **2017**, *56*, 523–532. [CrossRef]
40. Han, Q.; Gui, C.; Xu, J.; Lacidogna, G. A generalized method to predict the compressive strength of high-performance concrete by improved random forest algorithm. *Constr. Build. Mater.* **2019**, *226*, 734–742. [CrossRef]
41. Azimi-Pour, M.; Eskandari-Naddaf, H.; Pakzad, A. Linear and non-linear SVM prediction for fresh properties and compressive strength of high volume fly ash self-compacting concrete. *Constr. Build. Mater.* **2020**, *230*, 117021. [CrossRef]
42. Zhang, J.; Ma, G.; Huang, Y.; Aslani, F.; Nener, B. Modelling uniaxial compressive strength of lightweight self-compacting concrete using random forest regression. *Constr. Build. Mater.* **2019**, *210*, 713–719. [CrossRef]
43. Yaseen, Z.M.; Deo, R.C.; Hilal, A.; Abd, A.M.; Bueno, L.C.; Salcedo-Sanz, S.; Nehdi, M.L. Predicting compressive strength of lightweight foamed concrete using extreme learning machine model. *Adv. Eng. Softw.* **2018**, *115*, 112–125. [CrossRef]
44. Iqtidar, A.; Bahadur Khan, N.; Kashif-ur-Rehman, S.; Faisal Javed, M.; Aslam, F.; Alyousef, R.; Alabduljabbar, H.; Mosavi, A. Prediction of compressive strength of rice husk ash concrete through different machine learning processes. *Crystals* **2021**, *11*, 352. [CrossRef]
45. Amin, M.N.; Iqtidar, A.; Khan, K.; Javed, M.F.; Shalabi, F.I.; Qadir, M.G. Comparison of machine learning approaches with traditional methods for predicting the compressive strength of rice husk ash concrete. *Crystals* **2021**, *11*, 779. [CrossRef]
46. Shaik, S.B.; Karthikeyan, J.; Jayabalan, P. Influence of using agro-waste as a partial replacement in cement on the compressive strength of concrete–A statistical approach. *Constr. Build. Mater.* **2020**, *250*, 118746. [CrossRef]
47. Asteris, P.G.; Kolovos, K.G. Self-compacting concrete strength prediction using surrogate models. *Neural Comput. Appl.* **2019**, *31* (Suppl. S1), 409–424. [CrossRef]
48. Getahun, M.A.; Shitote, S.M.; Gariy, Z.C.A. Artificial neural network based modelling approach for strength prediction of concrete incorporating agricultural and construction wastes. *Constr. Build. Mater.* **2018**, *190*, 517–525. [CrossRef]
49. Behnood, A.; Golafshani, E.M. Predicting the compressive strength of silica fume concrete using hybrid artificial neural network with multi-objective grey wolves. *J. Clean. Prod.* **2018**, *202*, 54–64. [CrossRef]
50. Golafshani, E.M.; Behnood, A.; Arashpour, M. Predicting the compressive strength of normal and High-Performance Concretes using ANN and ANFIS hybridized with Grey Wolf Optimizer. *Constr. Build. Mater.* **2020**, *232*, 117266. [CrossRef]
51. Shariati, M.; Mafipour, M.S.; Mehrabi, P.; Bahadori, A.; Zandi, Y.; Salih, M.N.; Nguyen, H.; Dou, J.; Song, X.; Poi-Ngian, S. Application of a hybrid artificial neural network-particle swarm optimization (ANN-PSO) model in behavior prediction of channel shear connectors embedded in normal and high-strength concrete. *Appl. Sci.* **2019**, *9*, 5534. [CrossRef]
52. Han, B.; Wu, Y.; Liu, L. Prediction and uncertainty quantification of compressive strength of high-strength concrete using optimized machine learning algorithms. *Struct. Concr.* **2022**, *23*, 3772–3785. [CrossRef]
53. Yan, F.; Lin, Z.; Wang, X.; Azarmi, F.; Sobolev, K. Evaluation and prediction of bond strength of GFRP-bar reinforced concrete using artificial neural network optimized with genetic algorithm. *Compos. Struct.* **2017**, *161*, 441–452. [CrossRef]
54. Tien Bui, D.; Abdullahi MA, M.; Ghareh, S.; Moayedi, H.; Nguyen, H. Fine-tuning of neural computing using whale optimization algorithm for predicting compressive strength of concrete. *Eng. Comput.* **2021**, *37*, 701–712. [CrossRef]
55. Huang, X.Y.; Wu, K.Y.; Wang, S.; Lu, T.; Lu, Y.F.; Deng, W.C.; Li, H.M. Compressive Strength Prediction of Rubber Concrete Based on Artificial Neural Network Model with Hybrid Particle Swarm Optimization Algorithm. *Materials* **2022**, *15*, 3934. [CrossRef]
56. Andalib, A.; Aminnejad, B.; Lork, A. Compressive Strength Prediction of Self-Compacting Concrete-A Bat Optimization Algorithm Based ANNs. *Adv. Mater. Sci. Eng.* **2022**, *2022*, 8404774. [CrossRef]
57. Iftikhar, B.; Alih, S.C.; Vafaei, M.; Elkotb, M.A.; Shutaywi, M.; Javed, M.F.; Deebani, W.; Khan, M.I.; Aslam, F. Predictive modeling of compressive strength of sustainable rice husk ash concrete: Ensemble learner optimization and comparison. *J. Clean. Prod.* **2022**, *348*, 131285. [CrossRef]
58. Li, C.; Dias, D. Assessment of the Rock Elasticity Modulus Using Four Hybrid RF Models: A Combination of Data-Driven and Soft Techniques. *Appl. Sci.* **2023**, *13*, 2373. [CrossRef]
59. Li, J.; Li, C.; Zhang, S. Application of Six Metaheuristic Optimization Algorithms and Random Forest in the uniaxial compressive strength of rock prediction. *Appl. Soft Comput.* **2022**, *131*, 109729. [CrossRef]
60. Zhou, J.; Li, C.; Asteris, P.G.; Shi, X.; Armaghani, D.J. Chart-Based Granular Slope Stability Assessment Using the Modified Mohr–Coulomb Criterion. *Arab. J. Sci. Eng.* **2022**, *48*, 5549–5569. [CrossRef]
61. Abualigah, L.; Abd Elaziz, M.; Sumari, P.; Geem, Z.W.; Gandomi, A.H. Reptile Search Algorithm (RSA): A nature-inspired meta-heuristic optimizer. *Expert Syst. Appl.* **2022**, *191*, 116158. [CrossRef]
62. Zhou, J.; Dai, Y.; Du, K.; Khandelwal, M.; Li, C.; Qiu, Y. COSMA-RF: New intelligent model based on chaos optimized slime mould algorithm and random forest for estimating the peak cutting force of conical picks. *Transp. Geotech.* **2022**, *36*, 100806. [CrossRef]
63. Zawbaa, H.M.; Emary, E.; Grosan, C. Feature selection via chaotic antlion optimization. *PLoS ONE* **2016**, *11*, e0150652. [CrossRef] [PubMed]
64. Varol Altay, E.; Alatas, B. Bird swarm algorithms with chaotic mapping. *Artif. Intell. Rev.* **2020**, *53*, 1373–1414. [CrossRef]
65. Li, C.; Zhou, J.; Armaghani, D.J.; Li, X. Stability analysis of underground mine hard rock pillars via combination of finite difference methods, neural networks, and Monte Carlo simulation techniques. *Undergr. Space* **2021**, *6*, 379–395. [CrossRef]
66. Li, C.; Zhou, J.; Khandelwal, M.; Zhang, X.; Monjezi, M.; Qiu, Y. Six novel hybrid extreme learning machine–swarm intelligence optimization (ELM–SIO) models for predicting backbreak in open-pit blasting. *Nat. Resour. Res.* **2022**, *31*, 3017–3039. [CrossRef]

67. Li, C.; Zhou, J.; Tao, M.; Du, K.; Wang, S.; Armaghani, D.J.; Mohamad, E.T. Developing hybrid ELM-ALO, ELM-LSO and ELM-SOA models for predicting advance rate of TBM. *Transp. Geotech.* **2022**, *36*, 100819. [CrossRef]
68. Zhou, J.; Dai, Y.; Huang, S.; Armaghani, D.J.; Qiu, Y. Proposing several hybrid SSA—Machine learning techniques for estimating rock cuttability by conical pick with relieved cutting modes. *Acta Geotechnica* **2022**, *18*, 1431–1446. [CrossRef]
69. Zhao, Y.; Hu, H.; Song, C.; Wang, Z. Predicting compressive strength of manufactured-sand concrete using conventional and metaheuristic-tuned artificial neural network. *Measurement* **2022**, *194*, 110993. [CrossRef]
70. Tipu, R.K.; Panchal, V.R.; Pandya, K.S. An ensemble approach to improve BPNN model precision for predicting compressive strength of high-performance concrete. In *Structures*; Elsevier: Amsterdam, The Netherlands, 2022; Volume 45, pp. 500–508.
71. Zhang, J.; Dias, D.; An, L.; Li, C. Applying a novel slime mould algorithm-based artificial neural network to predict the settlement of a single footing on a soft soil reinforced by rigid inclusions. *Mech. Adv. Mater. Struct.* **2022**, 1–16. [CrossRef]
72. Abdalla, A.; Mohammed, A.S. Hybrid MARS-, MEP-, and ANN-based prediction for modeling the compressive strength of cement mortar with various sand size and clay mineral metakaolin content. *Arch. Civ. Mech. Eng.* **2022**, *22*, 194. [CrossRef]
73. Gupta, T.; Rao, M.C. Prediction of compressive strength of geopolymer concrete using machine learning techniques. *Struct. Concr.* **2022**, *23*, 3073–3090. [CrossRef]
74. Nasir, V.; Dibaji, S.; Alaswad, K.; Cool, J. Tool wear monitoring by ensemble learning and sensor fusion using power, sound, vibration, and AE signals. *Manuf. Lett.* **2021**, *30*, 32–38. [CrossRef]
75. Zhou, J.; Qiu, Y.; Zhu, S.; Armaghani, D.J.; Li, C.; Nguyen, H.; Yagiz, S. Optimization of support vector machine through the use of metaheuristic algorithms in forecasting TBM advance rate. *Eng. Appl. Artif. Intell.* **2021**, *97*, 104015. [CrossRef]
76. Nasir, V.; Kooshkbaghi, M.; Cool, J.; Sassani, F. Cutting tool temperature monitoring in circular sawing: Measurement and multi-sensor feature fusion-based prediction. *Int. J. Adv. Manuf. Technol.* **2021**, *112*, 2413–2424. [CrossRef]
77. Pianosi, F.; Wagener, T. A simple and efficient method for global sensitivity analysis based on cumulative distribution functions. *Environ. Model. Softw.* **2015**, *67*, 1–11. [CrossRef]
78. Pianosi, F.; Wagener, T. Distribution-based sensitivity analysis from a generic input-output sample. *Environ. Model. Softw.* **2018**, *108*, 197–207. [CrossRef]
79. Amin, M.N.; Iftikhar, B.; Khan, K.; Javed, M.F.; AbuArab, A.M.; Rehman, M.F. Prediction model for rice husk ash concrete using AI approach: Boosting and bagging algorithms. In *Structures*; Elsevier: Amsterdam, The Netherlands, 2023; Volume 50, pp. 745–757.

materials

Article

Experimental Investigation on the Influence of Crack Width of Asphalt Concrete on the Repair Effect of Microbially Induced Calcite Precipitation

Ling Fan, Jinghong Zheng, Shuquan Peng *, Zhize Xun and Guoliang Chen

School of Resources and Safety Engineering, Central South University, Changsha 410083, China; pqrfanlinger@csu.edu.cn (L.F.); jinghong.zheng@csu.edu.cn (J.Z.); xunzhize@csu.edu.cn (Z.X.); chenguoliang0216@163.com (G.C.)

* Correspondence: pqr97linger@csu.edu.cn

Abstract: The appearance of cracks is one of the reasons that affect the performance of asphalt pavement, and traditional repair methods have the potential problem of causing adverse effects on the environment. In this paper, an environmentally friendly method for asphalt concrete crack repair was investigated using microbially induced calcite precipitation (MICP) for asphalt concrete cracks of different widths (0.5 mm, 1.0 mm, 1.5 mm, and 3 mm), and the effectiveness of repair was evaluated using nondestructive and destructive experiments. A varied ultrasonic pulse velocity was used to evaluate the healing process, and it was found that the samples with an initial crack width of 0.5 mm showed the most significant increase in wave velocity of 18.06% after repair. The results also showed that the uniaxial compressive strength and indirect tensile strength of the MICP-repaired samples recovered up to 47.02% and 34.68%. Static creep test results showed that MICP-repaired samples with smaller width cracks had greater resistance to permanent deformation. The results of uniaxial compressive strength tests on larger width (3 mm) cracks repaired by MICP combined with fibers showed that the strength of the samples was significantly increased by the addition of fibers. In addition, the SEM/EDS results showed that the MICP products were spherical calcite particles with a particle size distribution from 0 to 10 μm. This study shows that MICP has some potential for repairing cracks in asphalt concrete of different widths within the range investigated.

Keywords: microbially induced calcite precipitation; asphalt concrete; cracks; uniaxial compressive strength; fibers; spherical calcite

Citation: Fan, L.; Zheng, J.; Peng, S.; Xun, Z.; Chen, G. Experimental Investigation on the Influence of Crack Width of Asphalt Concrete on the Repair Effect of Microbially Induced Calcite Precipitation. *Materials* **2023**, *16*, 3576. https://doi.org/10.3390/ma16093576

Academic Editor: Giovanni Polacco

Received: 9 April 2023
Revised: 29 April 2023
Accepted: 5 May 2023
Published: 7 May 2023

1. Introduction

Asphalt pavements are widely used in traffic engineering all over the world because of the advantages of a short construction period, a smooth surface, and comfortable driving [1]. There are two main types of asphalt pavement cracks: load and non-load cracks. Cracks are almost inevitable in asphalt pavement structures, and long-term traffic loading is an important cause of cracks [2]. Studies have shown that pavement deterioration and cracking can lead to a reduced engineering application performance of asphalt pavements, such as indirect tensile strength [3] and permanent deformation [4].

There are generally two types of solutions for this pathology. One acts before the development of cracks and the other after the appearance of cracks, the former being a proactive way to reduce the possibility of cracks and the latter being a reactive repair after the cracks appear. The first method is to enhance the crack resistance of the asphalt mix, with the effect of preventing the extension of cracks or reducing the likelihood of cracks appearing. This approach includes enhancing the bonding effect of the binder (with the addition of asphalt mastic, rubber, etc. [5–7]) and enhancing the properties of the asphalt concrete aggregate (with the addition of polyester, lignin, glass, basalt fibers, etc. [8–12]).

When cracks appear in asphalt concrete, the traditional methods of treating cracks are heat-induced self-healing, crack sealing, or filling with organic asphalt-based materials [13,14]. Crack sealers are hardly cost-effective because of their poor treatment performance, high maintenance costs, and high environmental risks [15,16].

Recently, promising results have been achieved in eco-friendly fracture repair using biocolloid technology. Inorganic materials precipitated by microbially induced calcite precipitation (MICP) for the effective repair of cracks in cement concrete, brick, mortar, and rock have been extensively investigated due to their environmental friendliness and low-carbon nature [17–19]. The basic principle is to drop a biological agent into an already existing crack in the substrate material. After the biological agent enters the crack, the bacteria contained in the biological agent solution induce the formation of calcium carbonate precipitation by metabolizing the calcium ions and urea in the surrounding environment (cementing solution) [20]. However, the application of MICP in asphalt concrete crack filling has rarely been reported so far.

It has been shown that pure inorganic materials can be used to rapidly repair asphalt concrete cracks in municipal pavement manhole structures [16]. In addition, there have been attempts to combine MICP and asphalt mixtures. The application of microbial technology in asphalt mixtures was first used for modified cold asphalt emulsion mixes (CAEMs) [21]. By comparing the MICP method for CAEMs through two scenarios, it was found that the MICP method could significantly improve the mechanical properties of CAEMs. Among the studies on the moisture sensitivity of CAEMs, it was also found that the resistance to moisture damage was significantly improved after treatment with the MICP technique [22]. The above studies showed that MICP products have some compatibility with asphalt mixtures. Theoretically, it is possible to repair cracks in asphalt concrete using the MICP method.

In this paper, the effect of MICP on the repair effect of cracked asphalt concrete of different widths was investigated experimentally, and the repair performance of cracked asphalt concrete, including uniaxial compressive strength, indirect tensile strength, and permanent deformation, was evaluated. In addition, the physical morphology and chemical components were analyzed using scanning electron microscopy (SEM) and energy dispersive spectrometer (EDS) analysis.

2. Material and Sample Preparation

2.1. Sample Preparation

The binder used for the asphalt concrete samples was cationic emulsified asphalt (52 mm needle penetration, 66 cm ductility, and 50.5 °C softening point) produced by Dongguan Dongjiao Asphalt Co. The aggregates for the asphalt concrete samples were crushed gravel, mineral powder, and coarse and medium-fine sand, with 10% asphalt content for each sample. Based on the Talbot classification method [23], the mass ratio of five sizes of particles P_i ($i = 1, \ldots, 5$) is as shown in Equation (1):

$$P_i = \frac{M_i}{M} = \left(\frac{d_i}{d_{\max}}\right)^n \quad (1)$$

where $d_{\max}$ is the maximum size of the particle and n is the Talbot index ($n < 1$).

The total mass of each sample was 800 g. The grading of the aggregates at a Talbot index of 0.5 is shown in Table 1.

Table 1. Aggregate grading.

particle size (mm)	0–0.125	0.125–0.5	0.5–3	3–6	6–9
Mass fraction (%)	11.8	11.8	34.1	23.9	18.4

To investigate the effect of pre-crack width on MICP repair, intact samples and samples with crack widths of 0.5 mm, 1.0 mm, and 1.5 mm were designed for direct MICP repair, as well as samples with crack widths of 3 mm for MICP repair with fibers. It has been shown that the maximum crack depth of asphalt concrete in the field is less than 50 mm [24]. In this paper, for small-width (≤1.5 mm) cracks, the crack depth was set to 30 mm. For cracks of large width (3 mm), the crack depth was set to 40 mm. The length of the precast cracks used was fixed at 40 mm.

NCHRP Report No. 425 [25] was used to prepare the sample. First, the aggregates were weighed with an electronic scale (according to Table 1) and stirred clockwise and counterclockwise for 30 s, and then emulsified asphalt was added to the aggregates and stirred with the same mixing method. The mixture was then poured into a steel mold with 70.7 mm sides and the samples were compacted with a hammer. Next, a PVC hard plastic sheet was embedded in the top center of each sample and the PVC sheet was pulled out with tweezers after 1 h. After 48 h, the samples were de-molded and dried in a programmable temperature and humidity tester (TH-80CH) at 60 °C for 24 h. In this study, 15 samples of each crack width (0.5 mm, 1.0 mm, 1.5 mm, and 3 mm) were prepared, for a total of 60 samples with cracks. In addition, 12 complete samples without cracks were used as control.

2.2. Bacterial Culture and Cementing Solution

The bacteria used in this experiment were Bacillus pasteurii because of its significant advantages, such as easy extraction and nonbiological pathogenicity. The composition of the liquid nutrient medium used in the experiment was urea (20 g/L), soy protein (5 g/L), casein (15 g/L), and sodium chloride (5 g/L), and the pH of the medium was adjusted to 7.3 with NaOH. After the nutrient medium was prepared, the bacteria were inoculated into the culture medium at a rate of 1%. Finally, the inoculated bacterial culture was put into a constant temperature shaker and incubated at 30 °C and 130 rpm for 24 h. The OD600 and conductivity were measured every eight hours until their OD600 and conductivity reached about 2 and 20, respectively. An equimolar urea–calcium chloride solution with a concentration of 1 M was used as the MICP cementing solution.

2.3. Crack Repair Methods

The repair method varies depending on the width of the crack.

For small-width cracks (≤1.5 mm), precast cracked asphalt concrete samples were repaired by peristaltic pump drip injection. A BT100M peristaltic pump was used to simultaneously pump 10 mL of bacterial solution and cementing solution into the cracks at a rate of 3.0 rpm for eight days. The crack repair is shown schematically in Figure 1.

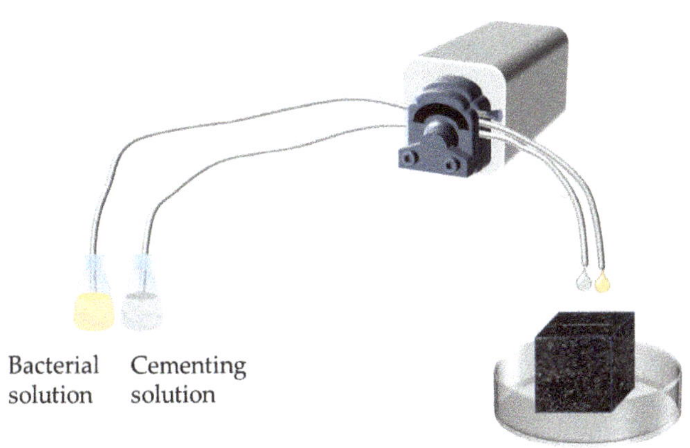

Figure 1. Schematic diagram of MICP grouting.

For samples with large-width cracks (3 mm), the MICP repair process used a peristaltic pump to simultaneously pump 20 mL of bacterial solution and 20 mL of the cementing solution into the cracks at 20.0 rpm, and the fibers (basalt fibers, polyester fibers, glass fibers, and steel fibers), as shown in Figure 2, were added separately to the cracks, along with the slurry solution, while grouting. In order to ensure the fibers were evenly distributed in the cracks, the fibers were slowly moved in place using a fine iron rod.

Figure 2. Fibers.

3. Evaluation of Repair Effect

Table 2 shows the experimental procedures of this study.

Table 2. Procedure for each group of tests.

Crack Widths		Whether to Add Fiber	Nondestructive Testing UPV	Destructive Testing		
				UCS	ITS	Creep
Small widths	0.5 mm	/	√	√	√	√
	1.0 mm	/	√	√	√	√
	1.5 mm	/	√	√	√	√
Large width	3.0 mm	√	/	√	/	/

Note: Three replicate trials were conducted for each group. '√' means carried out, '/' means not carried out.

3.1. Compressive Strength

The effect of MICP repair on the strength of small-width asphalt concrete cracked samples was determined using indirect and direct methods. For the indirect method, a conventional nondestructive test method called ultrasonic pulse velocity (UPV) was used to assess the degree of internal damage [26]. Prior to testing, petroleum jelly was applied to two opposing planes of the sample parallel to the fracture. A pair of transmitter and receiver transducers of the rock acoustic parameter tester (HS-YS4A) were placed close to these two planes for transmitting and receiving ultrasonic pulses. The ultrasonic velocity is the length of the path between the two transducers divided by the time for the ultrasonic pulse to pass through that path length. For the direct method, the UCS of the sample was tested with a servo tester (WHY-200/10). It is worth noting that during the test, the samples were loaded with cracks of different widths in different directions, as shown in Figures 3 and 4b.

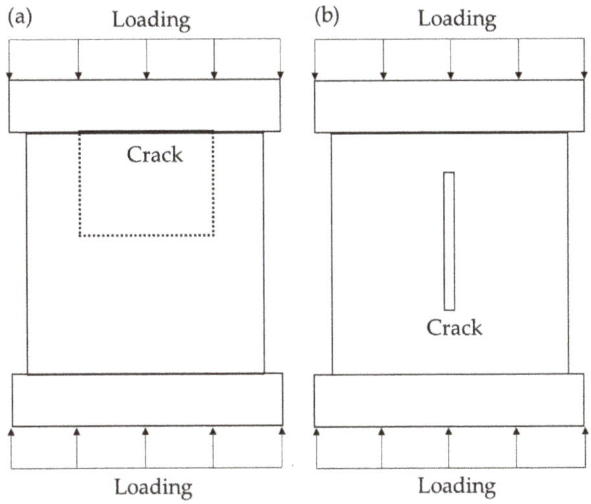

Figure 3. Schematic of loading method: (**a**) small-width crack (≤1.5 mm); and (**b**) large-width crack (3 mm).

Figure 4. Test procedures: (**a**) ultrasonic pulse velocity; (**b**) UCS test; (**c**) ITS test; and (**d**) static creep test.

3.2. Indirect Tensile Strength

An indirect tensile strength test was used to characterize the asphalt mixture in the tensile state [21], which can estimate the possibility of low temperature and fatigue cracking of asphalt concrete material under the action of thermal load and wheel load. Asphalt concrete samples underwent indirect tensile testing according to test method T0716-2011 of the technical specification [27]. A thin iron bar was placed parallel to the crack on the top surface of the sample, and a uniaxial compressive load was applied to the sample until the sample broke (Figure 4c). By applying a vertical load to the sample, this test produces a relatively uniform tensile along the plane where the crack is located, and fracture usually occurs in the same plane. The indirect (splitting) tensile strength is calculated as follows:

$$S_t = \frac{2P}{\pi A} \tag{2}$$

where P is the load when the sample is damaged and A is the cross-sectional area of the sample parallel to P.

3.3. Static Creep Test

Creep stiffness modulus is an important parameter for evaluating the deformation performance of asphalt concrete. According to NCHRP Project 9-6(1) [28], the test procedure was carried out using a servo tester (WHY-200/10) under a fixed loading compressive stress (σ_c) of 0.12 MPa (5–25% of the compressive strength without lateral limit), as shown in Figure 4d. The creep stiffness modulus is calculated using Equations (3) and (4):

$$E_{cq(t)} = \frac{\sigma_c}{\varepsilon_{c(t)}} \tag{3}$$

$E_{cq(t)}$ is the creep modulus at time t; σ_c is the compressive stress (Pa) applied to the sample; and $\varepsilon_{(c(t))}$ is the strain in the sample at time t (Calculation of creep stiffness modulus at 1, 10, 100, 500, 1000, 1500, 2000, 2500, 3000, 3500, and 3600 s).

$$\varepsilon_{c(t)} = \frac{\Delta_v(t)}{\ell} \tag{4}$$

where $\Delta_v(t)$ is the uniaxial deformation of the sample in time t and ℓ is the average height of the sample.

3.4. SEM/EDS

The split samples were analyzed using a scanning electron microscope (SEM) and energy dispersive spectroscopy (EDS) to obtain the microscopic morphological characteristics and chemical composition of microbial mineralization products.

The electron microscope and energy dispersive spectrometer used in this experiment were both from Hunan Kewei Testing Technology Co., Ltd., and the parameters were as follows: the model of the electron microscope was JSM-7610FPlus with a magnification of 500~10,000 times; the model of energy dispersive spectrometer was ULTIM MAX 40, and the detection elements were mainly C, O, and Ca.

4. Analysis and Evaluation of the Effect of Repairing the Width of Small Cracks (≤1.5 mm)

4.1. Observation of Repair Effect

Figure 5 shows the top surface (cross section) of the asphalt concrete samples after one day and 5 to 8 days of MICP treatment. It can be observed that in the MICP-treated samples, the cracks gradually healed as the number of days of treatment increased. After five days of repair, the samples with crack widths of 0.5 mm and 1.0 mm (Figures 3a and 5b) were basically healed, and visible cracks were still evident in the samples with crack widths of 1.5 mm (Figure 5c). After eight days of repair, the surface cracks of the samples were almost completely healed or sealed. In addition, the entire crack section of the sample was covered with precipitated calcium carbonate produced by the MICP process in the vicinity. The reason for the different distribution of white precipitates at the crack openings is that some organic matter produced during bacterial metabolism adheres the mineralization product particles together, thus forming dense, cohesive larger particles.

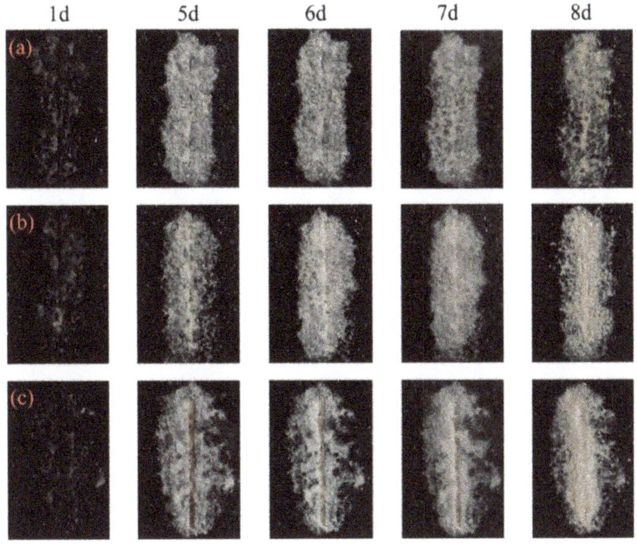

Figure 5. Filling situation of crack surface: (**a**) 0.5 mm; (**b**) 1.0 mm; and (**c**) 1.5 mm.

As Figure 6 shows the changing pattern of the surface repair rate and the fitted straight line in the repair process for different-width cracks. From the slope of the fitted straight line, it can be seen that the surface repair rate increases more rapidly for the sample with a crack width of 0.5 mm; the surface repair rate is the slowest for the sample with a crack width of 1.5 mm.

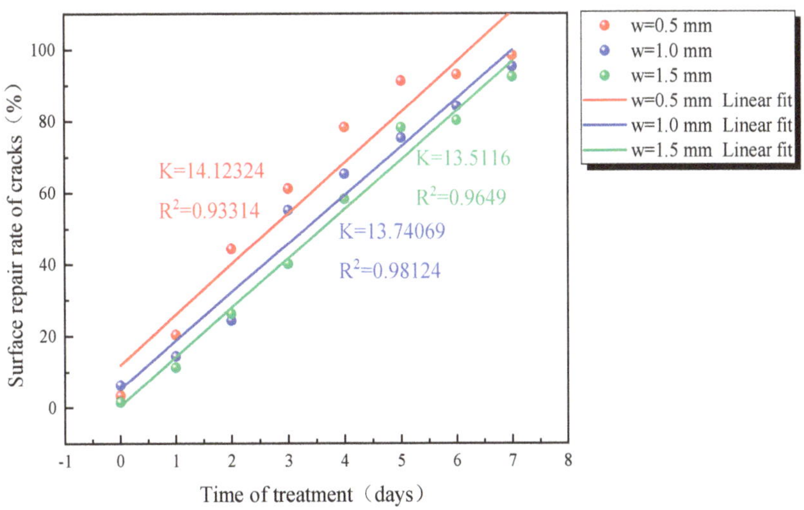

Figure 6. Percentage of fracture filling for different durations.

4.2. UPV

The greater the ultrasonic velocity, the better the repairing effect of asphalt concrete cracks. Therefore, the ultrasonic detection method can indirectly reflect the healing efficiency of asphalt concrete cracks. The results of the ultrasonic method for asphalt concrete before and after crack repair are shown in Figure 7a. After 8 d of healing, the acoustic wave velocity increased from 1545.36, 1536.96, and 1528.65 m/s to 1824.52, 1734.97, and 1683.33 m/s for initial crack widths of 0.5 mm, 1.0 mm, and 1.5 mm, respectively.

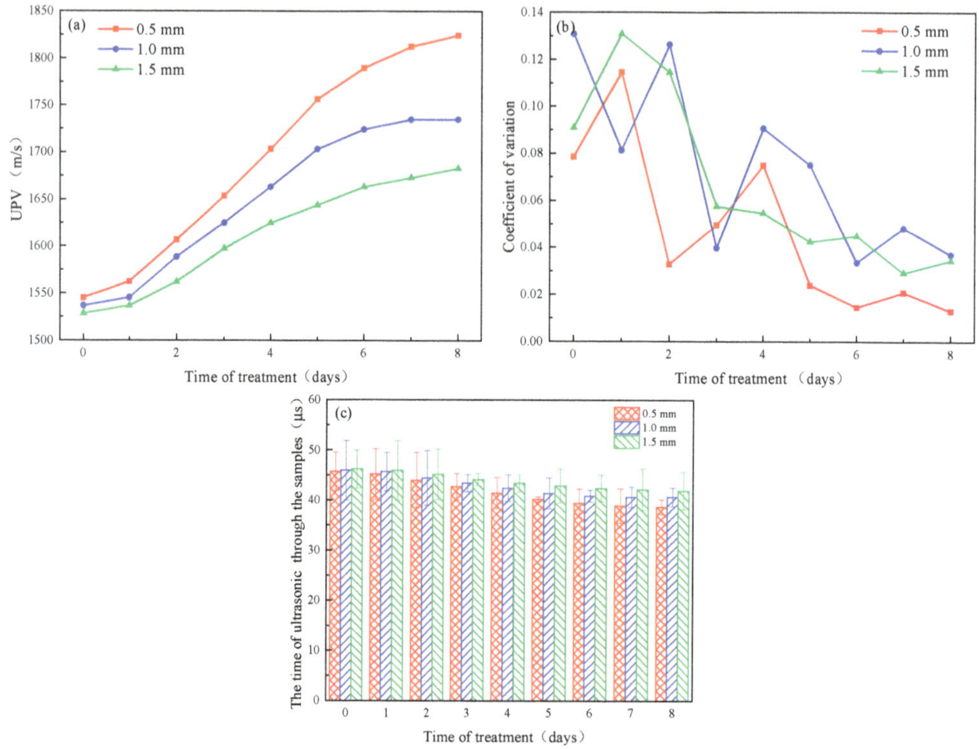

Figure 7. Ultrasonic velocity test results: (**a**) wave speed variation; (**b**) dispersion coefficient; and (**c**) sample ultrasonic treatment duration.

It can be seen that the ultrasonic velocity coefficient of variation for the three crack widths does not change much before and after the 8-day healing time (Figure 7b), and the changes are all within the range of 0.1. It is noteworthy that the coefficient of variation shows a decreasing pattern in general. The coefficient of variation for the cracks at a later stage of repair was lower than those for the cracks before repair, which may be due to the fact that the cracks were all better filled at the later stage of repair than at the earlier stage of repair, and thus the measured average wave velocities were more representative.

The asphalt concrete samples consisted of a solid and a gas phase because the asphalt concrete samples were sufficiently dry prior to the ultrasonic pulse velocity test. Due to the speed of propagation of the sound in solids being greater than that in air, when the cracks are filled, the solid phase of the asphalt concrete increases, the gas phase decreases, and the wave velocities show an increasing trend. The structural integrity of the 0.5 mm crack-width samples is greater than that of the 1.0 mm and 1.5 mm crack-width samples, and it is less difficult to form an effective 'bridging' effect on both sides of the crack than in the 1.0 mm and 1.5 mm crack-width samples, so the increase in wave velocity is most significant. The speed of the ultrasonic pulse increases as the crack width decreases, which is consistent with experimental expectations, and is in line with the results of Jongvivatsakul et al. [29] who showed the same pattern for the ultrasonic pulse velocities for the MICP repair of mortar cracks of different widths.

The above analysis shows that microorganisms can produce mineralization products for repairing asphalt concrete cracks of different widths, and the repairing effect is better for samples with smaller crack widths than those with larger crack widths.

4.3. UCS

The compressive strength of the mortar samples is shown in Figure 8. The presence of cracks in the samples was the main reason why the compressive strength of asphalt concrete was lower than that of intact samples. The compressive strengths of cracked asphalt concrete samples were 0.80, 0.70, and 0.55 MPa when the crack widths were 0.5, 1.0, and 1.5, respectively, which were 39.84%, 47.37%, and 58.65% lower than those of the control. This is due to the high-stress concentration caused by cracks, which resulted in damage at lower stresses compared to the control samples. After treatment with bacterial healing agents, the available bearing area increased with the formation of $CaCO_3$, thus allowing load transfer. The results showed that the stress concentration at the crack tip was reduced. As a result, the compressive strength of the cracked asphalt concrete increased to 1.05, 0.90, and 0.64 MPa. In addition, the compressive strength ratios were up to 78.94%, 67.67%, and 48.12% compared to the control samples. The increase in strength with crack healing is consistent with the increase in pulse velocity. Zheng et al. [30] and Qian et al. [31] confirmed that the higher the velocity of acoustic waves through the sample, the stronger the material.

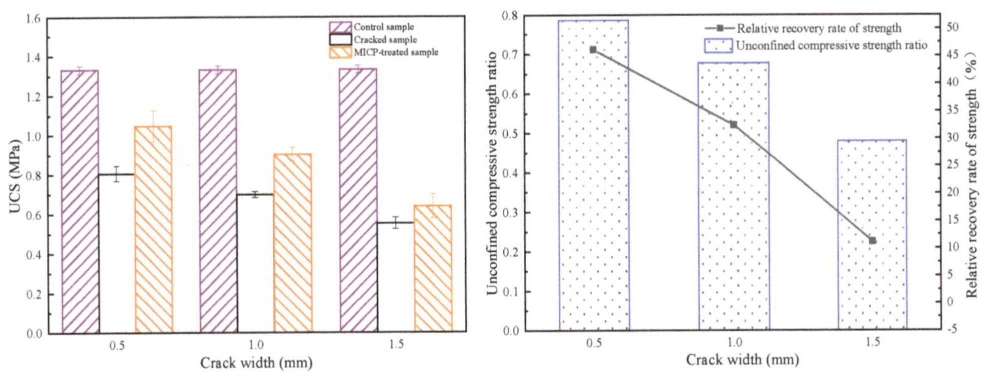

Figure 8. Compressive strength relationship for different crack widths under MICP treatment: (**left**) uniaxial compressive strength; and (**right**) strength ratio and recovery rate.

The asphalt has a strong bond between the aggregate and the calcium carbonate, which helps to form a bridge between the two sides of the crack [32]. The UCS decreases with increasing crack width, which indicates that strength loss increases with cracking. In addition, smaller crack widths exhibit greater microbial utilization during MICP repair [33], such that samples with a crack width of 0.5 mm show a larger UCS after MICP repair.

4.4. ITS

The indirect tensile strength (ITS) of MICP-repaired cracks of different widths is shown in Figure 9. It can be seen that the ITS values of unrepaired samples decreased with the increase in crack width. The ITS values of the samples repaired by MICP increased from 0.091, 0.070, and 0.055 MPa to 0.122, 0.092, and 0.063 MPa, respectively, and the recovery of tensile strength reached about 34.68%, 32.19%, and 13.64%.

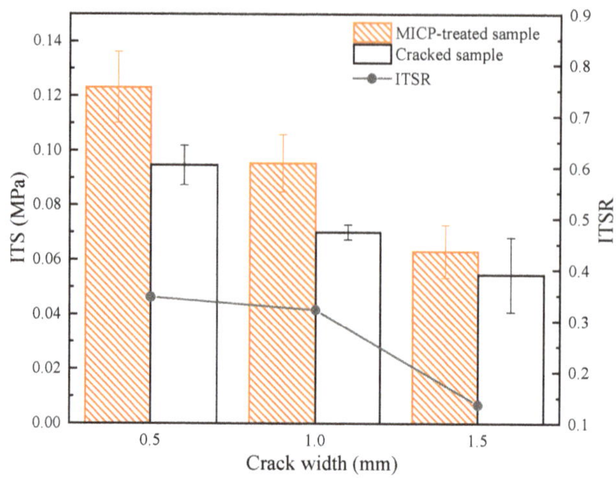

Figure 9. Indirect tensile strength relationship for different crack widths under MICP treatment.

It is speculated that the lower value of ITS may be due to insufficient crack healing and the formation of an imperfect bond between cracked samples. The samples with larger crack widths repaired by MICP showed smaller ITS. One reason for this may be the lack of bonding between the cracks and the smaller size and microstructure of the $CaCO_3$ distribution. The researchers found similar effects for potential applications in terms of indirect tensile testing following the bacterial repair of cracks in cement mortars [34]. Therefore, it can be concluded that the repair of asphalt concrete samples by MICP does increase the indirect tensile strength of the samples [35]. The mineralization products act as cementitious fillers in the cracks, and the ITS increase is more pronounced for small crack widths.

4.5. Permanent Deformation

The results of the static creep test are shown in Figure 10. It can be seen that the creep stiffness modulus has a large abrupt change within 100 s of loading, and after 500 s, the value of the creep stiffness modulus gradually stabilizes. The creep stiffness modulus decreases as the crack width increases. Notably, the creep stiffness modulus of asphalt concrete samples with crack widths w = 1.0 mm and 1.5 mm is slightly more prominent in the former than in the latter. In general, it is still consistent with the fact that the repaired creep stiffness modulus of the large crack width is smaller than that of the small crack width. This is because samples with large crack widths may have uneven filling as the cracks at the same depth are filled with mineralization products differently during microbial grouting.

The researcher, Manfro [36], found that the addition of calcium carbonate provides a resistance gain at the asphalt binder–aggregate interface, resulting in a modified asphalt mixture with greater resistance to permanent deformation. Samples with small crack widths and small voids between the two sides of the crack make it easier for MICP to achieve a relatively complete cementitious filling in the voids. They are easier to 'bridge', and the $CaCO_3$ bridge formed is less likely to break in the middle, so it has a greater ability in preventing permanent deformation.

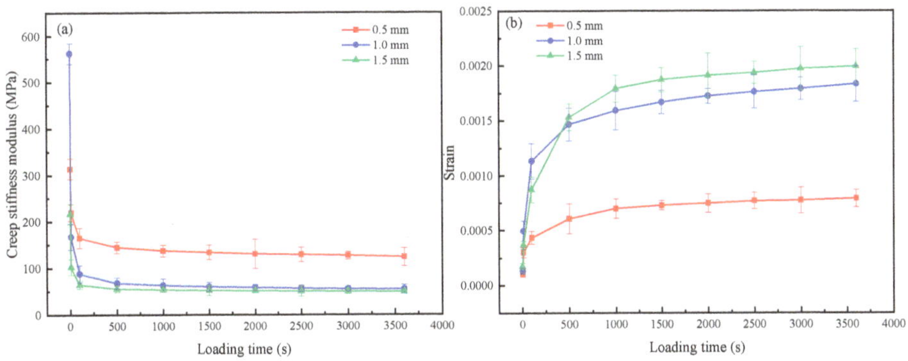

Figure 10. Static creep test: (**a**) creep stiffness modulus; and (**b**) strain-loading time curve.

4.6. SEM/EDS Analysis

SEM scans (Figure 11) show that the MICP mineralization products of this test have about four morphologies of crystallization. They are triangular conical, parallel hexahedral, prismatic, and spherical. Among them, the most common crystalline morphology is spherical.

Figure 11. Scanning results of MICP product electron microscopy: (**a**) triangular cone; (**b**) parallel hexahedron; (**c**) prism; and (**d**) spherical-shaped.

The particle size distribution of the particles in Figure 12 was measured using Image J2 software, and the selected particles are the particle sizes observable on the image surface, as shown in Figure 12a. Figure 12b shows the distribution range of the particle size, from which it can be seen that the majority of the MICP products have a particle size distribution in the range of 0 to 10 μm, and the mean value of the diameter of the 170 particles selected was 4.54 μm.

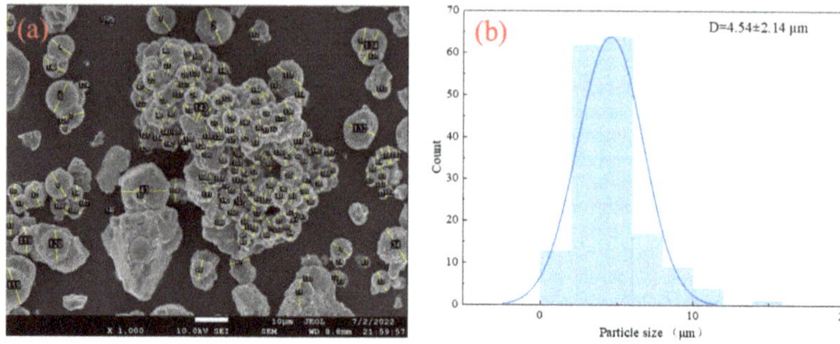

Figure 12. MICP product diameter measurement: (**a**) selected particles; and (**b**) particle size distribution.

A total of three EDS scans were performed in this experiment, and the analytical results are shown in Table 3 and Figure 13. The results showed that the precipitates were mainly composed of C, O, and Ca, with an atomic number approximately close to 1:3:1 and an atomic mass ratio of approximately 1:3:3 (Figure 13). It can be inferred that the MICP production is calcite.

Table 3. EDS test results.

Element	The First Test			The Second Test			The Third Test		
	Wt %	Wt % Sigma	Atomic %	Wt %	Wt % Sigma	Atomic %	Wt %	Wt % Sigma	Atomic %
C	15.09	0.09	24.83	13.47	0.10	22.41	13.59	0.15	20.71
O	44.83	0.12	55.40	45.95	0.14	57.37	42.97	0.17	55.23
Ca	40.08	0.10	19.77	40.58	0.12	20.22	43.44	0.12	24.06
Total	100	/	100	100	/	100	100	/	100

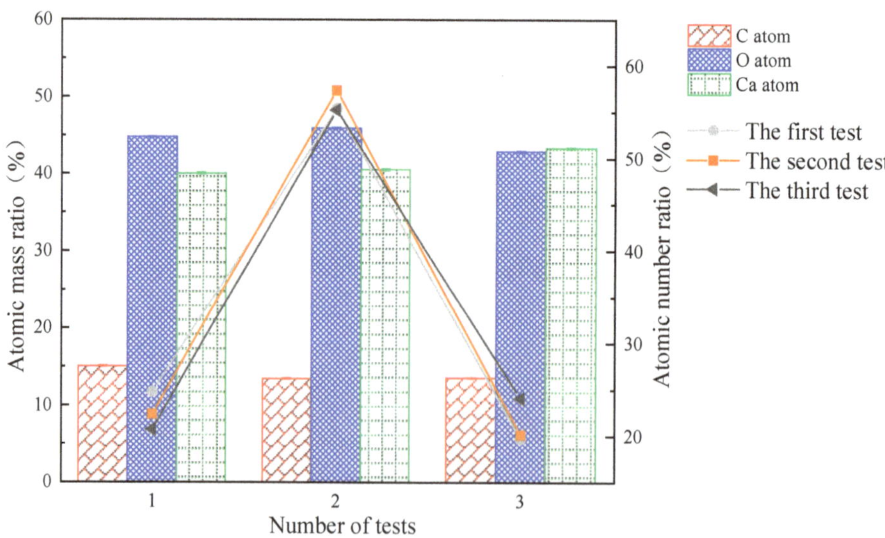

Figure 13. EDS scan results.

5. Analysis and Evaluation of the Effect of Repairing Large-Width Cracks (3 mm)

5.1. Effect of MICP Repair for Different Types of Fibers

Figure 14 shows the stress–strain curves of the samples with different fiber admixtures for repairing large-width asphalt concrete cracks. As a control, the stress–strain curves of the samples with 3 mm cracks without repair and without any fiber admixture are also shown in the figure.

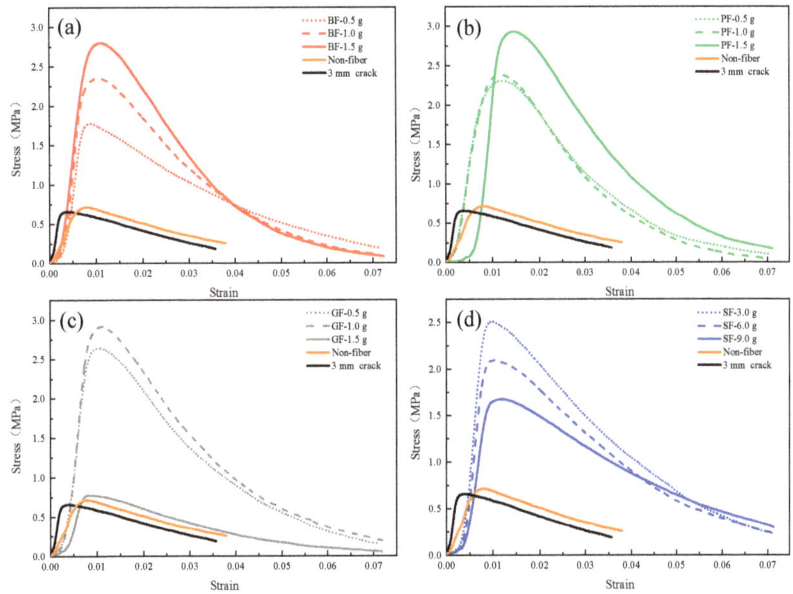

Figure 14. Stress–strain curves of repaired samples with different levels of fiber doping. (**a**) Basalt fibers (**b**) Polyester fibers (**c**) Glass fibers (**d**) Steel fibers.

Figure 14a illustrates that when the basalt fiber (BF) doping is increased from 0.5 g to 1.5 g, the strength increases, and the magnitude of the increase is close to about 0.45 MPa. In addition, the strain increases when the peak strength is reached. This indicates that the strength and toughness of the samples increased with the increase in basalt fiber doping during the MICP repair. Figure 14b demonstrates that when the doping of polyester fiber (PF) was increased from 0.5 g to 1.0 g, the strength increased, but not significantly, while the fiber doping was further increased from 1.0 g to 1.5 g, the strength increased relatively more significantly by about 0.54 MPa. This indicates that the small amount of basalt fiber (BF) and polyester fiber (PF) doping during MICP remediation is beneficial for bacteria to remain on them, and as a connecting bridge for MICP products, the fibers can form a bridging effect within the cracks for the integrity of the repaired samples.

Figure 14c shows that the strength of glass fiber (GF) increased by 0.27 MPa when increasing from 0.5 g to 1.0 g. In contrast to the pattern shown in Figure 14a,b, the strength did not increase but showed a significant decrease when increasing the fiber doping from 1.0 g to 1.5 g. The decreased strength was comparable to that of the MICP repair without any fiber doping. The reason for this phenomenon can be explained by the fact that the glass fiber admixture does help to improve the strength of asphalt concrete to some extent, the space inside the cracks of the samples is limited, and it is difficult for the fungus solution and the cementing solution to contact with it fully when the fiber admixture is too high, i.e., the MICP mineralization effect is negatively affected.

Figure 14d shows that during the MICP repair process, the strength gradually decreased as the steel fiber (SF) admixture increased from 3.0 g to 9.0 g, and the gradient of decrease was similar, about 0.4 MPa. This indicates that when the steel fiber admixture was too large, it squeezed too much space inside the cracks, and the effective bridging effect between the fiber and MICP products could not be formed.

Figure 15 visualizes the compressive strength of each sample with different fiber types and fiber mass for MICP repair. It can be seen that the measured compressive strengths differed from the previous sections when conducting the experimental study of repairing asphalt concrete samples with large-width cracks and were generally greater than those of the previous samples repairing cracks of different depths and small widths, which can be explained by the different placement of the samples when measuring the compressive strengths in this subsection compared to the previous subsections (shown in Figure 3). In addition, it is worth noting that the MICP repair without fiber doping showed an increase in strength compared to the samples with 3 mm cracks, but the increase was not significant, only 11.22%, indicating that when the crack width is greater than 3 mm, it is no longer appropriate to use a direct MICP repair. Scholars [37–40] have found similar effects when studying the coupling of MICP with each fiber.

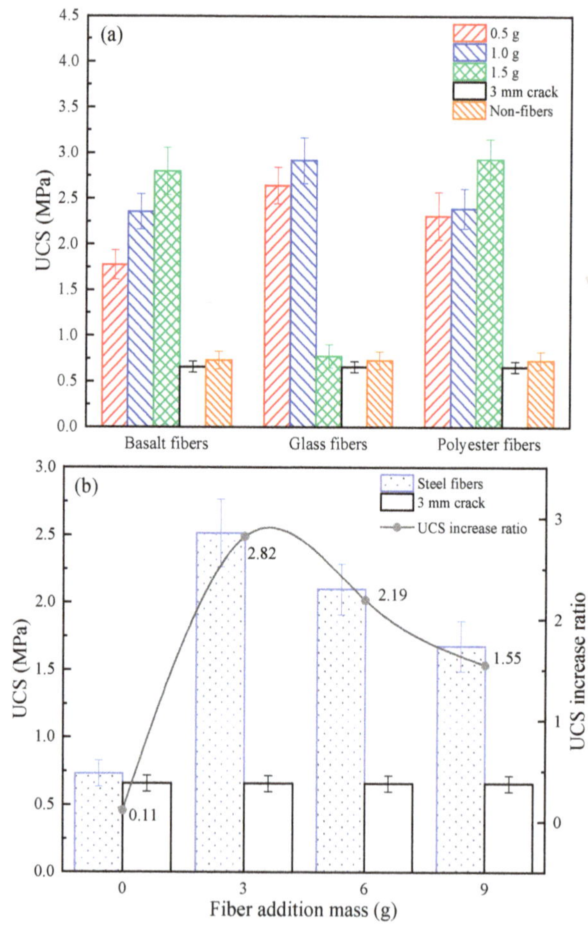

Figure 15. UCS with different fiber types and fiber doping: (**a**) BF, GF, and PF; and (**b**) SF.

The reasons for the above results can be inferred as follows:

(1) In the MICP repair process, the different fibers added in the cracks, even though they have the same quality, show different repair effects due to their different softness and volume, resulting in different distribution patterns in the cracks and different interaction relationships with the asphalt binder, aggregates, and MICP products.
(2) In the case of basalt fiber, its texture is hard, and it plays the role of strengthening the interconnection between calcium carbonate particles, and also provides the landing point for bacteria; in the case of polyester fiber, its texture is soft and it is more aggregated and distributed under the infiltration of slurry, which plays the local geotextile effect; for glass fiber, its softness and hardness are between basalt fiber and polyester fiber, and due to its longer length, it becomes entangled in the cracks. This winding effect can also strengthen the cementation between calcium carbonate particles, aggregates, and fibers to a certain extent; in the case of steel fibers, it has the hardest texture and the largest mass in the same volume, and the interlocking between fibers can also lead to the local siltation of calcium carbonate when repairing cracks via MICP due to the sinking through gravity, resulting in the effect with steel fibers where the MICP repair only occurs in a certain depth range of the cracks.

5.2. Investigation of Optimal Fiber Doping

Figure 16 shows the fitted curves of the doping mass of basalt fiber, glass fiber, polyester fiber, and steel fiber in the MICP process versus UCS, from which the fiber doping mass that has the best effect on UCS improvement can be deduced. The amount of addition for each fiber in this paper can be calculated using the following equation:

$$D = \frac{m}{V_c} \quad (5)$$

where m is the mass of fiber added and V_c is the crack volume.

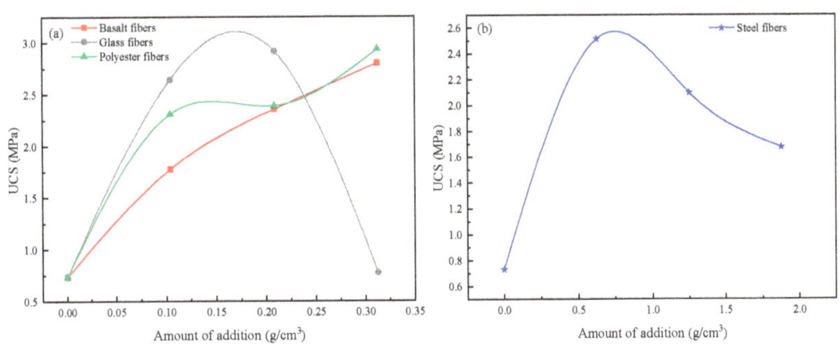

Figure 16. UCS–fiber doping fitting curve. (**a**) Basalt fibers, Glass fibers, Polyester fibers (**b**) Steel fibers.

It can be seen in the figure that the best doping mass of glass fiber is about 0.18 g/cm^3, the best doping mass of steel fiber is about 0.83 g/cm^3, and the best doping mass of polyester fiber and glass fiber is more than 0.31 g/cm^3. The specific values are subject to further experimental study.

6. Discussion

The spherical calcium carbonate particles of MICP mineralization products have better fluidity and dispersion [41], which are easily attached near the crack surface and clustered with each other from particle to particle, allowing the cracks to be fully cemented and filled.

Since the propagation speed of acoustic waves in solids is greater than that in air, the denser the mineralization quantities in products, the greater the measured wave speed, and the better the filling effect of cracks [42]. After the cracks are filled by MICP products, the bridging effect is produced on both sides of the cracks [43]. The degree of bridging effect is different due to the different widths of the cracks. The direct manifestation of the bridging effect is the ultrasonic wave velocity, and the ultrasonic wave velocity of the sample with large crack width after MICP repair is smaller than that of the smaller crack width, indicating that the large crack width is not easy to form a significant bridging effect. The indirect manifestation of the bridging effect is UCS, and the study showed that adding fibers to the cracks could enhance this bridging effect [37]. Corresponding to the ultrasonic wave velocity test results, the UCS of the MICP-repaired samples decreased as the crack width increased.

The adhesion between the microbial mineralization products and the cracks of the asphalt concrete samples depended upon the compatibility between the roughness of the crack surface and the mineralization products, while the strength of the mineralization products depended on their own properties [44]. Since the crack surface roughness was almost the same for all samples, the tensile strength depended more on the direct adhesion between the mineralization products and the cracks for the samples with small-width cracks. The indirect tensile strength of the MICP-repaired samples increased when the crack width decreased, which can be explained by the stronger adhesion between the MICP mineralized products and the cracks for small-width cracks. In addition, the stronger adhesion also resists a greater resistance to permanent deformation under this adhesion in the static creep test.

In order to facilitate quantitative research, all of the repair work in this paper was carried out on regular cracks. However, in practice, regular cracks are almost nonexistent, and, therefore, the effect of irregular cracks on the repair effect needs to be considered. Irregular crack extensions make it difficult to control the percolation path of the biological slurry within the crack and make the repair more difficult and complex. In future research, the percolation channels of MICP slurry under complex fracture extension and the mechanism of MICP product attachment under such complex percolation can be further explored.

7. Conclusions

In this study, the effectiveness and mechanism of the (MICP) technique for repairing asphalt concrete with different crack widths were investigated through a series of experiments. Several conclusions were drawn as follows:

(1) The wave speed increased from 1545.36, 1536.96, and 1528.65 m/s to 1824.52, 1734.97, and 1683.33 m/s for crack widths of 0.5 mm, 1.0 mm, and 1.5 mm, respectively, and the uniaxial compressive strength for the MICP-repaired samples also increased compared to the unrepaired samples by 78.94%, 67.67%, and 48.12%, respectively. In addition, the recovery rates of indirect tensile strength were 34.68%, 32.19%, and 13.64%, respectively.

(2) The static creep test results showed that the creep modulus of the MICP samples was finally stabilized at 123.85 MPa, 54.83 MPa, and 49.83 MPa after 3600 s of constant pressure loading, indicating that the repaired MICP with a crack width of 0.5 mm had greater resistance to deformation.

(3) SEM and EDS images showed that the MICP product of this study was $CaCO_3$, which was mainly in the form of spherical calcite. the particle size of $CaCO_3$ was mostly concentrated within 10 μm, with an average value of 4.54 μm.

(4) The results of MICP combined with different fibers to repair large-width cracks showed that the addition of fibers contributed to the filling effect of the MICP product in the cracks of the as-built concrete. the UCS results showed that the optimum dose was about 0.18 g/cm^3 for glass fibers, about 0.83 g/cm^3 for steel fibers, and over 0.31 g/cm^3 for polyester and basalt fibers.

The above findings indicate that the asphalt concrete samples with single cracks repaired with MICP showed some degree of recovery in performance. In general, the repair effect increased significantly as the crack width decreased. When the crack width was more extensive, the combined repair effect of MICP and fiber was better than the direct MICP repair.

Author Contributions: Methodology, S.P. and L.F.; software, J.Z.; validation, J.Z. and S.P.; formal analysis, L.F. and G.C.; data curation, J.Z. and Z.X.; writing—original draft preparation, J.Z.; writing—review and editing, S.P. and L.F.; project administration, S.P. and L.F.; funding acquisition, S.P. and L.F. All authors have read and agreed to the published version of the manuscript.

Funding: This project is financially supported by the National Natural Science Foundation of China (grant # 52174100 and # 51674287), and the National Science Foundation of Hunan Province, China (grant # 2021JJ30834).

Institutional Review Board Statement: Not applicable.

Informed Consent Statement: Not applicable.

Data Availability Statement: The data that support the findings of this study are available upon request from the authors.

Conflicts of Interest: The authors declare no conflict of interest.

Abbreviations

Abbreviation	Definition
MICP	microbially induced calcium precipitation
SEM	scanning electron microscope
EDS	energy dispersive spectrometer
CAEM	cold asphalt emulsion mixes
UPV	ultrasonic pulse velocity
UCS	uniaxial compressive strength
ITS	indirect tensile strength
BF	basalt fiber
PF	polyester fiber
GF	glass fiber
SF	steel fiber

References

1. Yang, Z. Study on Multi-Scale Behavioral Characteristics of Asphaltbefore and after Aging. Ph.D. Thesis, South China University of Technology, Guangzhou, China, 2018.
2. Kim, H.; Wagoner, M.P.; Buttlar, W.G. Simulation of Fracture Behavior in Asphalt Concrete Using a Heterogeneous Cohesive Zone Discrete Element Model. *J. Mater. Civ. Eng.* **2008**, *20*, 552–563. [CrossRef]
3. Ma, Z.; Liu, L.; Sun, L. Investigation of top-down cracking performance of in-situ asphalt mixtures based on accelerated pavement testing and laboratory tests. *Constr. Build. Mater.* **2018**, *179*, 177–184. [CrossRef]
4. Cascione, A.A.; Williams, R.C.; Yu, J. Performance testing of asphalt pavements with recycled asphalt shingles from multiple field trials. *Constr. Build. Mater.* **2015**, *101*, 628–642. [CrossRef]
5. Zhang, Z.; Sun, J.; Huang, Z.; Wang, F.; Jia, M.; Lv, W.; Ye, J. A laboratory study of epoxy/polyurethane modified asphalt binders and mixtures suitable for flexible bridge deck pavement. *Constr. Build. Mater.* **2021**, *274*, 122084. [CrossRef]
6. Shafabakhsh, G.; Ahmadi, S. Reflective cracking reduction by a comparison between modifying asphalt overlay and sand asphalt interlayer: An experimental evaluation. *Int. J. Pavement Eng.* **2019**, *22*, 192–200. [CrossRef]
7. Moreno-Navarro, F.; Sol-Sánchez, M.; Rubio-Gámez, M.C. The effect of polymer modified binders on the long-term performance of bituminous mixtures: The influence of temperature. *Mater. Design* **2015**, *78*, 5–11. [CrossRef]
8. Fan, T.; Si, C.; Zhang, Y.; Zhu, Y.; Li, S. Optimization Design of Asphalt Mixture Composite Reinforced with Calcium Sulfate Anhydrous Whisker and Polyester Fiber Based on Response Surface Methodology. *Materials* **2023**, *16*, 594. [CrossRef]
9. Xue, X.; Gao, J.; Wang, J.; Chen, Y. Evaluation of High-Temperature and Low-Temperature Performances of Lignin-Waste Engine Oil Modified Asphalt Binder and Its Mixture. *Materials* **2021**, *15*, 52. [CrossRef]
10. Li, H.; Yu, J.; Wu, S.; Liu, Q.; Li, Y.; Wu, Y.; Xu, H. Investigation of the Effect of Induction Heating on Asphalt Binder Aging in Steel Fibers Modified Asphalt Concrete. *Materials* **2019**, *12*, 1067. [CrossRef]
11. Khater, A.; Luo, D.; Abdelsalam, M.; Ma, J.; Ghazy, M. Comparative Life Cycle Assessment of Asphalt Mixtures Using Composite Admixtures of Lignin and Glass Fibers. *Materials* **2021**, *14*, 6589. [CrossRef]
12. Wang, W.; Cheng, Y.; Tan, G. Design Optimization of SBS-Modified Asphalt Mixture Reinforced with Eco-Friendly Basalt Fiber Based on Response Surface Methodology. *Materials* **2018**, *11*, 1311. [CrossRef] [PubMed]
13. Awuah, F.K.A.; Garcia-Hernández, A. Machine-filling of cracks in asphalt concrete. *Automat. Constr.* **2022**, *141*, 104463. [CrossRef]
14. Yin, J.; Pang, Q.; Wu, H.; Song, W. Using a polymer-based sealant material to make crack repair of asphalt pavement. *J. Test. Eval.* **2018**, *46*, 20170041. [CrossRef]

15. Gnatenko, R.; Tsyrkunova, K.; Zhdanyuk, V. Technological Sides of Crack Sealing in Asphalt Pavements. *Transport. Res. Procedia* **2016**, *14*, 804–810. [CrossRef]
16. Wang, Y.; Kong, L.; Chen, Q.; Lau, B.; Wang, Y. Research and application of a black rapid repair concrete for municipal pavement rehabilitation around manholes. *Constr. Build. Mater.* **2017**, *150*, 204–213. [CrossRef]
17. Sun, X.; Miao, L.; Wu, L.; Wang, H. Theoretical quantification for cracks repair based on microbially induced carbonate precipitation (MICP) method. *Cem. Concr. Comp.* **2021**, *118*, 103950. [CrossRef]
18. Cheng, L.; Kobayashi, T.; Shahin, M.A. Microbially induced calcite precipitation for production of "bio-bricks" treated at partial saturation condition. *Constr. Build. Mater.* **2020**, *231*, 117095. [CrossRef]
19. Peng, S.; Zhang, K.; Fan, L.; Kang, J.; Peng, K.; Wang, F.; Chen, Z. Permeability Reduction and Electrochemical Impedance of Fractured Rock Grouted by Microbial-Induced Calcite Precipitation. *Geofluids* **2020**, *2020*, 8876400. [CrossRef]
20. Iqbal, D.M.; Wong, L.S.; Kong, S.Y. Bio-Cementation in Construction Materials: A Review. *Materials* **2021**, *14*, 2175. [CrossRef]
21. Attaran Dovom, H.; Mohammadzadeh Moghaddam, A.; Karrabi, M.; Shahnavaz, B.; Attaran Dowom, S. Investigation of the mechanical and physical properties of bio-modified cold asphalt emulsion mixtures by microbial carbonate precipitation. *Int. J. Pavement Eng.* **2019**, *22*, 404–417. [CrossRef]
22. Attaran Dovom, H.; Mohammadzadeh Moghaddam, A.; Karrabi, M.; Shahnavaz, B. Improving the resistance to moisture damage of cold mix asphalt modified by eco-friendly Microbial Carbonate Precipitation (MCP). *Constr. Build. Mater.* **2019**, *213*, 131–141. [CrossRef]
23. Xin, J.; Pei, J.; Akiyama, M.; Li, R.; Zhang, J.; Shao, L. A Study on the Design Method for the Material Composition of Small Particle-Size Asphalt Mixture for Controlling Cracks in Asphalt Pavement. *Appl. Sci.* **2019**, *9*, 1988. [CrossRef]
24. Uhlmeyer, J.S.; Willoughby, K.; Pierce, L.M. Top-Down Cracking in Washington State Asphalt Concrete Wearing Courses. *Transport. Res. Rec.* **2000**, *1730*, 110–116. [CrossRef]
25. Brown, E.R. Designing stone matrix asphalt mixtures for rut-resistant pavements (no. 425–430). In *Transportation Research Board*; Transportation Research Board, National Research Council: Washington, DC, USA, 1999.
26. Bani Baker, M.I.; Abendeh, R.M.; Khasawneh, M.A. Freeze and Thaw Effect on Asphalt Concrete Mixtures Modified with Natural Bentonite Clay. *Coatings* **2022**, *12*, 1664. [CrossRef]
27. Ministry of Transport of the People' Republic of China. *Standard Test Methods of Bitumen and Bituminous Mixtures for Highway Engineering JTG E20–2011*; China Communication Press: Beijing, China, 2011.
28. Quintus, H.V.; Hughes, C.S.; Scherocman, J.A. NCHRP asphalt-aggregate mixture analysis system. *Transport. Res. Rec.* **1992**, *1353*, 90–99.
29. Jongvivatsakul, P.; Janprasit, K.; Nuaklong, P. Investigation of the crack healing performance in mortar using microbially induced calcium carbonate precipitation (MICP) method. *Constr. Build. Mater.* **2019**, *212*, 737–744. [CrossRef]
30. Wang, Z.; Ning, J.; Ren, H. Frequency characteristics of the released stress wave by propagating cracks in brittle materials. *Theor. Appl. Fract. Mech.* **2018**, *96*, 72–82. [CrossRef]
31. Qian, C.; Zheng, T.; Zhang, X.; Su, Y. Application of microbial self-healing concrete: Case study. *Constr. Build. Mater.* **2021**, *290*, 123226. [CrossRef]
32. Pannem, R.; Chintalapudi, K. Evaluation of Strength Properties and Crack Mitigation of Self-healing Concrete. *Jordan J. Civ. Eng.* **2019**, *13*, 386–393.
33. Sun, X.H.; Miao, L.C. Application of Bio-remediation with Bacillus megaterium for Crack Repair at Low Temperature. *J. Adv. Concr. Technol.* **2020**, *18*, 307–319. [CrossRef]
34. Kulkarni, P.B.; Nemade, P.D.; Wagh, M.P. Healing of Generated Cracks in Cement Mortar Using MICP. *Civ. Eng. J.* **2020**, *6*, 679–692. [CrossRef]
35. Lu, C.; Ge, H.; Li, Z.; Zheng, Y. Effect evaluation of microbial mineralization for repairing load-induced crack in concrete with a cyclic injection-immersion process. *Case Stud. Constr. Mater.* **2022**, *17*, e01702. [CrossRef]
36. Manfro, A.L.; Melo, J.V.S.; Carpio, J.A.V. Permanent deformation performance under moisture effect of an asphalt mixture modified by calcium carbonate nanoparti-cles. *Constr. Build. Mater.* **2022**, *342*, 128087. [CrossRef]
37. Li, H.F.; Li, Z.; Liu, Y. Effect of basalt fibers on the mechanical and self-healing properties of expanded perlite solid-loaded microbial mortars. *J. Build. Eng.* **2022**, *62*, 105201. [CrossRef]
38. Yang, D.F.; Xu, G.B.; Duan, Y. Self-healing cement composites based on bleaching earth immobilized bacteria. *J. Clean Prod.* **2022**, *358*, 132045. [CrossRef]
39. Zhang, D.; Shahin, M.A.; Yang, Y. Effect of microbially induced calcite precipitation treatment on the bonding properties of steel fiber in ultra-high performance concrete. *J. Build. Eng.* **2022**, *50*, 104132. [CrossRef]
40. Zhao, J.T.; Tong, H.W.; Shan, Y. Effects of Different Types of Fibers on the Physical and Mechanical Properties of MICP-Treated Calcareous Sand. *Materials* **2021**, *14*, 268. [CrossRef]
41. Song, C.; Elsworth, D.; Zhi, S.; Wang, C. The influence of particle morphology on microbially induced $CaCO_3$ clogging in granular media. *Mar. Georesour. Geotec.* **2019**, *39*, 74–81. [CrossRef]
42. Zheng, T.; Su, Y.; Zhang, X.; Zhou, H.; Qian, C. Effect and Mechanism of Encapsulation-Based Spores on Self-Healing Concrete at Different Curing Ages. *ACS. Appl. Mater. Interfaces* **2020**, *12*, 52415–52432. [CrossRef]

43. Lin, H.; Suleiman, M.T.; Brown, D.G. Investigation of pore-scale $CaCO_3$ distributions and their effects on stiffness and permeability of sands treated by microbially induced carbonate precipitation (MICP). *Soils. Found.* **2020**, *60*, 944–961. [CrossRef]
44. Choi, S.-G.; Wang, K.; Wen, Z.; Chu, J. Mortar crack repair using microbial induced calcite precipitation method. *Cem. Concr. Comp.* **2017**, *83*, 209–221. [CrossRef]

materials

MDPI

Article

Numerical Investigation on the Compressive Behavior of Desert Sand-Based Backfill Material: Parametric Study

Haitian Yan [1], Honglin Liu [1,2,*], Guodong Li [1,2,3], Xiangyu Wang [3] and Yinjian Hang [4]

[1] School of Geology and Mining Engineering, Xinjiang University, Urumqi 830046, China; 107552101622@stu.xju.edu.cn (H.Y.); cklgd2011@xju.edu.cn (G.L.)

[2] Key Laboratory of Environmental Protection Mining for Minerals Resources at Universities of Education Department of Xinjiang Uygur Autonomous Region, Xinjiang University, Urumqi 830047, China

[3] School of Mines, China University of Mining and Technology, Xuzhou 221116, China; wangxiangyu79@126.com

[4] Xinjiang Sail Energy Co., Xuzhou Coal Mining Group, Tacheng 834700, China; hangyinjian@126.com

* Correspondence: liuhonglin@xju.edu.cn; Tel.: +86-139-9982-2448

Abstract: As a key node in the promotion of the "Western Development" strategy in Xinjiang, China, the large-scale mining of coal resources is bound to cause a series of ecological and environmental problems, such as surface subsidence. Desert areas are widely distributed in Xinjiang, and from the perspective of reserves and sustainable development, it is crucial to fully utilize desert sand to make filling materials and predict its mechanical strength. In order to promote the application of High Water Backfill Material (HWBM) in mining engineering, a modified HWBM doped with Xinjiang Kumutage desert sand was used to prepare a desert sand-based backfill material, and its mechanical properties were tested. The discrete element particle flow software PFC3D is used to construct a three-dimensional numerical model of desert sand-based backfill material. The parameters such as sample sand content, porosity, desert sand particle size distribution, and model size are changed to study their impact on the bearing performance and scale effect of desert sand-based backfill materials. The results indicate that a higher content of desert sand can effectively improve the mechanical properties of HWBM specimens. The stress–strain relationship inverted by the numerical model is highly consistent with the measured results of desert sand-based backfill materials. Improving the particle size distribution of desert sand and reducing the porosity of filling materials within a certain range can significantly improve the bearing capacity of desert sand-based backfill materials. The influence of changing the range of microscopic parameters on the compressive strength of desert sand-based backfill materials was analyzed. This study provides a desert sand-based backfill material that meets the requirements of mine filling, and predicts its strength through numerical simulation.

Keywords: desert sand; high-water backfill material; PFC3D; parametric study

Citation: Yan, H.; Liu, H.; Li, G.; Wang, X.; Hang, Y. Numerical Investigation on the Compressive Behavior of Desert Sand-Based Backfill Material: Parametric Study. *Materials* **2023**, *16*, 3878. https://doi.org/10.3390/ma16103878

Academic Editor: Krzysztof Schabowicz

Received: 25 April 2023
Revised: 14 May 2023
Accepted: 16 May 2023
Published: 22 May 2023

1. Introduction

High-water backfill materials (HWBMs) have high early strength, a short gelling time, a simple construction technology, and other advantages [1–3]. Many mines have used HWBMs for goaf filling and roadway side support [4–6]. However, filling materials with a high-water content are expensive and decrease in strength with the passage of time. With the promotion of gob-side entry retaining technology and the expansion of the application scope of the HWBM in mine engineering [7–9], the performance of HWBMs in engineering practice is further improved.

In fact, a modified HWBM has the same mechanical properties as pure high-water backfill material. Many scholars use different materials, such as fly ash, river sludge, silica fume, and polyethylene plastics, to modify high-water backfill materials and analyze the changes in the properties of HWBMs [10–14]. On adding fly ash, the uniaxial compressive strength and the elastic modulus of the HWBM decrease, the cost decreases, and the residual strength increases [15]. The influence of lithium carbonate and aluminum sulfate on the properties of HWBM was analyzed via a hydration heat test, X-ray diffraction, a scanning electron microscope, and other microscopic tests. It was found that the addition of lithium carbonate and aluminum sulfate can promote HWBM hydration, shorten the solidification time, and improve the early and late compressive strength of the hydration [16,17].

However, most of the coal mines in northwest China are located in arid and semi-arid areas that are short of water. Water resources are precious. It is a potential method to improve its performance to make desert sand-based backfill material with abundant desert sand. The use of desert sand as fine aggregate for concrete production has become a hot research issue, and relevant studies have proved that desert sand concrete has good mechanical properties. Reasonable grading of desert sand aggregate can make concrete denser, reduce the water consumption and cement consumption of concrete mixture per unit volume, and also make the skeleton and stability of aggregate reach a good state [18–21].

The discrete element method was first proposed by Cundall [22], and the PFC particle flow method, as a kind of discrete element method, is mostly used to study the meso-mechanical properties of granular medium materials. Discrete element particle flow software PFC3D is based on the discrete element method DEM code to track the trajectory of each particle. At present, many scholars build discrete element numerical models by different methods. YingYan et al. [23] established the DEM numerical model by using the quaternion method and simulated the direct shear test of irregular limestone gravel. Y.H. et al. [24], combined with the improved bond particle model, considered the physical characteristics of solid bonds when measuring the compressive strength in the unconfined compaction process, proving that the solid bond model must be considered when using the DEM method to simulate the particle compaction process. Ahmadi et al. [25] conducted a numerical study on the scaling method of polygon particles in DEM and calibrated the model according to the shear test. Peña et al. [26] studied the mechanical response of non-viscous particle materials under monotone loading through molecular dynamics simulation. Tanoli et al. [27] used the zoning method to conduct numerical analysis on deep foundation pit engineering in Shanghai, providing a comprehensive solution for predicting the displacement related to deep excavation in soft clay. Numerical simulation of PFC3D can simulate mechanical parameters and failure forms of research objects more accurately [28,29], but most studies mainly focus on the mechanical properties of research objects, and studies using PFC3D to study the microscopic parameters of models are relatively few. Based on previous studies, this study used discrete element particle flow software PFC3D to build a parallel bond model, study the influence of some parameters on desert sand-based backfill materials, and predict its compressive strength.

The main factors affecting the mechanical properties of desert sand-based backfill materials are the properties of desert sand, such as the content of desert sand, particle size distribution of desert sand, etc. [30]. In addition, porosity is also one of the factors affecting the mechanical properties of rocks and building materials [31]. Therefore, in this paper, we mainly use the discrete element particle flow software PFC3D to carry out numerical inversion. By adjusting the range of the mesoscopic parameters, it analyzes the relationship between the mesoscopic parameters and the macro mechanical properties of the desert sand-based backfill material and further discusses the influence of the sand content, porosity, desert sand particle size distribution range, and model size on the compressive strength of the desert sand-based backfill material model. The research results are of great significance to the practical application of desert sand-based filling materials in mining filling.

2. Experimental Study

Previous studies have shown that the water–cement ratio and the content of impurities influence HWBMs [32,33]. Guodong Li et al. [34,35] used the HWBM from Yangzhou China Mining Construction New Material Technology Co., Ltd. (Yangzhou, China), selected the water–cement ratio of the HWBM to be 1, 1.5 and 2, set the desert sand doping amount to be 0, 30%, and 60%, and made seven types of desert sand-based filling material samples.

Through a series of tests, the influence of desert sand-based backfill material on its initial setting time, mechanical properties, microstructure, and the effect of water–cement ratio on the performance of HWBM was studied.

See Table 1 for specific specimen information, and see Figure 1 for desert sand particle size grading curve. The test piece is numbered according to the different parameters of the test piece, and the test piece is numbered with H. The first number represents the water–cement ratio, and the last number represents the sand content of the test piece. For example, the water–cement ratio is 1, the sand content is 0%, and the test piece number is H-1-00.

Table 1. Specific parameters of test pieces.

Specimen	W/C	Sc	Width (mm)	Height (mm)
H-1-00	1	0%	50.20	100.40
H-1-30	1	30%	50.10	100.60
H-1-60	1	60%	50.50	101.20
H-1.5-00	1.5	0%	50.30	100.20
H-1.5-60	1.5	60%	50.20	100.50
H-2-00	2	0%	50.60	100.80
H-2-60	2	60%	50.10	100.40

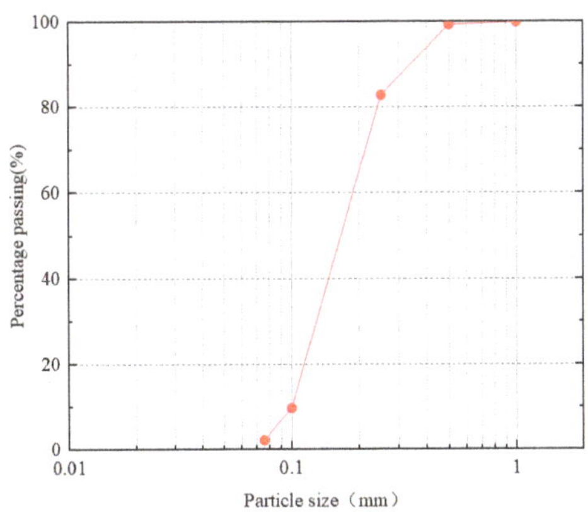

Figure 1. Grain size gradation curve of desert sand.

The test results show that the initial setting time of the HWBM modified by desert sand is reduced, the high content of desert sand can effectively improve the compressive strength and elastic modulus of the HWBM, the water–cement ratio increases, and the compressive strength of the specimen decreases. By adding desert sand to modify the HWBM, the compressive performance of the test piece has changed. When the sand content of the test piece reaches 60%, its compressive performance has been significantly improved. According to the calculation, the uniaxial compressive strength of specimens H-1-60, H-1.5-60, and H-2-60 increased by 46.63%, 78.04%, and 113.23%, respectively, compared with that without desert sand. Therefore, high content of desert sand can effectively improve the compressive strength of HWBMs. At the micro level, the addition of desert sand changes the microstructure of the HWBM, generates more hydrated calcium silicate (C-S-H) to fill in the gap between ettringite crystals, and its compactness increases, thus improving the mechanical properties of the HWBM.

Based on the test results, the stress–strain curves of different samples under uniaxial compression when the water–cement ratio is 1 are drawn, as shown in Figure 2. It can be seen that, during compression failure, as the compressive stress increases, the specimen roughly undergoes four deformation stages: (1) In the initial compression stage, under the influence of the compressive stress, the cracks in the specimen gradually close, the stress increases slowly, and the strain increases rapidly, which is the compaction stage. (2) With the increase in compressive stress, the stress and strain increase linearly and uniformly with time, which is a linear elastic failure stage. At this stage, the pores and cracks in the specimen are gradually compacted under the axial stress, and a new fracture surface begins to appear at the same time. (3) When the stress is close to the peak value, the specimen is damaged many times in a short period, and the stress fluctuates many times and reaches the peak value, which is the yield stage. At this stage, the internal structure of the specimen is rapidly destroyed, and the cracks are interconnected to form a macro fracture surface. (4) When the peak stress is reached, the specimen collapses, and the compressive strength rapidly decreases to a lower value, which is the residual stress stage. However, due to the friction between the fracture surfaces, the specimens can still be loaded by friction.

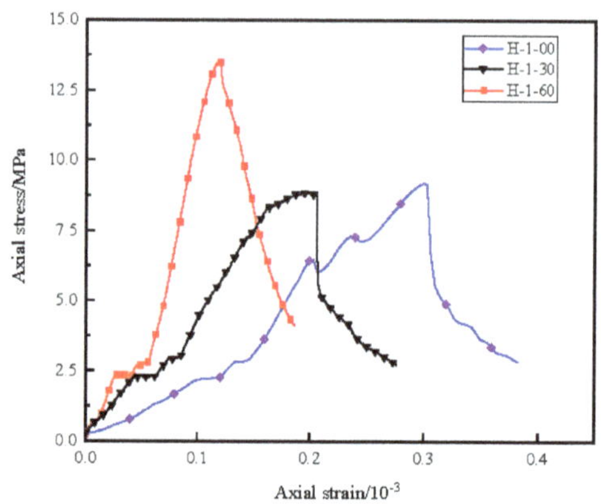

Figure 2. Stress–strain curves of different specimens under uniaxial compression.

3. PFC3D Numerical Model Establishment

3.1. Contact Constitutive Model

In PFC3D software, the constitutive model between discrete elements is the basis for realizing the macro mechanical behavior simulation. PFC3D software includes the Hertz contact model, linear contact model, contact bond model, and parallel bond model. Among them, the parallel bond model is used to bond the mechanical behavior of materials, such as the bond of aggregate between cement. Its bonding component is parallel to the linear element and establishes elastic interaction between contacts. The existence of parallel keys does not exclude the possibility of sliding. Parallel bonding can transfer forces and moments between different entities.

The linear parallel bonding model provides the behavior of two interfaces: the friction interface with finite size and the moment bonding interface (Figure 3). The friction interface is equivalent to a linear model: it does not resist relative rotation and adjusts the sliding force by applying the Coulomb limit to the shear force. The moment bonding interface is called a parallel bond because it is parallel to the friction interface during bonding. When the moment bond interface is bonded, it can resist relative rotation, and its behavior is linear elastic until it exceeds the strength limit, and the bond breaks, making it lose its bond. When the second interface is unbound, it will not carry any load. The unbonded linear parallel bonded model is equivalent to the linear model [36].

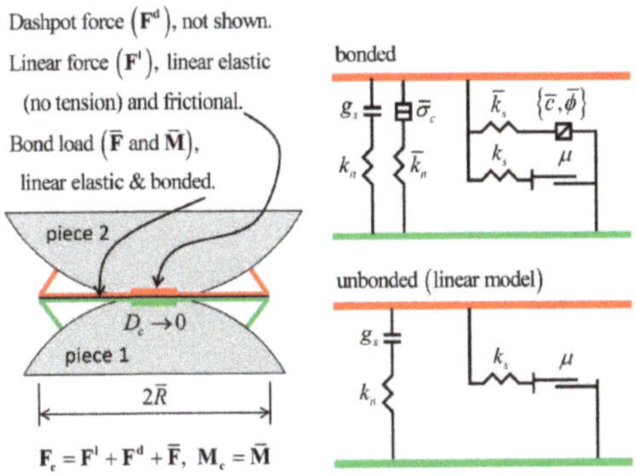

Figure 3. Linear parallel bond model.

3.2. Model Construction

In this paper, three different proportions of desert sand-based backfill material with a water–cement ratio of 1 are used as the research object. The discrete element particle flow software PFC3D was used for the simulation, and the influence of sand content, porosity, desert sand particle size, and model size on the performance of the test pieces was analyzed. Particle flow analysis usually regards a discrete medium as the aggregate of particle units. Each particle unit moves independently and is in contact with, and interacts with, other particle units. The particles themselves are rigid bodies that can overlap each other but cannot deform. The interaction between particles is realized through different contact models.

We obtained the rock strength, elastic modulus, and other parameters through a uniaxial compression test in the laboratory and conducted the rock uniaxial numerical simulation test using PFC3D. The numerical model adopts the parallel bonding contact model and the linear model. The parallel bonding model is used between the material particles, and the linear model is used between the sphere and the wall. First of all, the domain is created to determine the limits of the area where the model is to be built, and then three walls are set up, which are the loading and download boards of infinite size and the side confining plate, respectively. After that, the particles are generated. First, a region is created using geometry, and then spherical particles are generated in the geometry region by using the ball command. At this point, it will be found that the locations of some particles are beyond the range of geometry and wall; in this case, it is likely to appear that the particles are extruded from the wall, so we can reduce the size of the sphere in the X, Y, and Z directions to make the particles shrink, so as to avoid the bad effect of this situation on the simulation [29].

The structural system of a desert sand-based backfill material is composed of skeleton particles, and the actual sample particles are in thousands, so a calculation model is difficult to create. To reduce the calculation amount, we adopted the radius amplification method. Thus, the minimum particle radius was set to 1 mm, and the ratio of the maximum radius to the minimum radius was 1.66. The grain unit of the HWBM was randomly generated according to the particle size range of 2~3.32 mm in the designated space according to a uniform distribution, while the grain unit of desert sand was randomly generated according to the particle size range of 1~1.66 mm in the designated space according to a uniform distribution. The size of the model was 50 × 100 mm (width × height), the porosity was 0.3, the density of the HWBM was 1800 kg/m^3, and the density of desert sand was 1500 kg/m^3. The generation model is shown in Figure 4, where green particles represent the HWBM and blue particles represent desert sand.

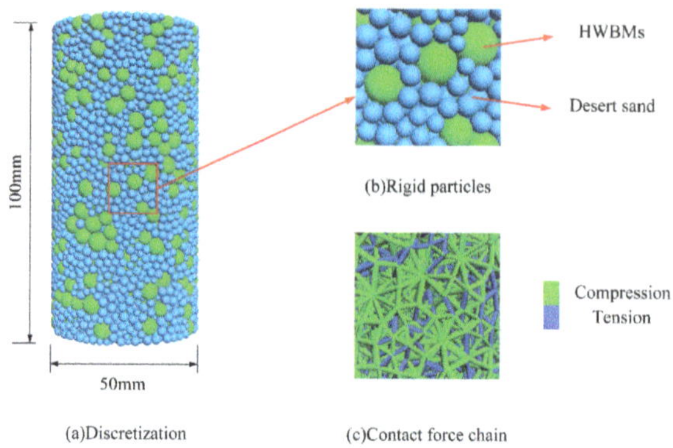

Figure 4. Discrete element modeling desert sand-based backfill material.

3.3. Parameter Calibration

To obtain a macroscopic response consistent with the indoor experimental results in the particle flow program, it is necessary to adjust the microscopic parameters to obtain the macroscopic parameters in line with the actual conditions. When building the PFC3D numerical model, the trial-and-error method is generally used to calibrate the mesoscopic parameters. Adjust the microscopic parameters of particles and bonds multiple times through trial and error until these parameters better reflect the mechanical properties of real rocks [36–39]. Firstly, the particle flow model is given the estimated mesoscopic parameters, and the numerical simulation results are compared with those of the laboratory tests. If the difference is large, the microparameters of the particle flow model are adjusted continuously by the trial-and-error method, and the model parameters are adjusted many times until the numerical simulation results are basically consistent with the indoor test results.

The specific numerical model test particle mesoscopic parameters are shown in Table 2. The stress–strain curve of the sample calculated using the particle mesoscopic parameters shown in Table 2 is shown in Figure 5. The mechanical parameters and failure patterns of indoor tests and numerical simulations are shown in Table 3. Through comparison, it can be seen that the uniaxial compression stress–strain curve of numerical simulation is relatively close to the uniaxial compression stress–strain curve of the indoor test, and the mechanical parameters of the indoor test and numerical simulation in Table 3 are similar. However, the curves of specimens H-1-00 and H-1-30 after failure are slightly different from the actual situation. The selected particle size is different from the actual particle size of the sand used, so the internal connections and actual cementation are different. This situation can be well simulated before the destruction; after the destruction, it cannot be consistent with reality. The uniaxial compression test is also a process of energy exchange to destroy the desert sand-based backfill material specimens: part of the energy is used for pore crack compaction and crack extension, and the remaining part is stored as elastic strain energy. Observing its failure, a desert sand-based backfill material specimen absorbs a large amount of energy under axial pressure and converts it into elastic properties. When the pressure it bore exceeded its yield limit, a fracture occurred at one end of the specimen first, and cracks occurred inside the specimen. The cracks gradually developed, increased, and penetrated the whole specimen, and some materials fell off, damaging the specimen. The failure results of numerical simulation are relatively close to the actual situation, so the numerical simulation can more accurately simulate the mechanical parameters and the failure mode of the specimen.

Table 2. Micromechanical parameters of the numerical model.

Specimen	emod/Pa	kratio	pb_kratio	pb_emod/Pa	pb_ten/Pa	pb_coh/Pa	fric
HWBM	18×10^9	1.0	1.0	18×10^9	4.0×10^6	3.0×10^6	0.35
Sand	200×10^9	1.0	1.0	200×10^9	25.0×10^6	21.0×10^6	0.55

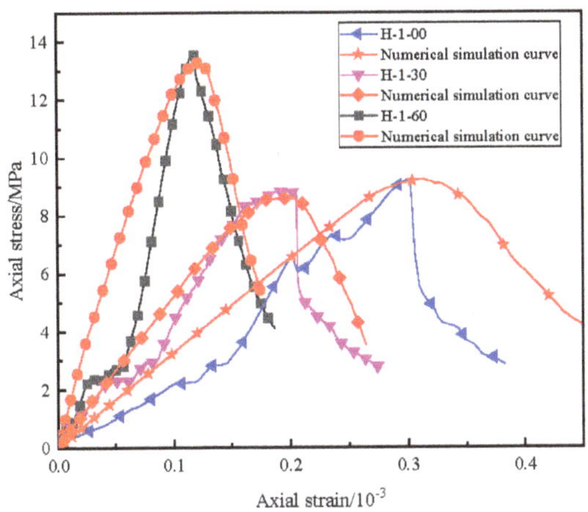

Figure 5. Comparison of stress–strain curves of the laboratory test and the numerical simulation uniaxial compression test.

Table 3. Mechanical parameters and failure modes of the laboratory test and the numerical simulation.

Variable	Indoor Test		Numerical Simulation		Specimen
	Mechanical Parameters	Failure Form	Failure Form	Mechanical Parameters	
Peak intensity/MPa	9.20			9.24	H-1-00
Elastic modulus/GPa	0.57			0.56	
Peak strain/10^{-3}	0.30			0.29	
Peak intensity/MPa	9.04			9.09	H-1-30
Elastic modulus/GPa	0.82			0.78	
Peak strain/10^{-3}	0.20			0.21	
Peak intensity/MPa	13.52		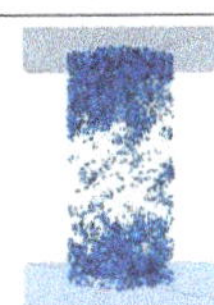	13.55	H-1-60
Elastic modulus/GPa	1.28			1.19	
Peak strain/10^{-3}	0.12			0.13	

4. Analysis of Influencing Factors

The influence of sand content, porosity, desert sand size distribution, and model size on the compressive properties of the desert sand-based backfill material was studied by taking the values obtained from indoor results as basic calculation parameters.

4.1. Effect of Sand Content on Uniaxial Compressive Strength

The sand content of the specimen was adjusted to 10%, 20%, 30%, 40%, 50%, and 60%, respectively; its uniaxial compression test was simulated, and a stress–strain curve was drawn, as shown in Figure 6. It can be seen that when the sand content is lower than 40%, the compressive performance of the specimen decreases slightly, while when the sand content is higher than 40%, the uniaxial compressive strength of the specimen has significantly improved, increasing to 13.3 MPa when the sand content is 60%.

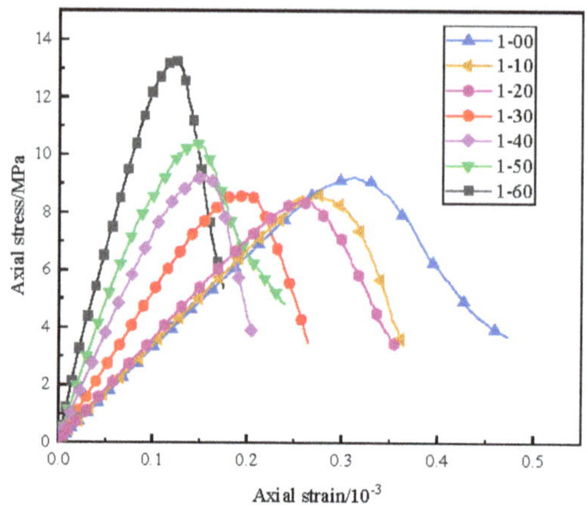

Figure 6. Stress–strain curve of uniaxial compression test for models with different sand content.

Research on desert sand concrete has shown that the optimal sand content of desert sand is between 20 and 60%, and the numerical simulation results are approximately the same as the experimental results [30,35]. When the desert sand content is low, the compressive strength of the specimen decreases slightly. This is because when there is less desert sand, the specimen skeleton is restructured, and there are more large pores between the particles of high-water filling materials, which cannot be completely filled in the desert sand concrete. The average particle size of desert sand is small, and when the sand content is high, the tiny pores between aggregates can be filled in the production of filling materials, forming a dense skeleton and thus improving the compressive strength of the specimen. Unlike other tests, the indoor test in this study uses a high-water backfill material, not ordinary concrete. Material A of the HWBM is composed of sulfoaluminate cement and a suspension agent, while material B is generally composed of gypsum, lime, and an early strength agent. This material has a high slurry volume after adding water, enabling the desert sand to be better embedded in it. From the microstructure, it can also be seen that when high content of desert sand is added, the compactness of the desert sand-based backfill material is more compact [34]. Therefore, the high content of desert sand can effectively improve the compressive strength of desert sand-based backfill materials.

4.2. Effect of Porosity on Uniaxial Compressive Strength

The porosity was set at 0.26, 0.28, 0.30, 0.32, and 0.34. Figure 7 displays the uniaxial compression test stress–strain curves of models with different porosities. While other mesoscopic parameters remained unchanged, with the increase in porosity, the numerical simulation uniaxial compressive strength of different samples decreased gradually, and the peak strength also increased gradually as strain peaked. When the porosity increased from 0.3 to 0.34, the values of uniaxial compressive strength of specimens H-1-00, H-1-30, and H-1-60 decreased by 16.90%, 24.30%, and 28.01%, respectively.

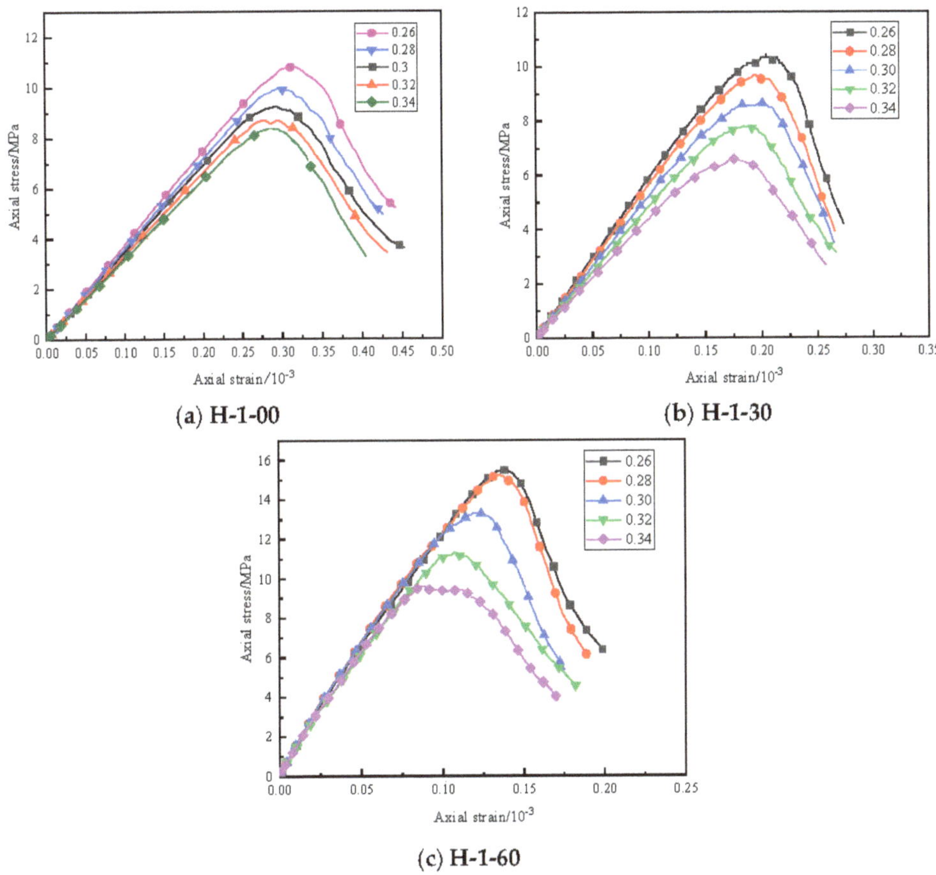

Figure 7. Stress–strain curve of the model uniaxial compression test under different porosity values.

The mesoscopic parameters of different specimen models under different porosity values are summarized in Table 4. With an increase in porosity, the number of particles in the model, the number of bonds between particles, the number of fractures, and the uniaxial compressive strength showed a downward trend. The number of particles contained in the model itself decreased, and so did the number of contact bonds. The contact bonding between particles corresponded to the microbonding surface or bonding point formed by the slurry cementation of the HWBM between desert sand aggregates. These microbonding surfaces and bonding points are important factors for maintaining the strength of desert sand-modified HWBM. Therefore, with the increase in porosity, the decrease in the number of contact bonds between particles was the main reason for the decrease in the uniaxial compressive strength. In addition, with the decrease in porosity, the particles are more closely arranged, and the intergranular embedding is stronger. Under the action of external forces, the position adjustment between particles is not easy to occur, and the resistance of particle clusters to deformation is enhanced, which shows that the compressive strength increases with the decrease in porosity.

Table 4. Macroscopic parameters and compressive strength of the numerical model.

Specimen	Porosity	Balls	Contacts	Fractures	f_c/MPa
H-1-00	0.26	3302	11,957	5602	10.80
	0.28	3217	11,435	5333	9.94
	0.30	3129	10,925	4948	9.24
	0.32	3041	10,333	4635	8.68
	0.34	2956	9650	4420	8.35
H-1-30	0.26	5804	24,044	12,470	10.30
	0.28	5622	22,522	11,703	9.67
	0.30	5436	20,633	10,426	8.68
	0.32	5258	19,113	8735	7.79
	0.34	5137	17,442	7598	6.57
H-1-60	0.26	9814	42,359	10,800	15.47
	0.28	9478	39,403	10,394	15.23
	0.30	9226	37,230	9566	13.33
	0.32	8974	34,786	8296	11.33
	0.34	8720	31,631	8248	9.59

4.3. Effect of Desert Sand Particle Size Distribution on Uniaxial Compressive Strength

Keeping other parameters unchanged, we changed the distribution range of desert sand particle size to 0.8~1.33 mm, 0.9~1.49 mm, 1~1.66 mm, 1.1~1.83 mm, and 1.2~1.99 mm and conducted a uniaxial compression simulation experiment. Figure 8 shows the stress–strain curve obtained from the experimental results. According to the results, the size distribution of desert sand had a great impact on the uniaxial compressive strength of the specimen. When the size distribution of desert sand increased from 0.8~1.33 mm to 1.2~1.99 mm, the uniaxial compressive strength of specimen H-1-30 increased from 7.99 MPa to 9.12 MPa, and the uniaxial compressive strength of specimen H-1-60 increased from 12.60 MPa to 14.29 MPa, increasing by 14.14% and 13.41%, respectively.

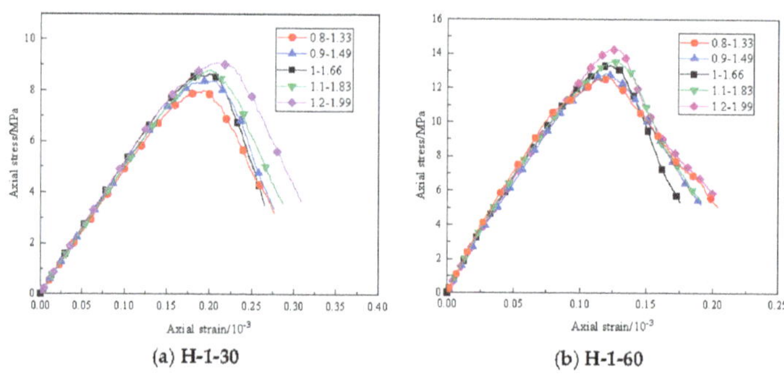

Figure 8. Stress–strain curve of the model uniaxial compression test under different desert sand particle size distributions.

Therefore, increasing the size distribution of desert sand in a certain range can improve the compressive performance of desert sand-modified HWBMs. This is because when the other conditions are the same, the smaller the desert sand particles, the larger the specific surface area of desert sand per unit weight, and its water demand will also increase. Under the same amount of cementitious material, size distribution plays a role in reducing the strength of the specimen. However, when the particle size of desert sand increases, the specific surface area of desert sand per unit weight is larger, the particles of high-water backfill material in the contact area of particles are more, and the cohesive force is larger, which leads to the enhancement of the compressive strength of the specimen of desert sand-based backfill material. However, the strength of the specimen will be larger without blindly increasing the particle size of desert sand. To improve the compression resistance of desert sand-modified HWBM, the particle size distribution of desert sand can be improved within a certain range.

4.4. Effect of Model Size on Uniaxial Compressive Strength

The effect of model size on the specimen was studied by increasing the difference between 25 mm and 50 mm. The size effect degree is used to quantitatively analyze the size effect of compressive strength of desert sand-based filling material specimens [40]. Taking the 50 × 100 mm specimen as the benchmark specimen, the size effect degree γ_{75} and γ_{100} of the cylindrical desert sand-based backfill material specimen were defined as follows:

$$\gamma_{75} = \frac{f_{c,50} - f_{c,75}}{f_{c,50}} \times 100\%$$

$$\gamma_{100} = \frac{f_{c,50} - f_{c,100}}{f_{c,50}} \times 100\%$$

where $f_{c,50}$, $f_{c,75}$, and $f_{c,100}$, respectively, represent the compressive strength of the sample with a model size of 50 × 100 mm, 75 × 150 mm, and 100 × 200 mm of the desert sand-based filling material.

Figure 9 shows the stress-strain curves of specimens with different model sizes. It can be seen that with the increase in the model size, the uniaxial compressive strength of the specimen decreased. Detailed parameters are shown in Table 5. When the model size was from 50 × 100 mm to 100 × 200 mm, the uniaxial compressive strength of specimen H-1-00 decreased from 9.24 MPa to 8.75 MPa, the uniaxial compressive strength of specimen H-1-30 decreased from 8.68 MPa to 8.20 MPa, and the uniaxial compressive strength of specimen H-1-60 decreased from 13.33 MPa to 12.03 MPa. According to the calculation, when the model size is 75 × 150 mm, the size effect degree of the three specimens is 2.92%, 3.00%, and 3.60%, respectively. When the model size is 100 × 200 mm, the size effect degree of the three specimens is 5.30%, 5.53%, and 9.75%, respectively.

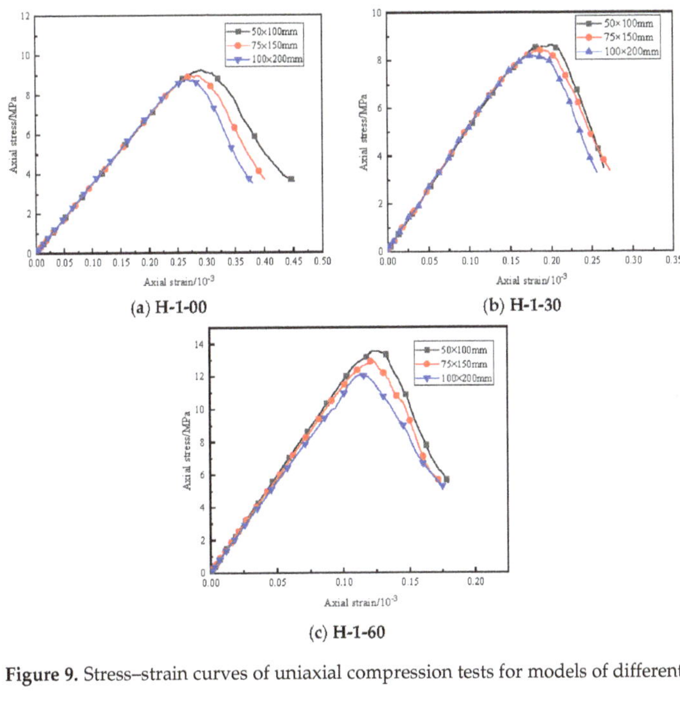

Figure 9. Stress–strain curves of uniaxial compression tests for models of different sizes.

Table 5. Uniaxial test results of models of different sizes.

Specimen	Size/mm (Width × Height)	f_c/MPa	$\varepsilon_u/10^{-3}$
H-1-00	50 × 100	9.24	0.29
	75 × 150	8.97	0.27
	100 × 200	8.75	0.26
H-1-30	50 × 100	8.68	0.20
	75 × 150	8.42	0.18
	100 × 200	8.20	0.17
H-1-60	50 × 100	13.33	0.13
	75 × 150	12.85	0.12
	100 × 200	12.03	0.11

According to the size effect theory of Bazant [41] fracture mechanics, the instability and expansion of fracture zones cause the failure of quasi-brittle materials such as concrete, and the development of fracture zones formed by micro-cracks is related to energy consumption and the release of stored energy. There is a fracture process zone composed of micro-cracks in the front end of concrete cracks. The expansion of cracks in the fracture process zone involves the release and consumption of energy. Concrete fracture is caused by the mismatch between the release and consumption of stored energy in size.

Bazant adopted the method of dimensional analysis, combined with experiments, and made use of energy balance to deduce the formula of size effect, as shown below.

$$\sigma_N = \frac{Bf_t}{\sqrt{1+\beta}}, \beta = \frac{D}{D_0}$$

where f_t, B, and D, respectively, represent the tensile strength, infinite programmatic number, and constant dependent on the structural geometry of the material.

It can be seen that the compressive strength of the specimen decreases with the increase in size. The compressive strength of desert sand-based backfill material specimens exhibits a relatively obvious size effect. However, the increased size of desert sand-based backfill material still has good compressive performance.

5. Conclusions

In this paper, through a series of experiments, the effects of water–cement ratio and desert sand doping amount on the properties of desert sand-based backfill material were studied. At the same time, based on the discrete element particle flow software PFC3D, the three-dimensional structure of the desert sand-based backfill material model was established, and the influence of sand content, porosity, desert sand particle size distribution, and model size on the performance of the specimen was analyzed. Conclusions are summarized as follows:

(1) The initial setting time of the high-water filling material modified by desert sand is reduced; the compressive strength and elastic modulus of the high-water filling material can be effectively improved by the high-water sand mixed with high content, and the compressive strength of the specimen decreases with the increase in the water–cement ratio;

(2) With the increase in desert sand content, the compressive strength of desert sand-based filling materials decreases first and then increases, and all specimens with sand content between 40 and 60% have good compressive performance;

(3) The uniaxial compressive strength of numerical simulation is related to the particle size distribution of desert sand. In a certain range, increasing the particle size distribution of desert sand can improve the compressive performance of desert sand-based backfill material specimens;

(4) The specimen size effect of desert sand-based backfill material is significant, and its compressive performance decreases with the increase in model size.

The discrete element particle flow software PFC3D can simulate and predict the mechanical properties and failure process of desert sand-based backfill materials. Due to the problems of the PFC3D software and algorithm, the calculation efficiency is not high, which greatly limits the calculation scale. The desert sand-based backfill material contains two kinds of particles with different sizes and properties, namely desert sand and high-water backfill material. In order to simplify the complexity of the problem and improve the computational efficiency, the effect of gradation effect was not considered in this paper. Therefore, it is necessary to consider the effect of gradation effect on the macro and micro mechanical properties of desert sand-based backfill materials in the future, and carry out in-depth research.

Author Contributions: Conceptualization, H.Y., H.L. and G.L.; methodology, H.L., H.Y. and G.L.; software, H.Y. and H.L.; validation, H.Y., Y.H. and X.W.; investigation, H.Y., Y.H. and X.W.; data curation, H.Y. and H.L.; writing—original draft preparation, H.Y. and Y.H.; writing—review and editing, H.Y., H.L. and Y.H. All authors have read and agreed to the published version of the manuscript.

Funding: This research was funded by the National Natural Science Foundation of China, grant numbers 51964043 and U1903216, and also sponsored by the Natural Science Foundation of Xinjiang Uyghur Autonomous Region (2022D01E31), Xinjiang Uygur Autonomous Region "Tianshan Talent Training" Program (2022TSYCCX0037), Xinjiang Uygur Autonomous Region Special Program for Key R&D Tasks (2022B01034,2022B01051), and the National College Student Innovation Project (202210755007).

Institutional Review Board Statement: Not applicable.

Informed Consent Statement: Not applicable.

Data Availability Statement: The research data used to support the findings of this study are currently under embargo while the research findings are commercialized. Requests for data 12 months after the publication of this article will be considered by the corresponding author.

Acknowledgments: The author would like to thank H.L. and G.L. for their continuous support in the data and writing of the paper.

Conflicts of Interest: The authors declare no conflict of interest.

References

1. Bai, E.; Guo, W.; Tan, Y.; Huang, G.; Guo, M.; Ma, Z. Roadway Backfill Mining with Super-High-Water Material to Protect Surface Buildings: A Case Study. *Appl. Sci.* **2019**, *10*, 107. [CrossRef]
2. Péra, J.; Ambroise, J. New applications of calcium sulfoaluminate cement. *Cem. Concr. Res.* **2004**, *34*, 671–676. [CrossRef]
3. García-Maté, M.; Santacruz, I.; Ángeles, G.; León-Reina, L.; Aranda, M.A. Rheological and hydration characterization of calcium sulfoaluminate cement pastes. *Cem. Concr. Compos.* **2012**, *34*, 684–691. [CrossRef]
4. Liu, H.; Zhang, B.; Li, X.; Liu, C.; Wang, C.; Wang, F.; Chen, D. Research on roof damage mechanism and control technology of gob-side entry retaining under close distance gob. *Eng. Fail. Anal.* **2022**, *138*, 106331. [CrossRef]
5. Zhao, H. State-of-the-art of standing supports for gob-side entry retaining technology in China. *J. S. Afr. Inst. Min. Metall.* **2019**, *119*, 891–906. [CrossRef]
6. Buzylo, V.; Pavlychenko, A.; Borysovska, O. Ecological aspects of filling of worked-out area during underground coal mining. In *E3S Web of Conferences*; EDP Sciences: Ulis, France, 2020; Volume 201, p. 01038.
7. Chang, Q.; Tang, W.; Xu, Y.; Zhou, H. Research on the width of filling body in gob-side entry retaining with high-water materials. *Int. J. Min. Sci. Technol.* **2018**, *28*, 519–524. [CrossRef]
8. Zhao, H.; Ren, T.; Remennikov, A. A hybrid tubular standing support for underground mines: Compressive behaviour. *Int. J. Min. Sci. Technol.* **2021**, *31*, 215–224. [CrossRef]
9. Shaat, A.; Fam, A. Axial loading tests on short and long hollow structural steel columns retrofitted using carbon fibre reinforced polymers. *Can. J. Civ. Eng.* **2006**, *33*, 458–470. [CrossRef]
10. Zhang, L.; Liu, C.; Wu, F.; Lu, Y.; Zhou, X. Study on the characteristics of compressive strength and hydration mechanism of high-water-content materials modified by furnace slag and silica fume. In *IOP Conference Series Earth and Environmental Science*; IOP Publishing: Bristol, UK, 2019; Volume 330, p. 042028.

11. García-Maté, M.; De la Torre, A.G.; León-Reina, L.; Aranda, M.A.G.; Santacruz, I. Hydration studies of calcium sulfoaluminate cements blended with fly ash. *Cem. Concr. Res.* **2013**, *54*, 12–20. [CrossRef]
12. Jiang, Y.; Liu, C.; Xie, H.; Li, J.; Luo, P. Test and analysis of main physical and mechanical parameters of modified high-water-concretion. *Sichuan Daxue Xuebao (Gongcheng Kexue Ban)/J. Sichuan Univ. (Eng. Sci. Ed.)* **2015**, *47*, 65–69.
13. Diao, Z.F.; Liu, C.W.; Zhang, L.W.; Sun, W.; Feng, B. Deformation Characteristics and Mechanical Response of Modified High-Water Material Mixed with Sludge under Typical Loading Methods. *Sci. Technol. Eng.* **2018**, *18*, 262–266.
14. Zhang, Z.; Liu, C.W.; Ye, D.Y. Experimental study on modification of polyethylene plastic for high water filling materials. *Min. Res. Dev.* **2020**, *40*, 50.
15. Feng, B.; Liu, C.; Xie, H.; Meisu, W.U.; Jiang, Y. Experimental Study on the Size and the Shape of High-water-content Material that Modified. *Adv. Eng. Sci.* **2017**, *49*, 121–127.
16. Zhang, Y.; Wang, Y.; Li, T.; Xiong, Z.; Sun, Y. Effects of lithium carbonate on performances of sulphoaluminate cement-based dual liquid high-water material and its mechanisms. *Constr. Build. Mater.* **2018**, *161*, 374–380. [CrossRef]
17. Wang, Y.; Yu, J.; Wang, J.; Guan, X. Effects of Aluminum Sulfate and Quicklime/Fluorgypsum Ratio on the Properties of Calcium Sulfoaluminate (CSA) Cement-Based Double Liquid Grouting Materials. *Materials* **2019**, *12*, 1222. [CrossRef]
18. Al-Harthy, A.S.; Halim, M.A.; Taha, R.; Al-Jabri, K.S. The properties of concrete made with fine dune sand. *Constr. Build. Mater.* **2007**, *21*, 1803–1808. [CrossRef]
19. Benabed, B.; Azzouz, L.; Kadri, E.H.; Kenai, S.; Belaidi, A.S.E. Effect of fine aggregate replacement with desert dune sand on fresh properties and strength of self-compacting mortars. *J. Adhes. Sci. Technol.* **2014**, *28*, 2182–2195. [CrossRef]
20. Bosco, E.; Claessens, R.J.M.A.; Suiker, A.S. Multi-scale prediction of chemo-mechanical properties of concrete materials through asymptotic homogenization. *Cem. Concr. Res.* **2020**, *128*, 105929. [CrossRef]
21. Zhao, H.; Sun, W.; Wu, X.; Gao, B. The effect of coarse aggregate gradation on the properties of self-compacting concrete. *Mater. Des.* **2012**, *40*, 109–116. [CrossRef]
22. Cundall, P.A.; Strack, O.D. A discrete numerical model for granular assemblies. *Geotechnique* **1979**, *29*, 47–65. [CrossRef]
23. Yan, Y.; Ji, S. Discrete element modeling of direct shear tests for a granular material. *Int. J. Numer. Anal. Methods Geomech.* **2009**, *34*, 978–990. [CrossRef]
24. He, Y.; Evans, T.J.; Yu, A.B.; Yang, R.Y. Numerical Modelling of Die and Unconfined Compactions of Wet Particles. *Procedia Eng.* **2015**, *102*, 1390–1398. [CrossRef]
25. Ahmadi, A.; Larsson, S.; Wersäll, C. Scaling granular material with polygonal particles in discrete element modeling. *Particuology* **2023**, *75*, 151–164. [CrossRef]
26. Pena, A.A.; Lizcano, A.; Alonso-Marroquin, F.; Herrmann, H.J. Biaxial test simulations using a packing of polygonal particles. *Int. J. Numer. Anal. Methods Geomech.* **2008**, *32*, 143–160. [CrossRef]
27. Tanoli, A.Y.; Yan, B.; Xiong, Y.L.; Ye, G.L.; Khalid, U.; Xu, Z.H. Numerical analysis on zone-divided deep excavation in soft clays using a new small strain elasto–plastic constitutive model. *Undergr. Space* **2022**, *7*, 19–36. [CrossRef]
28. Li, Z.; Rao, Q.H. Quantitative determination of PFC3D microscopic parameters. *J. Cent. South Univ.* **2021**, *28*, 911–925. [CrossRef]
29. Lopez, R.; Larsson, S.; Silfwerbrand, J. A discrete element material model including particle degradation suitable for rockfill embankments. *Comput. Geotech.* **2019**, *115*, 103166. [CrossRef]
30. Abu Seif, E.S.S.; Sonbul, A.R.; Hakami, B.A.H.; El-Sawy, E.K. Experimental study on the utilization of dune sands as a construction material in the area between Jeddah and Mecca, Western Saudi Arabia. *Bull. Eng. Geol. Environ.* **2016**, *75*, 1007–1022. [CrossRef]
31. Pavičić, I.; Briševac, Z.; Vrbaški, A.; Grgasović, T.; Duić, Ž.; Šijak, D.; Dragičević,, I. Geometric and fractal characterization of pore systems in the upper triassic dolomites based on image processing techniques example from umberak mts nw croatia. *Sustainability* **2021**, *13*, 7668. [CrossRef]
32. Sun, Y.; Xie, S.; Wu, C.; Cui, J.; Chen, D.; Guo, F.; Jiang, Z.; Ren, Y.; Lu, W. Influence of the ultra-fine fly ash dosages on the mechanical properties of the sulfoaluminate cement-based high water backfilling material. *Sci. Rep.* **2023**, *13*, 1564. [CrossRef]
33. Chen, D.; Sun, Y.; Xie, S.; Wang, E.; Wu, X.; Qi, P. Change in the Mechanic Properties and Electrical Resistivity of High Water Backfill Material under Electrochemical Treatment. *Adv. Mater. Sci. Eng.* **2020**, *2020*, 3248294. [CrossRef]
34. Li, G.; Liu, H.; Deng, W.; Wang, H.; Yan, H. Behavior of CFRP-Confifined Sand-Based Material Columns under Axial Compression. *Polymers* **2021**, *13*, 3994. [CrossRef]
35. Li, G.; Wang, H.; Liu, Z.; Liu, H.; Yan, H.; Liu, Z. Effects of Aeolian Sand and Water−Cement Ratio on Performance of a Novel Mine Backfill Material. *Sustainability* **2023**, *15*, 569. [CrossRef]
36. Salazar, A.; Saez, E.; Pardo, G. Modeling the direct shear test of a coarse sand using the 3D Discrete Element Method with a rolling friction model. *Comput. Geotech.* **2015**, *67*, 83–93. [CrossRef]
37. Lee, H.; Jeon, S. An experimental and numerical study of fracture coalescence in pre-cracked specimens under uniaxial compression. *Int. J. Solids Struct.* **2010**, *48*, 979–999. [CrossRef]
38. Smolnik, G. DEM (PFC3D) numerical simulation of the influence of grain orientation on the strength of the Kimachi sandstone. In *IOP Conference Series: Earth and Environmental Science*; IOP Publishing: Bristol, UK, 2019.
39. Ren, J.; Xiao, M.; Liu, G. Rock Macro–Meso Parameter Calibration and Optimization Based on Improved BP Algorithm and Response Surface Method in PFC 3D. *Energies* **2022**, *15*, 6290. [CrossRef]

40. Liu, J.; Min, D.U.; Xiu-li, D.U.; Zhen-bao, L.I. Size effect on the failure of stirrp-confined reinforced concrete columns under axial compression. *Eng. Mech.* **2018**, *35*, 93–101.
41. Bažant, Z.P. Size effect in blunt fracture: Concrete, rock, metal. *J. Eng. Mech.* **1984**, *110*, 518–535. [CrossRef]

materials

MDPI

Article

Simplified Evaluation of Shear Stiffness Degradation of Diagonally Cracked Reinforced Concrete Beams

Kaiqi Zheng [1,2,3], Siyuan Zhou [1], Yaohui Zhang [2,*], Yang Wei [1,*], Jiaqing Wang [1], Yuxi Wang [1] and Xiaochuan Qin [3]

[1] College of Civil Engineering, Nanjing Forestry University, Nanjing 210037, China; k.zheng@njfu.edu.cn (K.Z.); 8210610638@njfu.edu.cn (S.Z.); jiaqingw@njfu.edu.cn (J.W.); wyxwyx11@126.com (Y.W.)
[2] State Key Laboratory of Mechanical Behavior and System Safety of Traffic Engineering Structures, Shijiazhuang Tiedao University, Shijiazhuang 050043, China
[3] State Key Laboratory of High Performance Civil Engineering Materials, Jiangsu Research Institute of Building Science, Nanjing 210008, China; qinxiaochuan@cnjsjk.cn
* Correspondence: sjzzhangyh@126.com (Y.Z.); wy78@njfu.edu.cn (Y.W.)

Citation: Zheng, K.; Zhou, S.; Zhang, Y.; Wei, Y.; Wang, J.; Wang, Y.; Qin, X. Simplified Evaluation of Shear Stiffness Degradation of Diagonally Cracked Reinforced Concrete Beams. *Materials* **2023**, *16*, 4752. https://doi.org/10.3390/ma16134752

Academic Editor: Loic Hilliou

Received: 19 May 2023
Revised: 23 June 2023
Accepted: 27 June 2023
Published: 30 June 2023

Abstract: Shear cracking in concrete box-girder bridges, which could cause excessive deflection during the serviceability limit state, cannot be effectively avoided by code-guided design. While elastic shear deformation only accounts for a small proportion of total deformation for un-cracked reinforced concrete (RC) beams, the magnitude of after-cracking shear deformation becomes comparable to flexural deformation for RC beams. However, there is still a lack of practical models to predict the after-cracking shear deformation of RC beams. First, six thin-webbed I beams were tested to investigate the shear stiffness degradation mechanism and the decrease ratio. Then, a very simple truss strut angle formula, which is the crucial parameter for shear stiffness, was established. Furthermore, a stiffness degradation rule for partially cracked beams was proposed considering the influence of concrete tension stiffening, which is essential for predicting the development process of after-cracking shear deformation. Finally, directly measured shear strains were used to validate the proposed shear stiffness model. The results showed that the shear stiffness drops to about 30~40% of the original stiffness after the first diagonal crack, and the remaining shear stiffness is only about 10% of the original one when the stirrup yields. Increasing the stirrup ratio is a more effective method to control shear stiffness degradation for diagonally cracked RC beams. Also, the proposed shear stiffness model well captures the main features of the shear stiffness degradation, and it provides a relatively accurate prediction of the equivalent shear stiffness at the post-cracking stage.

Keywords: reinforced concrete beam; diagonal cracking; shear deformation; shear stiffness; strut angle; tension stiffening; stiffness degradation

1. Introduction

As is well known, the deformation of concrete beams mainly consists of two parts: bending deformation and shear deformation [1]. Generally, the deformation of beams is mainly bending deformation, and the magnitude of shear deformation is small, which can be ignored [2,3]. However, for long-span concrete box-girder bridges commonly used in bridge engineering, this assumption may cause deviations [4]. Especially for the thin-webbed box girder with diagonal cracking, its shear deformation may be equivalent to the bending deformation [5].

According to a finite element analysis of a thin-webbed box-girder bridge with large web height [4], the magnitude of elastic shear deformation under load is considerably large and may reach a level that cannot be ignored. Scholars' experiments on thin-webbed concrete beams [1,6] have shown that: after the appearance of diagonal cracks, the shear deformation of the beam will significantly increase, with a shear deformation ratio of over 30% of the total deformation. Further, according a survey on an existing concrete box-girder bridge conducted by the Research Institute of Highway Ministry of Transport of China [7], diagonal web cracking occurs in more than 90% of this type of bridge. This indicates that with the development of diagonal cracks in the web, the impact of shear deformation on the deflection of thin-webbed beam bridges will be greater and should be taken seriously. Considering that the influence mechanism of diagonal cracks on deformation is extremely complex, conducting a shear test on thin-webbed concrete beams is the most effective way to discover and summarize its laws of influence [8,9].

So far, although thousands of shear tests have been conducted on concrete beams, the vast majority of specimens have beam heights not exceeding 500 mm, and only a few beams without web reinforcement have reached heights exceeding 1000 mm [10–20]. Due to the influence of size effects, these specimens cannot truly reflect the actual stress behavior of large-sized beams widely used in practical engineering. And, most of the existing experiments have focused on shear strength and failure mode, with less attention paid to the contribution of shear deformation after shear cracking.

As RC beams are subjected to the combined action of the shear and bending moment, the corresponding shear and flexural deformation are coupled with each other. It is difficult to measure the shear deformation separately. Experimental research on post-crack deformation of RC beams is also limited to the comparison of total deformation [21–23]. Although there are a few quantitative experimental studies focusing on direct shear deformation of concrete beams after shear cracking, in which a decoupling technique of shear deformation and bending deformation has been proposed [24,25], the data are limited and are not compatible with variable depth beams. Based on the iso-parametric concept from the finite element method, a theoretical deformation decoupling method for variable depth beams was proposed [26], which provides a new approach for the direct measurement of shear deformation.

In another aspect, researchers proposed calculating methods for after-cracking shear deformation. Based on the assumption of homogeneous characteristics for cracked concrete, the modified compression field theory (MCFT) could provide the whole load-deformation curve for RC beams [27–30]. However, iterative calculations are required, which are not suitable for the rapid evaluation of existing bridges. The variable angle truss model (VATM)-based calculating theory for shear deformation prediction is proposed by Pan et al. [7], whereby constant tangent stiffness degradation rules are suggested for diagonally cracked RC beams, which is very simple and conservative for engineering practice. However, efforts are still needed to achieve maximum optimization in computational accuracy and simplicity [31,32].

This study aims to obtain the shear stiffness degradation law of large-scale thin-webbed concrete beams and to propose a simplified and practical prediction method, which can be further applied to concrete box-girder bridges with diagonal web cracks. The degradation of after-cracking shear stiffness is studied by experimental tests. Further, a simplified and practical model is proposed to depict the degradation of shear stiffness. And finally, shear deformation test data in this paper and in the literature are used to verify the accuracy and applicability of the proposed simplified shear stiffness degradation model.

2. Experimental Test of Shear Stiffness Degradation

2.1. Test Object and Design Concept

To study the impact of diagonal cracks on the degradation of shear stiffness and development of shear deformation of concrete thin web beams, a direct shear measurement test was conducted on six concrete thin-webbed constrained beams. The main experimental objectives include:

(1) Achieve continuous direct measurement of shear deformation before and after diagonal cracks in the concrete web;
(2) Analyze the amplitude of changes in shear deformation values before and after shear cracking and study the degree of influence of diagonal cracks on shear deformation;
(3) Study the degradation law of shear stiffness after the development of diagonal cracks.

In order to achieve the above experimental objectives, the main experimental ideas used include:

(1) Using large-scale thin-webbed I-shaped cross-section specimens to better simulate the stress behavior of thin web bridges, while facilitating the testing of web strain and the observation of diagonal cracks.
(2) Adopting a reinforcement design with "strong bending and weak shear" concept, ensuring the priority occurrence and full development of diagonal cracks, with a focus on observing the impact of diagonal cracks on shear deformation and shear stiffness.
(3) Constrained beams are used to investigate the shear performance of concrete beams under different combinations of bending and shear internal forces.
(4) The effects of inclined bottom chord on diagonal crack and shear strength were investigated by using two types of specimens, namely, equal-height beam and variable-height beam.
(5) Propose a strain-based shear deformation calculation method for arbitrary quadrilateral lattices, achieving direct peeling measurement of bending deformation and shear deformation.

2.2. Specimen Parameters and Setup

A total of 6 I-shaped cross-section specimens were made and divided into two groups: Group BC consisted of 2 beams of equal height; Group BV consisted of 4 variable-height beams, with the upper flange of the cross-section horizontal and the lower flange height varying in a parabolic manner. All the specimens were constructed in two batches on site (Figure 1). The width of the beam web is 100 mm, and the beam length is 5400 mm.

(**a**) Rebar binding (**b**) Strain gauge installation

(**c**) Formwork assembling (**d**) Concrete casting

Figure 1. Fabrication process of specimens.

All the specimens are simply supported with a cantilever, which were loaded with two concentrated loads at the cantilever end and within the simply supported span, respectively. The ratio of cantilever load to span load is 1:2 for BC specimens, and 1:1 for BV specimens. The detailed dimensions and elevation layout of the component are shown in Figures 2 and 3, The detailed arrangement of the measuring lattice and corresponding calculation method could refer to a pre-publication focusing on the measuring technology [26].

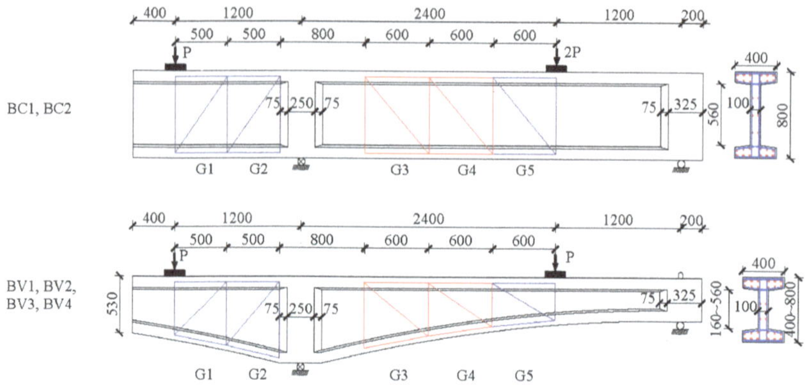

Figure 2. Beam dimension and strain measuring lattice arrangement (unit: mm).

(**a**) BC1, BC2

(**b**) BV1, BV2, BV3, BV4

Figure 3. Test setup.

For the specimens, D25 bars and D12 bars with a total cross section area of 3624 mm^2 and average yield strength of 497 MPa were used as longitudinal bars, which were arranged symmetrically in top and bottom flanges. D8 round bars (two legs) with average yield strength of 326 MPa were arranged as stirrups at a spacing of 200 mm (ρ_V = 0.5%) or 250 mm (ρ_V = 0.4%) for deferent beams along the whole span. The average concrete cylinder strengths were 39.0 MPa and 36.0 MPa for batch I (BC1, BV1, BV2) and batch II (BC1, BV1, BV2), respectively. It must be mentioned that excessive longitudinal reinforcement is arranged to prevent early flexural failure. The specimen number and main design parameters are shown in Table 1, with the changing parameters being concrete strength and reinforcement ratio.

Table 1. Details of specimens.

Resources	Specimen NO.	f'_c (MPa)	E_c (GPa)	d_v (mm)	b_w (mm)	f_{yv} (MPa)	ρ_v (%)	ρ_s (%)	$M/(Vd)$	θ_u (Degree)	λ_u
Author	BC1-G3	39.0	29.4	684	100	327	0.5	4.8	0.4	26.3	0.182
	BC1-G4	39.0	29.4	684	100	327	0.5	4.8	0.4	26.3	0.182
	BC2-G3	36.0	28.2	684	100	327	0.4	4.8	0.4	25.2	0.169
	BC2-G4	36.0	28.2	684	100	327	0.4	4.8	0.4	25.2	0.169
	BV1-G3	39.0	29.4	543.6	100	327	0.5	6.0	1.2	25.8	0.184
	BV1-G4	39.0	29.4	450.9	100	327	0.5	7.23	0.4	25.4	0.185
	BV2-G3	39.0	29.4	543.6	100	327	0.4	6.0	1.2	24.6	0.168
	BV2-G4	39.0	29.4	450.9	100	327	0.4	7.2	0.4	24.3	0.169
	BV3-G3	36.0	28.2	543.6	100	327	0.5	6.0	1.2	25.9	0.187
	BV3-G4	36.0	28.2	450.9	100	327	0.5	7.2	0.4	25.6	0.188
	BV4-G3	36.0	28.2	543.6	100	327	0.4	6.0	1.2	24.7	0.171
	BV4-G4	36.0	28.2	450.9	100	327	0.4	7.2	0.4	24.4	0.172
Hansapinyo et al. [24]	S1	33.0	27.0	320	150	370	0.47	4.26	2.6	26.4	0.179
	S2	33.0	27.0	320	150	370	0.47	4.26	3.5	26.4	0.179
	S3	33.0	27.0	320	150	370	0.47	2.13	2.6	28.5	0.170
	S4	33.0	27.0	320	150	370	0.31	2.13	2.6	26.1	0.142

2.3. Specimen Failure Modes

The final failure mode of all 6 specimens is shear failure, manifested as the yielding of web stirrups and concrete crushing at the bending reverse point, as shown in Figure 4.

The test reveals that the development of bending cracks was slow or almost non-existent during the loading process, and the strain increment of longitudinal reinforcement was small and did not reached yield [10]. With the appearance of diagonal cracks, their development is relatively rapid, quickly developing from the middle of the web to the upper and lower flanges, and gradually penetrating the entire web. With the formation of the main diagonal crack, the stirrups at the diagonal crack also yielded quickly. Afterwards, the number, width and range of diagonal cracks further expanded, and the stirrups of different shear span also yielded one after another. Finally, the concrete of the beam web at the bending reverse point collapsed, and the deformation increased sharply, declaring the failure of the specimen.

Though the first diagonal crack occurs at the cantilever span with maximum shear forces, all specimens failed in shear uniformly at the bending reverse point The positive and negative bending moment on both sides of the bending reverse point intensifies the shear deformation of the concrete web, ultimately leading to the tearing of the concrete web and the diagonal compression of the concrete web. The phenomenon reflects that the bending reverse point section is the weakest position for shear failure of continuous beams (constrained beams).

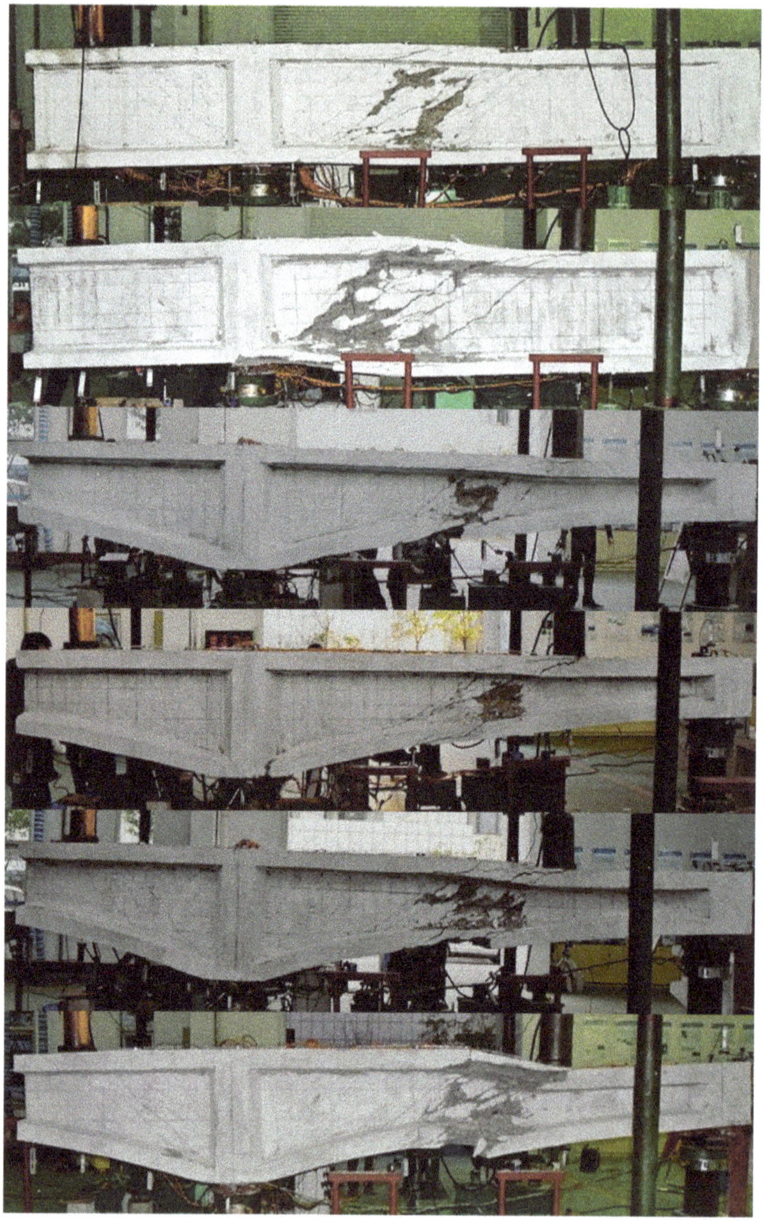

Figure 4. Failure mode of specimens.

2.4. Observed Shear Stiffness Degradation

Besides failure modes, the sequence of shear crack, strain development and yielding of stirrup, and failure load of each shear span can refer to an earlier publication [26]. Here, we only focus on the degradation of shear stiffness. Shear stiffness is the most important measure of structural shear deformation. In the elastic stress stage, the shear stiffness K_e of the section can be expressed as GA_v, and considering the Poisson's ratio, it becomes $EA_v/[2(1 + \mu)]$. After cracking, due to the destruction of structural continuity, the shear stiffness is no longer equal to elastic shear stiffness. Generally, the nominal shear stress of the test beam after cracking can be taken as τ, which is the ratio of shear force P to the shear cross-sectional area A_v. Based on the stress–strain relationship, the equivalent shear stiffness K_{eq} of the shear cracked specimen is defined by the following equation,

$$K_{eq} = \frac{\tau}{\gamma} A_v = \frac{P}{\gamma} \tag{1}$$

To study the variation of shear stiffness after the development of diagonal cracks, a shear stiffness degradation factor is defined as λ, namely, the ratio of equivalent shear stiffness K_{eq} to elastic shear stiffness K_e,

$$\lambda = K_{eq}/K_e \tag{2}$$

Considering that the tested shear strain in the elastic stage is very small, if the calculated $\lambda > 1$, take $\lambda = 1$. The shear stiffness reduction factor λ of each observation lattice can be plotted as a function of nominal shear stress τ, as shown in Figure 5. It can be seen that except for specimen BV1, the shear stress levels of all specimen frame G3 and G4 lattices with significantly reduced shear stiffness are around 2MPa, which is in good agreement with the initial shear crack load. This indicates that the shear stiffness reduction factor λ can properly reflect the impact of diagonal web cracks on shear stiffness.

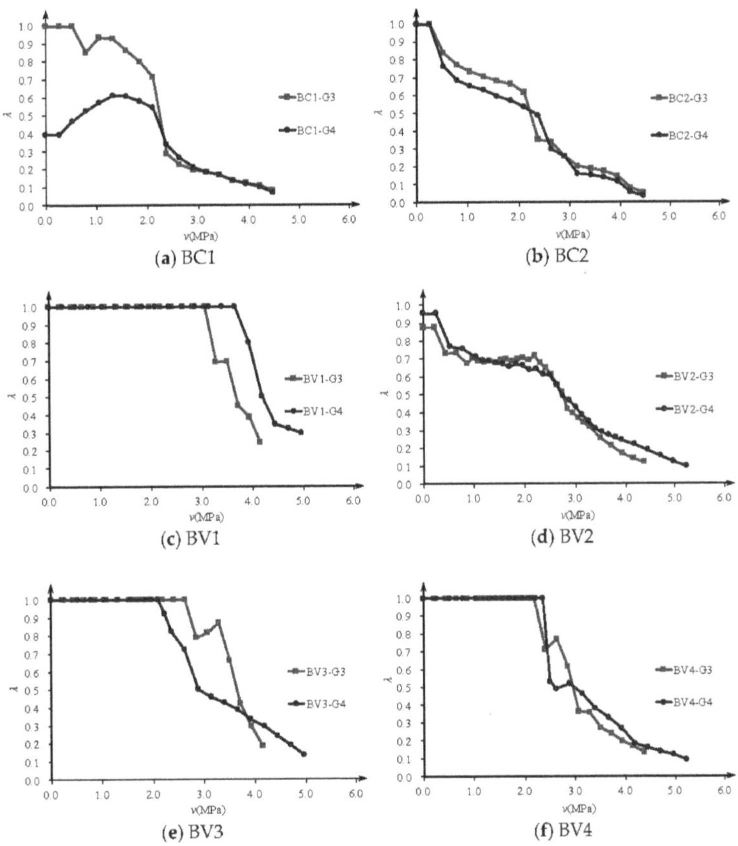

Figure 5. Degradation law of shear stiffness of specimens.

Comparing the degradation curves of shear stiffness of specimens with different concrete strength grades (BV1 and BV3, BV2 and BV4) and different reinforcement ratios (BV1 and BV2, BV3 and BV4), it was found that increasing the concrete strength grade (mainly the elastic modulus) and increasing the reinforcement ratio can both improve the shear stiffness after cracking. The contribution of increasing the reinforcement ratio to the remaining shear stiffness is more significant after shear cracking, and the suppression of shear deformation is more effective. Generally, after the diagonal cracking of the specimen, the shear stiffness is about 30~40% of the original, but when the stirrup yields, the remaining shear stiffness is only about 10%.

3. Proposed Shear Stiffness Degradation Model

3.1. Fully Diagonally Cracked Shear Stiffness

To evaluate the cracked shear stiffness, the truss model is recommended by scholars [33–36]. For slender beams, the inclined cracks are roughly parallel to each other. Therefore, a standard VATM can be used for analysis. As the truss model ignores the tensile stresses between cracked concrete, it is only suitable for calculating the shear stiffness of fully diagonally cracked RC beams. If the strut angle θ_u is determined, the corresponding shear stiffness K_u can be expressed as,

$$K_u = \frac{n\rho_v \cot^2\theta_u}{1 + n\rho_v \csc^4\theta_u} E_c A_v \tag{3}$$

where K_u is the fully diagonally cracked shear stiffness, n is the ratio of E_s to E_c, E_s is the modulus of elasticity of reinforcing steel, ρ_v is the stirrup ratio, and θ_u is the strut angle.

3.2. Ultimate Shear Stiffness Degradation Factor

Similarly to Equation (2), we can define ultimate shear stiffness degradation factor λ_u, which is equal to the ratio of K_u and K_e,

$$\lambda_u = \frac{K_u}{K_e} = \frac{2n(1+\mu)\rho_v \cot^2\theta_u}{1 + n\rho_v \csc^4\theta_u} \tag{4}$$

The factor λ_u is defined as the shear stiffness degradation factor, which reflects the decreasing magnitude of shear stiffness at the fully cracked stage. As shown in Equation (4), the main parameters that influence λ_u are the stirrup ratio ρ_v and the strut angle θ_u. If the only unknown parameter θ_u is determined, λ_u can be easily calculated by Equation (4).

3.3. Determination of Strut Angle θ_u

Scholars have already proposed various solving methods for the strut angle [33–35], most of which employed the minimum energy principle or plasticity theory. Their findings imply that the strut angle closely relates to the stirrup ratio, longitudinal reinforcement ratio, or concrete strength.

For calculating the shear stiffness calculation of fully diagonally cracked slender RC beams, Pan et al. [8] suggested that the strut angle can be calculated by Equation (5). It accounts for the influence of the web and longitudinal reinforcement.

$$\theta_u = \arctan\left[\left(\frac{1 + \frac{1}{n\rho_s}}{1 + \frac{1}{n\rho_v}}\right)^{0.25}\right] \tag{5}$$

In addition, He et al. [9] derived the strut angle of slender beams based on the lower-bound theorem of plasticity (Equation (6)). It accounts for the influence of the stirrup ratio and the concrete strength while assuming that the longitudinal reinforcement will not yield before shear failure. The key parameter for the equation ω is the mechanical web reinforcement ratio, and $\omega = \rho_v f_{yv}/f'_c, f'_c$ is the compressive strength of concrete.

$$\theta_u = \arcsin\left(\sqrt{0.23\left(\omega + \sqrt{\omega^2 + 13\omega}\right)}\right) \tag{6}$$

For simplification, compression field theory [27] is adopted for beam shear analysis to determine the strut angle (see Figure 6). The formulation process assumes that concrete beams are subject to service load, under which the steel bars and the inclined concrete struts behave linearly elastically.

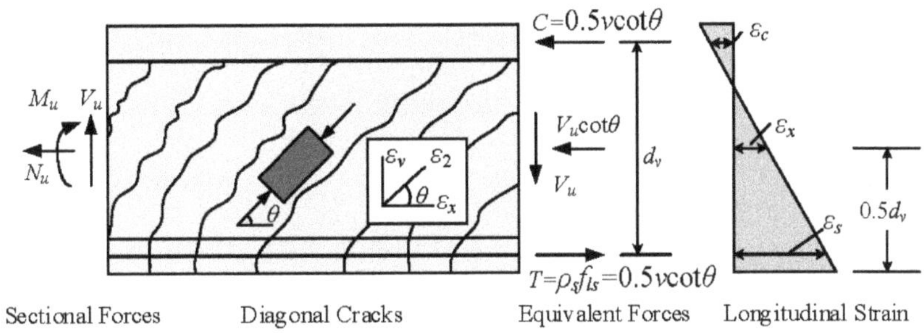

Figure 6. CFT adopted for shear analysis of RC beam.

The equilibrium of average stress and compatibility of average strain in a beam section are summarized in Equations (7)–(13), in which the longitudinal strain ε_x at the middle height is assumed to be 0.5 times the longitudinal strain ε_s at the center of longitudinal steel bars [3].

$$\rho_v f_v = v \tan\theta_u \tag{7}$$

$$\rho_s f_{ls} = 0.5v \cot\theta_u \tag{8}$$

$$f_2 = v(\tan\theta_u + \cot\theta_u) \tag{9}$$

$$\tan^2\theta_u = \frac{\varepsilon_x + \varepsilon_2}{\varepsilon_z + \varepsilon_2} \tag{10}$$

$$\varepsilon_2 = \frac{f_2}{E_c} \tag{11}$$

$$\varepsilon_x = \frac{\varepsilon_s}{2} = \frac{f_{ls}}{2E_s} = \frac{f_{ls}}{2nE_c} \tag{12}$$

$$\varepsilon_z = \frac{f_v}{E_s} = \frac{f_v}{nE_c} \tag{13}$$

where v, f_v and f_{ls} are the shear stress, stirrup stress and longitudinal reinforcement stress, respectively, and ε_x, ε_s, ε_z and ε_2 are the longitudinal strain, longitudinal strain at the center of longitudinal steel bars, vertical strain and main compression strain, respectively.

From Simultaneous Equations (7)–(13), we can obtain,

$$\theta_u = \arctan\left[\left(\frac{1 + \frac{1}{4n\rho_s}}{1 + \frac{1}{n\rho_v}}\right)^{0.25}\right] \tag{14}$$

Equation (14) is mainly influenced by the web and longitudinal reinforcement ratio.

3.4. Comparison of θ_u with Other Methods

To investigate the validity of angle prediction for slender beams, parameter analysis and comparison are performed according to Pan et al. [8], He et al. [9] and the proposed Equation (14). Parameter values of the reference specimen are ρ_v = 0.5%, ρ_s = 2%, f'_c = 50MPa and f_{yv} = 400MPa. Figure 6 shows how the strut angle changes as a function of stirrup ratio ρ_v, longitudinal reinforcement ratio ρ_s and concrete strength f'_c, respectively.

As is shown in Figure 7, the proposed equation gives an intermediate prediction of strut angle θ_u, while θ_{Pan} predicts the highest value, and θ_{He} predicts the lowest. All three angles are affected by stirrup ratio ρ_v, along with which the predicted angle grows (Figure 7a). Meanwhile, both θ_u and θ_{Pan} reflect the inverse relationship between longitudinal reinforcement ratio ρ_s and the strut angle (Figure 7b). θ_{He} ignores the influence of longitudinal reinforcement ratio ρ_s but emphasizes the importance of concrete strength f'_c (Figure 7c), and it may cause larger deviations for specimens with low longitudinal reinforcement ratio or high concrete strength.

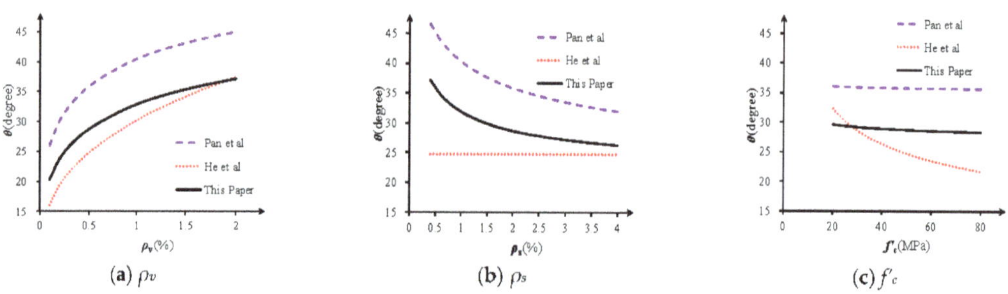

Figure 7. Comparison of different calculation methods of strut angle θ [8,9].

3.5. Proposed Degradation Rules

In spite of the fact that we have obtained the elastic shear stiffness and the fully cracked shear stiffness, it is still difficult to evaluate the effective shear stiffness K_{eff} of a partially diagonal cracked RC beam. As the transition from elastic stiffness to post-cracking stiffness is very complicated and is controlled by many parameters, establishing an exact and quantified expression for K_{eff} is almost impossible.

As is shown in Figure 8, the simplest and most ideal shear stiffness degradation model is the secant stiffness linear degradation model, which assumes that the post-cracking secant shear stiffness will degrade linearly with the shear force (Figure 8a), and it tends to give an unsafe prediction of shear deformation under service state. However, another shear stiffness model, namely, the constant tangent stiffness degradation model, which assumes that the post-cracking tangent shear stiffness will keep constant before stirrup yielding (Figure 8b), is only suitable for thin-webbed beams and tends to give a larger shear deformation prediction.

For RC beams, tension stiffening arises from tension carried by the concrete between the cracks (whether flexural or shear cracks). This contribution decreases with an increasing load after the member has cracked. To simulate the tension stiffening effect, the effective moment of inertia I_e approach introduced by Branson [37] facilitates a gradual transition from un-cracked to a fully cracked section as the ratio of service load moment M_a to cracking moment M_{cr} increases.

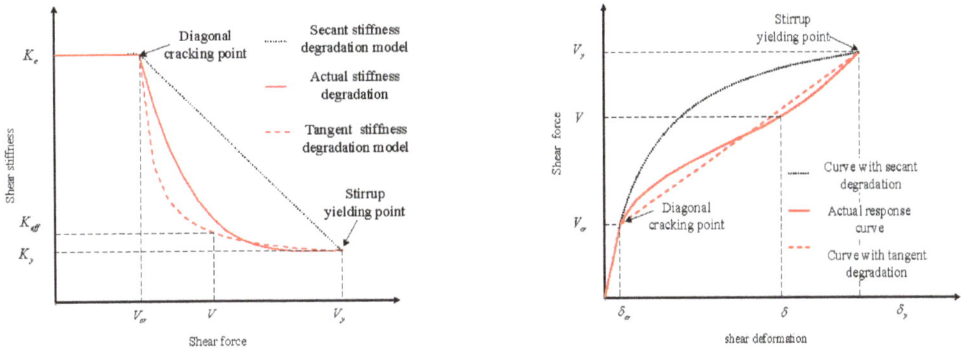

(**a**) Shear deformation curve (**b**) Shear stiffness degradation mode

Figure 8. The shear stiffness degradation of diagonally cracked RC beam.

The ACI 318 standard [2] adopts Branson's degradation method when calculating the bending stiffness after flexural cracking. To make the deformation calculation equation uniform and simple, we recommend using a degradation criterion similar to Branson to reflect the effect of tensile hardening. The recommended shear stiffness degradation formula is as follows,

$$K_{eff} = K_u + \left(\frac{V_u - V}{V_u - V_{cr}} \right)^3 (K_e - K_u) \leq K_e \tag{15}$$

where the diagonal cracking load V_{cr} and stirrup yielding load V_u can be calculated by Equations (16) and (17), respectively.

$$V_{cr} = 0.17\sqrt{f_c'}b_w d \tag{16}$$

$$V_u = V_{cr} + \rho_v f_{yv} b_w d_v \cot \theta_u \tag{17}$$

4. Test Verification and Discussion

4.1. Experiment Introduction

The shear deformation of 12 beam lattices in 6 thin web-restrained beams was directly measured and analyzed with self-designed strain-measuring lattices. Further, the shear deformation test of six beams conducted by Hansapinyo [24] was used to verify the proposed degradation model. Data for a total of 16 shear deformation measurement lattices are used for experimental verification. The main parameters of the specimens are listed in Table 1. Hansapinyo's test gives detailed shear strain test results, in which electronic transducers were also used to measure the normal and shear strains of each lattice based on the rosette concept.

As the stress distribution is disturbed by local point load in D regions (discontinuity regions, such as the lattice regions of G1, G2 and G5 shown in Figure 2), the direct strut component joins in the force transfer mechanisms in addition to the flexural and shear components. Consequently, the total deformation consists of not only flexural deformation and shear deformation but also the deformation of direct strut compression. Mean shear strain in these D regions becomes insignificant. Therefore, shear strain analyses are only performed on the lattice where the local point load disturbances are negligible (Lattice G3 and G4).

4.2. Comparing Results and Discussion

The measured and calculated shear strain of 16 zones of 10 beams are shown in Figures 9–11, respectively. Compared with the measured shear force–strain, the theoretical prediction results are in good agreement with the measured values. It can be concluded that the proposed shear stiffness degradation model simulates the degradation process of shear stiffness very well, and it tends to give a conservative prediction for the shear strain after diagonal cracking.

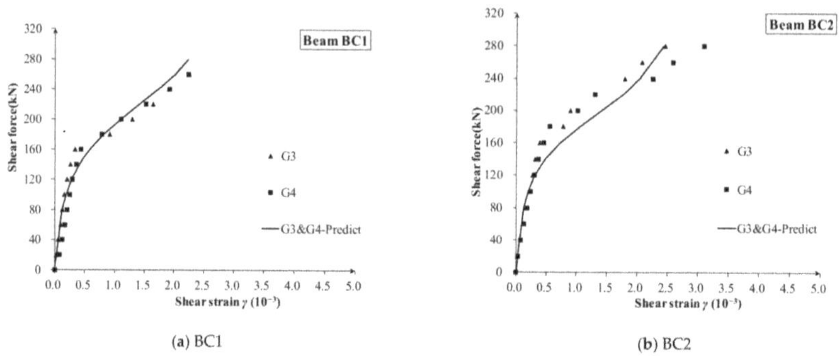

Figure 9. The shear strain of each measuring lattice in specimens BC1 and BC2.

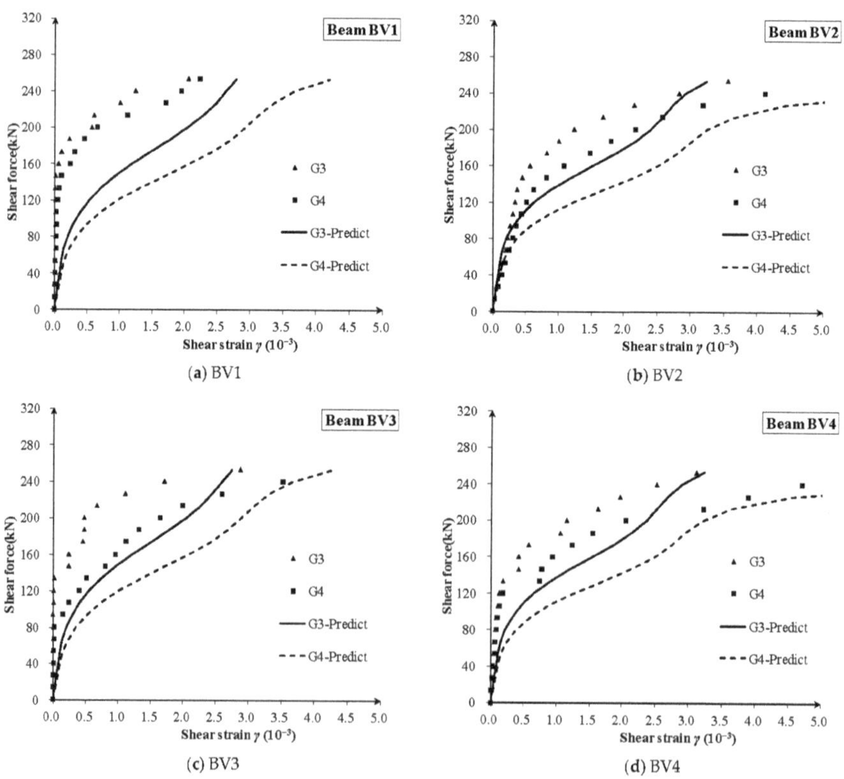

Figure 10. The shear strain of each measuring lattice in specimens BV1, BV2, BV3 and BV4.

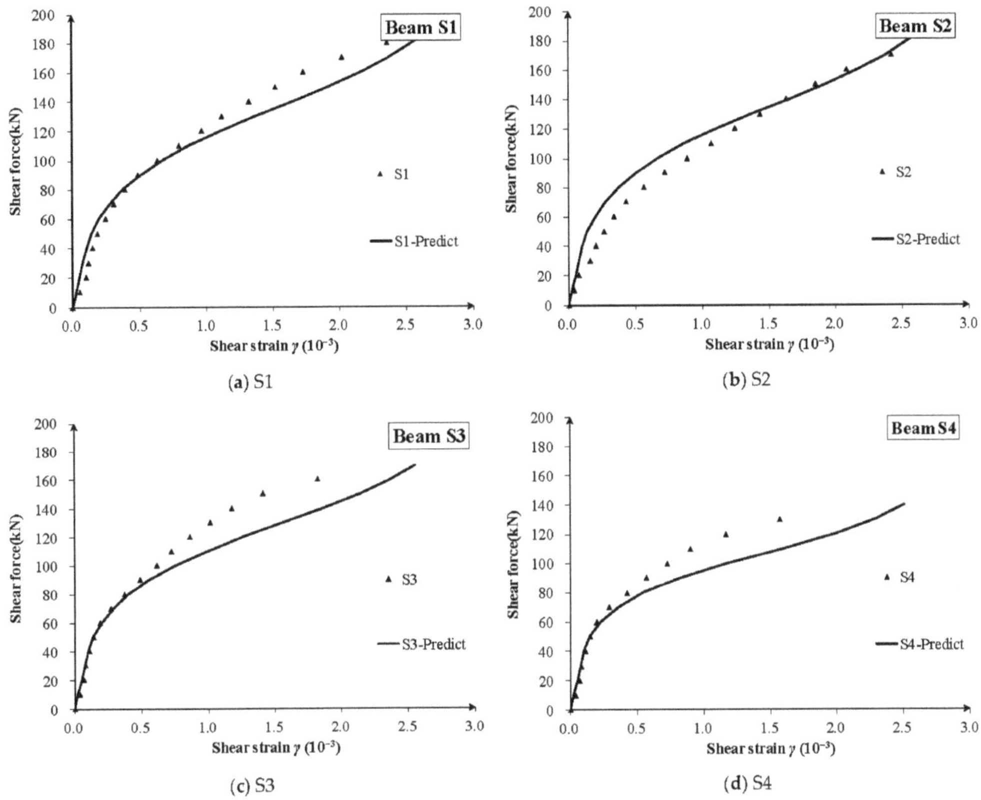

Figure 11. The shear strain of each measuring lattice in specimens S1, S2, S3 and S4.

For the specimen CV series (Figure 9) and S series (Figure 11) with constant depth, the proposed degradation model is a little conservative and gives very acceptable accuracy. Meanwhile, the degradation rule well captures the main characteristics of the shear strain curves, such as the turning point for the first diagonal crack and the gradual evolution from elastic stiffness to fully diagonally cracked stiffness.

For the specimen BV series with variable depth (Figure 10), it should be noted that the inclined lower chord bears part of the shear force, which is not completely consistent with the assumption for constant depth beam, for which the concrete web bears most of the shear force based on elastic beam theory. Therefore, it can be foreseen that the shear deformation prediction of the BV series specimens will be slightly larger than experimental results in the early loading stage, but the degradation law of shear stiffness is still in good agreement among each specimen.

In addition, although the shear span ratios of different observed beam lattices have large changes (such as lattices G3 and G4 of BV series in Figure 9, and lattices S1 and S2 in Figure 10), there is no obvious relationship between measured shear strain and shear depth-to-span ratio. This shows that the shear depth-to-span ratio has little effect on the degradation of the shear stiffness, and its effect on the after-cracking shear deformation is less.

As the beams crack randomly during the loading stage, they cannot get into a fully diagonally cracked status that matches the theoretical assumptions. The real shear stiffness of the beams might be larger than theoretical values, while the real shear deformations are just the opposite. In total, the proposed model will give a relatively conservative shear stiffness prediction for partially diagonal cracked beams.

Here, specimen S3 (ρ_s = 2.13%) is taken as an example to show the validation process. The calculated V_{cr} = 45kN, and the calculated V_y for the proposed method, Pan method and He method are 179.4 kN, 148.7 kN and 144.3 kN, respectively, while the values of λ_u are 0.169, 0.127 and 0.178, respectively. The predicted ultimate shear stiffness degradation reduction factor λ_u and shear strain calculated by three methods are compared to the measured data (Figure 12), which shows that the proposed method is a little better than others for the after-cracking shear stiffness evaluation.

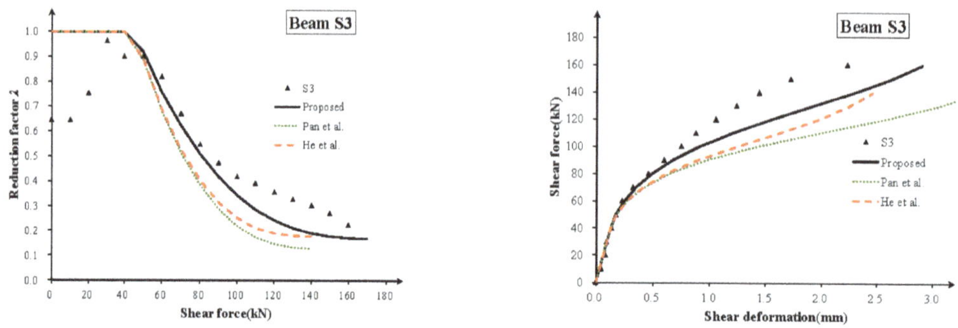

(**a**) Reduction factor (**b**) Shear deformation

Figure 12. Comparison of three methods in prediction of shear stiffness and shear strain [8,9].

In summary, the shear stiffness degradation model proposed in this paper can better evaluate the shear stiffness degradation of each beam lattice, and it can give a reasonable prediction of shear deformation, which can be used for the evaluation of the shear stiffness of diagonally cracked RC beams.

5. Conclusions

This study aims to obtain the shear stiffness degradation law and to propose a simplified and practical prediction method of thin-webbed beams, which can be further applied to existing concrete box-girder bridges with diagonal web cracks. Six large-scale thin-webbed concrete beams were tested in this paper to investigate the shear stiffness degradation of RC beams before and after shear cracking. Considering the effect of concrete tension stiffening, a practical shear stiffness degradation model was proposed and validated. The main conclusions can be drawn as follows:

1. The shear deformation test showed that the shear stiffness drops to about 30~40% of the original stiffness following the occurrence of the first main diagonal crack, and it further drops to only about 10% of the original stiffness when the stirrup yields.
2. The strut angle θ_u was deduced by combining CFT and elastic beam theory. Compared with two other methods from the literature, the proposed angle tends to give a moderate prediction of strut angles and shear deformation with higher accuracy.
3. Considering the tensioning stiffness effect, a simplified shear stiffness degradation rule was suggested for a diagonally cracked RC beam. A cubic form degradation equation consistent with the degradation form of flexural stiffness was established and validated.
4. Data for a total of 16 zones of lattice shear deformation from 10 beams were measured or collected for verification. The results showed that a turning point occurs in the shear deformation curve corresponding to the first diagonal crack. And, rather than the pre-cracking stage, the shear span-to-depth ratio has little effect on the shear deformation of RC beams in the post-cracking stage.
5. The results showed that the proposed method gives a good and consistent prediction of the effective shear stiffness and shear strain development. The proposed model

could capture the development characteristics of shear deformation curves. However, for the BV series, the bottom flanges bear part of the shear force, which will cause a larger predicted shear strain.

6. In general, the proposed simplified shear degradation model tends to give a conservative prediction of shear stiffness, and it is very practical for the early evaluation of diagonally cracked box-girder bridges in service.

Moreover, the contribution of the inclined flanges of variable depth specimens on shear capacity and shear deformation could be an interesting topic for further study. Also, the digital image correlation (DIC) technique maybe a good choice for shear deformation tests in future studies [38–41].

Author Contributions: Conceptualization, K.Z., Y.Z. and Y.W. (Yang Wei); data curation, S.Z. and Y.W. (Yuxi Wang); formal analysis, J.W.; funding acquisition, K.Z. and J.W.; methodology, K.Z.; project administration, Y.Z., Y.W. (Yang Wei) and X.Q.; resources, Y.W. (Yang Wei) and X.Q.; software, Y.W. (Yuxi Wang); supervision, Y.Z.; validation, S.Z.; visualization, X.Q.; writing—original draft, K.Z. and S.Z.; writing—review and editing, Y.Z. and J.W. All authors have read and agreed to the published version of the manuscript.

Funding: This research was funded by the National Natural Science Foundation of China (52108150, 52108408), China Postdoctoral Science Foundation (2021M691606), State Key Laboratory of Mechanical Behavior and System Safety of Traffic Engineering Structures (KF2020-17), State Key Laboratory of High-Performance Civil Engineering Materials (2020CEM006), Jiangsu Planned Projects for Postdoctoral Research Funds (2020Z349), the Natural Science Foundation of Jiangsu Province (BK20200793), the Natural Science Foundation of the Jiangsu Higher Education Institutions of China (19KJB560017) and Nanjing Forestry University Undergraduate Innovation Training Program (202310298024Z, 2022NFUSPITP0127).

Institutional Review Board Statement: Not applicable.

Informed Consent Statement: Not applicable.

Data Availability Statement: Data will be requested to the authors.

Acknowledgments: The discussion from Fenghui Dong is greatly appreciated.

Conflicts of Interest: The authors declare no conflict of interest.

References

1. Park, R.; Paulay, P. *Reinforcement Concrete Structures*; John Wiley & Sons: Christchurch, New Zealand, 1975; pp. 315–319.
2. *ACI318-19*; Building Code Requirements for Structural Concrete and Commentary. ACI: Farmington Hills, MI, USA, 2019.
3. AASHTO. *LRFD Bridge Design Specifications*, 9th ed.; AASHTO: Washington, DC, USA, 2020.
4. Li, H.; Wang, R.; Li, W.; Ye, J. Shear deformation analysis of large-span prestressed concrete box beams. *J. Appl. Fundam. Eng. Sci.* **2012**, *20*, 286–295. (In Chinese)
5. Zhou, M.; Zhang, Y.; Lin, P.; Zhang, Z. A new practical method for the flexural analysis of thin-walled symmetric cross-section box girders considering shear effect. *Thin-Walled Struct.* **2021**, *171*, 108710. [CrossRef]
6. Leonhardt, F.; Walther, R. *The Stuttgart Shear Tests*; Cement and Concrete Association: London, UK, 1964; Volume 111, pp. 1–134.
7. Wang, G.; Xie, J.; Fu, Y. Investigation and research on cracks of long-span prestressed concrete box girder bridge in use. *Highw. Transp. Technol.* **2008**, *25*, 52–56. (In Chinese)
8. Pan, Z.; Li, B.; Lu, Z. Effective shear stiffness of diagonally cracked reinforced concrete beams. *Eng. Struct.* **2014**, *59*, 95–103. [CrossRef]
9. He, Z.Q.; Liu, Z.; John Ma, Z. Shear deformations of RC beams under service loads. *J. Struct. Eng.* **2017**, *143*, 04016153. [CrossRef]
10. *ACI 445-R99*; Committee Recent Approaches to Shear Design of Structural Concrete. American Concrete Institute: Farmington Hills, MI, USA, 2009.
11. Li, H.; Wang, L.; Wei, Y.; Wang, B.J.; Jin, H. Bending and shear performance of cross-laminated timber and glued-laminated timber beams: A comparative investigation. *J. Build. Eng.* **2021**, *45*, 103477. [CrossRef]
12. Chen, S.; Wei, Y.; Ding, M.; Zhao, K.; Zheng, K. Combinatorial design and flexural behavior of laminated bamboo–timber composite beams. *Thin-Walled Struct.* **2022**, *181*, 109993. [CrossRef]
13. Wang, Z.; Wei, Y.; Hu, Y.; Chen, S.; Zhao, K. An investigation of the flexural performance of bamboo-concrete composite beams with precast light concrete slabs and dowel connectors. *J. Build. Eng.* **2021**, *41*, 102759. [CrossRef]

14. Ghahremannejad, M.; Abolmaali, A. Shear Capacity of Reinforced Concrete Box Culverts Compared with AASHTO Shear Equation. *J. Bridg. Eng.* **2019**, *24*, 04019032. [CrossRef]
15. El-Helou, R.G.; Graybeal, B.A. Shear Design of Strain-Hardening Fiber-Reinforced Concrete Beams. *J. Struct. Eng.* **2023**, *149*, 04022234. [CrossRef]
16. Gao, X.; Xiang, D.; Li, J.; Ren, X. Decomposition of the Shear Capacity of Steel Fiber–Reinforced Concrete Coupling Beams. *J. Struct. Eng.* **2021**, *147*, 04021176. [CrossRef]
17. Lee, J.Y.; Watanabe, F. Shear Design of Reinforced Concrete Beams with Shear Reinforcement Considering Failure Modes. *ACI Struct. J.* **2000**, *97*, 477–484.
18. Voo, Y.L.; Poon, W.K.; Foster, S.J. Shear Strength of Steel Fiber-Reinforced Ultrahigh—Performance Concrete Beams without Stirrups. *J. Struct. Eng.* **2010**, *136*, 1393–1400. [CrossRef]
19. Gérin, M.; Adebar, P. Simple Rational Model for Reinforced Concrete Subjected to Seismic Shear. *J. Struct. Eng.* **2009**, *135*, 753–761. [CrossRef]
20. Dönmez, A.A.; Carloni, C.; Cusatis, G.; Bažant, Z.P. Size Effect on Shear Strength of Reinforced Concrete: Is CSCT or MCFT a Viable Alternative to Energy-Based Design Code? *J. Eng. Mech.* **2020**, *146*, 04020110. [CrossRef]
21. Desalegne, A.S.; Lubell, A.S. Consideration of Shear Deformations for Slender Concrete Beams. *ACI Spec. Publ.* **2012**, *284*, 1–18. [CrossRef]
22. Rahal, K.N. Post-cracking shear modulus of reinforced concrete membrane elements. *Eng. Struct.* **2010**, *32*, 218–225. [CrossRef]
23. Zhu, R.; Hsu, T.; Lee, J.Y. Rational Shear Modulus for Smeared-Crack Analysis of Reinforced Concrete. *ACI Struct. J.* **2001**, *98*, 443–450.
24. Hansapinyo, C.; Pimanmas, A.; Maekawa, K.; Chaisomphob, T. Proposed model of shear deformation of reinforced concrete beam after diagonal cracking. *J. Div. Mater. Concr. Struct. Pavements* **2010**, *725*, 305–319. [CrossRef]
25. Debernardi, P.G.; Taliano, M. Shear deformation in reinforced concrete beams with thin web. *Mag. Concr. Res.* **2006**, *58*, 157–171. [CrossRef]
26. Zheng, K.Q.; Kuwornu, M.; Liu, Z. Shear Test of Variable Depth RC Beams with Inflection Point. *MATEC Web Conf.* **2019**, *275*, 02003. [CrossRef]
27. Vecchio, F.J.; Collins, M.P. The modified compression field theory for reinforced concrete elements subjected to shear. *ACI J.* **1986**, *83*, 219–231.
28. Wakjira, T.G.; Ebead, U.A. Simplified Compression Field Theory-Based Model for Shear Strength of Fabric-Reinforced Cementitious Matrix-Strengthened Reinforced Concrete Beams. *ACI Struct. J.* **2020**, *117*, 91–104.
29. Agarwal, A.; Foster, S.J.; Stewart, M.G. Model error and reliability of reinforced concrete beams in shear designed according to the Modified Compression Field Theory. *Struct. Concr.* **2021**, *22*, 3711–3726. [CrossRef]
30. Perera, S.V.T.J.; Mutsuyoshi, H. Shear Behavior of Reinforced High-Strength Concrete Beams. *ACI Struct. J.* **2013**, *110*, 43–52.
31. Suchorzewski, J.; Korol, E.; Tejchman, J.; Mróz, Z. Experimental study of shear strength and failure mechanisms in RC beams scaled along height or length. *Eng. Struct.* **2018**, *157*, 203–223. [CrossRef]
32. Golewski, G.L. The Phenomenon of Cracking in Cement Concretes and Reinforced Concrete Structures: The Mechanism of Cracks Formation, Causes of Their Initiation, Types and Places of Occurrence, and Methods of Detection—A Review. *Buildings* **2023**, *13*, 765. [CrossRef]
33. Wang, T.; Dai, J.-G.; Zheng, J.-J. Multi-angle truss model for predicting the shear deformation of RC beams with low span-effective depth ratios. *Eng. Struct.* **2015**, *91*, 85–95. [CrossRef]
34. Lee, J.D.; Mander, J.B. Unified Truss-Arch Model for the Analysis of Bending-Shear Interaction in Reinforced Concrete Members. *J. Struct. Eng.* **2022**, *148*, 04022074. [CrossRef]
35. Pang, X.B.; Hsu, T.T.C. Fixed-Angle Softened-Truss Model for Reinforced Concrete. *ACI Struct. J.* **1996**, *93*, 197–207.
36. Kim, J.H.; Mander, J.B. Influence of transverse reinforcement on elastic shear stiffness of cracked concrete elements. *Eng. Struct.* **2007**, *29*, 1798–1807. [CrossRef]
37. Branson, D.E. *Deformation of Concrete Structures*; McGraw-Hill Companies: Toronto, ON, Canada, 1977; pp. 545–546.
38. Duan, M.; Zou, X.; Bao, Y.; Li, G.; Chen, Y.; Li, Z. Experimental investigation of headed studs in steel-ultra-high performance concrete (UHPC) composite sections. *Eng. Struct.* **2022**, *270*, 114875. [CrossRef]
39. Golewski, G.L. Evaluation of fracture processes under shear with the use of DIC technique in fly ash concrete and accurate measurement of crack path lengths with the use of a new crack tip tracking method. *Measurement* **2021**, *181*, 109632. [CrossRef]
40. Chen, Y.; Li, H.; Xu, W.; Yuan, C.; Xue, X.; Corbi, O. Experimental study on Mode I fracture characteristics of laminated flattened-bamboo lumber com-bined with VIC-3D system. *Eng. Fract. Mech.* **2023**, *282*, 109195. [CrossRef]
41. Zhang, F.; Wang, C.; Liu, J.; Zou, X.; Sneed, L.H.; Bao, Y.; Wang, L. Prediction of FRP-concrete interfacial bond strength based on machine learning. *Eng. Struct.* **2023**, *274*, 115156. [CrossRef]

materials

MDPI

Article

Investigation on Buckling Performance of Prefabricated Light Steel Frame Materials under the Action of Random Defects during Construction

Gang Yao, Yuxiao Chen, Yang Yang *, Xinlong Ma and Wulei Men

Key Laboratory of New Technology for Construction of Cities in Mountain Area, School of Civil Engineering, Chongqing University, Chongqing 400045, China
* Correspondence: 20121601009@cqu.edu.cn

Abstract: This investigation proposes an analytical approach for analyzing the impact of random defects on light steel frame materials. The addition of random defects for the overall and the component units was achieved by integrating Matlab R2022a and Ansys R19.0 finite element software. Nonlinear analysis was conducted to calculate ultimate load factors and nodal ultimate displacements of the materials under various random defects at each stage of construction. A two-factor analysis was employed to investigate the effects of random defects on the calculation results during different construction stages. The investigation reveals that the response of the light steel frame materials to initial defects is more pronounced during the construction stage. Moreover, the construction stage is the main factor that affects the ultimate load factor and nodal ultimate displacement, compared with random defects. The influence of different random defects on structural displacements varies significantly. The displacement development of the light steel frame materials under the influence of component unit defects tends to be more rapid than that of the overall defects. However, their buckling critical loads are essentially similar.

Keywords: light steel frame materials; construction stages; random defects; two-factor analysis; numerical simulation

Citation: Yao, G.; Chen, Y.; Yang, Y.; Ma, X.; Men, W. Investigation on Buckling Performance of Prefabricated Light Steel Frame Materials under the Action of Random Defects during Construction. *Materials* **2023**, *16*, 5666. https://doi.org/10.3390/ma16165666

Academic Editor: Andrea Di Schino

Received: 20 July 2023
Revised: 15 August 2023
Accepted: 16 August 2023
Published: 17 August 2023

1. Introduction

Steel is an extensively used material in prefabricated buildings due to its lightweight, high strength, ease of production and processing, high construction efficiency, and strong assemblability [1]. Furthermore, under the promotion of the Chinese policy, assembled light steel materials have emerged as a preferred option for rural housing renovation and residential construction and have been built in large numbers [2,3]. Currently, common low-rise assembled building structural systems include the composite insulated reinforced welded mesh concrete shear wall system (CL building structural system), the assembled composite wall structural system, the prefabricated reinforced concrete hollow mould shear wall structural system, the assembled wood structural system, the dense-column-supported frame structural system, the modular structural system, and the low-rise assembled light steel framed structural system [4–7]. Among these, the low-rise assembled light steel frame structure system is the most favored for low-rise building structures due to its advantages in modularization, standardization, environmentalization, economization, factory assembly, and informationization production.

Congenital, processing, fabrication, connection, transportation, and installation defects in steel structures significantly impact their safety and applicability. Especially in rural construction, with many unique characteristics, light steel structures' production and installation stages are particularly susceptible to a decline in structural performance due to construction irregularities. Random defects can exacerbate this issue, potentially resulting in premature material buckling and posing a significant risk to the building's safety. Furthermore, improving the flexural properties of materials under the effect of defects not only contributes to the structure's safety but also helps save costs and improve construction efficiency. Therefore, the impact of structural defects on prefabricated steel materials cannot be overlooked [8].

Currently, research on the stability and bearing capacity of steel frame materials with defects mostly focuses on large and complex structures, such as mesh frames and mesh shell structures. The defects studied are primarily initial geometrical defects and residual stresses in the structure or members [9–13]. Usually, the initial geometric defects mainly refer to the buckling of the bar, increasing the perturbation and deformation of the material, and the defects' shape can be simulated by sinusoidal waves [14]. Residual stresses mainly refer to the stresses present in hot-rolled or welded steel members from processes such as rolling, welding, and cold forming, which may reduce the material's fatigue life and increase the risk of fracture [15]. Kani et al. [16] used the tangent stiffness matrix to solve the instability mode of the structure and found that node misalignment has a more significant effect on the structure than defects in the bars. Bielewicz et al. [17] conducted a stochastic analysis on the static response of a nonlinear model of a defective shell structure, combining the Monte Carlo method with finite element program analysis to analyze the nonlinear post-buckling behavior of shells and discuss the method's accuracy. Lauterbach et al. [18] used a stochastic method to simulate structural imperfections and found that the effect of geometrical imperfections can be neglected in areas where support exists in thin-walled members. Kala et al. [19] investigated the deformation of planar trusses with stochastic imperfections subjected to vertical loading and found that asymmetric defects have a detrimental effect on the load-carrying capacity of the trusses. Roy et al. investigated the flexural behavior of back-to-back built-up cold-formed steel and the axial strength of angle columns through tests and numerical simulations [20,21].

Research on defects of steel frame materials started earlier and developed faster, and the current focus is on constructing a database of material defects and studying the effects of defects on structures by probabilistic methods. Arrayago et al. [22] conducted a statistical analysis of the main parameters affecting the strength of steel based on data from the last decades. They considered factors such as steel type, cross-section geometry, defects, and residual stresses and proposed a compatible probabilistic model to provide a database for research on steel structure defects. Mirzaie [23] measured the geometry of steel tube defects, analyzed the characteristics of the defects caused by the manufacturing process and the errors in the measurement of the defects, and demonstrated the feasibility of using probabilistic methods to generate geometric defects consistent with the measurements. Fina et al. [24] used a probabilistic approach to establish a Gaussian random field for random defects, extending the classical probabilistic approach to the fuzzy-random approach, providing a more reasonable description of the inaccurate random defect sampling method and evaluating the simulation results of the fit.

In advanced structural analysis, there are three primary methods for considering structural defects: the direct analysis method [25], the equivalent nominal load method [26], and the reduced tangent modulus method [27]. The equivalent nominal load and reduced tangential modulus methods are approximate methods proposed when computer technology was not yet mature [28]. As hardware and software have developed, the direct analysis method of the overall structure has been adopted by design codes, and it is the most commonly used structural analysis method now. The direct analysis method involves introducing a definite defect value directly on the member in the structural analysis [29–31]. This method recognizes that the initial state of the member is no longer ideal and considers the direct effect of the defect on the light steel materials.

Research on defects mainly focuses on traditional steel structures, with a limited investigation on the working performance of low-rise light steel materials under defective states [32–34]. Additionally, relevant codes have not made provisions for this, and the steel structure design code is still adopted as the defect control standard in the design of light steel structures. Therefore, it is crucial to investigate the effect of initial defects on the stability performance of assembled lightweight steel buildings, given their widespread popularity [35,36]. To ensure the safety of light steel framed buildings and the sustainability of this material, this paper proposes an improved Monte Carlo-based method to analyze the buckling performance of low-rise assembled light steel frame materials under the separate effects of random overall geometrical defects and random component geometrical defects during the structure's service and construction stages.

2. Model of Low-Rise Assembled Light Steel Frames

2.1. Project Overview

This study focuses on a residential project in Shangxing Town of Liyang City, as shown in Figure 1. This low-rise steel residence is constructed with light steel frame materials. The project consists of three floors above ground and has a design service life of 50 years. The building is classified as structural safety class II, with a structural importance coefficient of 1.0, a seismic protection category of C, a seismic protection intensity of 7 degrees, and a design basic seismic acceleration of 0.10 g.

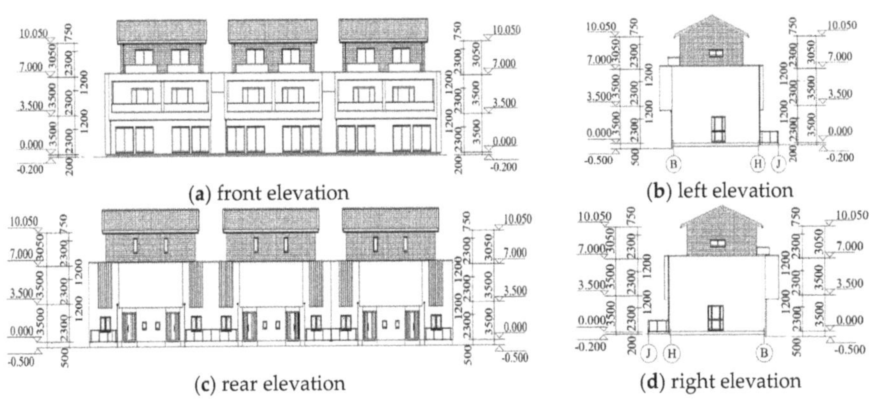

Figure 1. Elevation drawings of the light steel structure.

According to the Load Code for the Design of Building Structures [37] and the Code for Construction of Steel Structures [38], structural loads are assessed based on the construction and usage stages. For permanent loads, the value remains consistent during the construction and usage stages, whereas variable loads are determined by the actual load conditions of the structure in each stage. Load combinations are established as per the specifications, and due to the significant differences in horizontal stiffness, horizontal loads are combined based on the direction of the structure's smaller lateral stiffness. Moreover, given the short construction period for light steel structures, seismic action during the construction stage is not considered, and code-specified structural measures are employed to ensure load-carrying capacity. Table 1 displays the structural load conditions, while Table 2 presents the standard values of structural loads [37].

Table 1. Structural load combination.

Working Condition	Permanent Load	Variable Load	Wind Load	Snow Load
1	1.35	0.7	0.6	0.7
2	1.2	1.4	0.6	0.7
3	1.2	0.7	1.4	0.7
4	1.2	0.7	0.6	1.4

Table 2. Standard value for structural load.

Permanent Load		Variable Load	
Name	**Standard Value**	**Name**	**Usage (Construction) Value**
Reinforced concrete	2.5 kN/m^3	Floor live load	2.0 (0.6) kN/m^2
Steel	78.5 kN/m^3	Roof live load	0.5 (0.5) kN/m^2
C-type light steel keel partition wall	0.54 kN/m^2	Wind load	0.3 (0.3) kN/m^2
300 mm cement hollow block	9.6 kN/m^3	Snow load	0.25 (0.25) kN/m^2
Hardwood flooring	0.4 kN/m^3	/	/
Clay flat tile roofing	0.55 kN/m^3	/	/

2.2. Finite Element Modeling

The project's beams, columns, purlins, and braces are constructed from Q235-B steel, which satisfies the Von Mises yield criterion. The steel's modulus of elasticity is 2.06×10^5 MPa, with a Poisson's ratio of 0.24. The concrete has a modulus of elasticity of 3×10^4 MPa and a Poisson's ratio of 0.2. Both steel and concrete are isotropic materials.

The finite element analysis software Ansys is selected for the structural analysis. BEAM188 cell (3D current finite strain beam cell) was used to simulate the beams, columns, purlins, and bracing, with two integration points set along the length of the beams. The floors and roofs were simulated using the SHELL181 cell (4-node finite strain shell cell), with the cell integration option selecting the full integration of the uncoordinated mode and the bending and membrane stiffness considered. The horizontal support of the structure was simulated using the LINK180 cell (3D finite strain rod cell) [39]. Furthermore, the SURF154 cell (3D structural surface effect cell) overlays on the SHELL cell to apply various loads and surface effects. The components are interconnected through coupling that considers all degrees of freedom, mimicking welded or bolted connections. Moreover, the primary functions of the inner partition wall and exterior walls in the light steel frame synergistic system are space partitioning and enclosure. Consequently, during the modeling process, these elements are considered equivalent to homogeneous loads applied to the beam and column members of the system. The column and beam shapes and dimensions of the structure are shown in Table 3, and the model is shown in Figure 2.

The accuracy of finite element calculations is affected by the mesh size used. Generally, finer mesh sizes yield more accurate results but require longer calculation times, while larger mesh sizes may lead to inaccurate or incorrect results. For eigenvalue buckling analysis, Ansys does not distinguish between local and overall buckling, making it necessary to set the number of mesh divisions in the columns to one to obtain the overall buckling mode of the materials and determine their defective form. To facilitate subsequent analysis, the finite element model based on the ideal materials was meshed, and the maximum displacement and first-order buckling mode were calculated in the model under working condition 1. A suitable finite element model for adding random defects was then selected. The meshing method is shown in Table 4.

Table 3. Shapes and dimensions of column and beam (unit: mm).

Column Category	Section Size	Beam Category	Section Size
I-column Z1	259 × 107 × 6 × 9	I-beam B1	250 × 125 × 6 × 9
Square column Z2	75 × 75 × 4.5	I-beam B2	250 × 125 × 3.2 × 4.5
Square roof column Z3	60 × 60 × 2.5	C-beam B3	150 × 50 × 50 × 4.0
		Double C-type roof beam B4	100 × 500 × 20 × 2.5

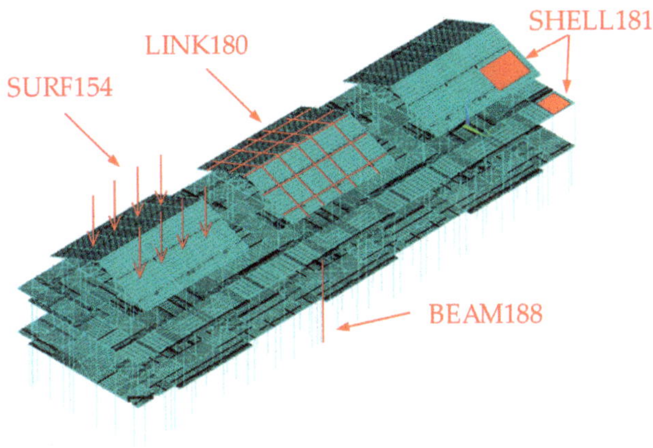

Figure 2. Schematic diagram of the finite element model.

Table 4. Meshing method of the ideal light steel frame materials.

Beam [1]	Column [1]	X-Direction Plate [2]	Y-Direction Plate [2]
2	1	0.5	0.5
4	5	1	1
6	10	/	/

[1] Indicates the number of meshing for the corresponding component; [2] Indicates the ratio between plate and beam nodes.

The meshing method utilized in this study generated 36 sets of data. Trial calculations revealed that the finite element results were more stable when the mesh was divided into smaller cells. Specifically, the first-order buckling factor obtained from the structural eigenvalue buckling analysis was slightly larger when the column mesh was divided into 1, compared to when it was divided into 5 or 10. This was because overall instability occurs when more than one local member is destabilized, and it is difficult to obtain the overall instability pattern of the structure if the column mesh is not divided into 1 [40]. The average deviation of the structural static analysis displacement solution due to changes in the column mesh size was approximately 1×10^{-5}, while the average deviation of the structural first-order buckling factor was about 1.2×10^{-3}, which was negligible. Similarly, when the beam mesh division was changed, the average deviation of the structural static analysis displacement solution was about 1.5%, and the average deviation of the structural first-order buckling factor was about 1.2×10^{-3}. However, when the plate mesh division was finer, the deviation of the two solutions was below 1×10^{-4}, and its effect was negligible. It is worth noting that the mesh division of the plate is associated with the beam, and the computational accuracy increases with finer divisions. However, the number of cells and the size of the model file also increase significantly. Considering the balance between computational accuracy and model file size, selected model number ZT1 for overall random defect analysis and model number JB5 for component unit random defects analysis. Table 5 and Figure 3 show the selected finite element model and calculation results.

Table 5. Calculated results after meshing.

Model	Beam [1]	Column [1]	X-Direction Plate [2]	Y-Direction Plate [2]	Max Displacement	First-Order Buckling Mode
ZT1	4	1	1	1	3.8904	2.8378
JB5	4	5	1	1	3.8966	2.8115

[1] Indicates the number of meshing for the corresponding component; [2] Indicates the ratio between plate and beam nodes.

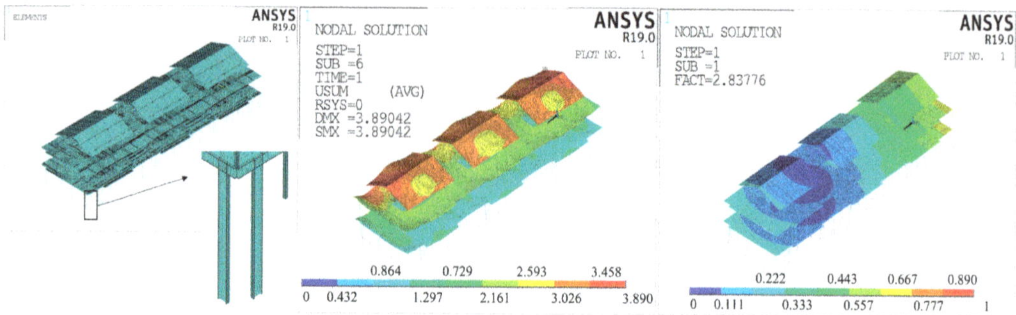

(**a**) ZT1

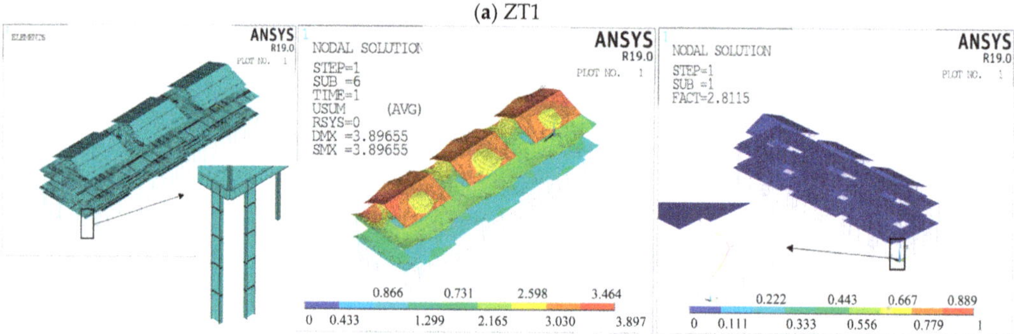

(**b**) JB5

Figure 3. Schematic diagram of the selected finite element model (from left to right: model, static analysis, and eigenvalue buckling analysis).

The simulation of the construction stage was conducted using the elastic-plastic stress analysis with the death/birth elements method, which utilizes the EKILL and EALIVE commands to activate or deactivate corresponding units [41]. However, in Ansys, the EKILL command does not delete the unit but reduces its stiffness to a small value. Consequently, the killed unit still participates in the calculation, necessitating further constraints on the nodes of the killed unit to ensure calculation accuracy. In this investigation, the light steel frame structure was divided into six construction stages based on the degree of completion, following the principle of constructing from the bottom to the top and from the middle to both sides. The construction stages were as follows: the bottom intermediate frame, the bottom frame, the two-story intermediate frame, the two-story frame, the roof beams, and the roof panels. The finite element model is shown in Figure 4, with the blue section representing the newly constructed materials in the current construction stage.

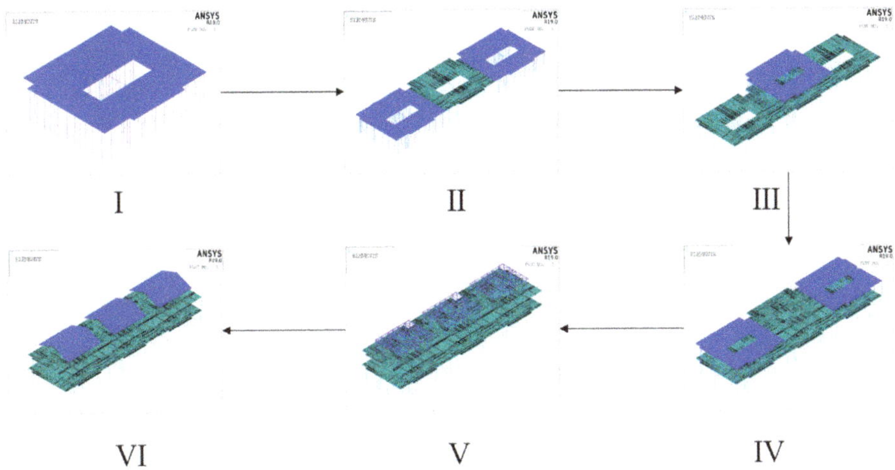

Figure 4. Schematic illustration of construction stages division.

2.3. Determination of the Most Unfavorable Working Conditions

To determine the most unfavorable working conditions under the four loads listed in Table 1, ZT1 and JB5 were calculated at six construction stages for each condition. The corresponding eigenvalues of the first-order buckling mode were obtained and are presented in Tables 6 and 7.

Table 6. Eigenvalues of first-order buckling mode of ZT1.

Working Condition	I	II	III	IV	V	VI
1	7.039	5.502	3.929	3.210	3.196	2.901
2	7.499	5.872	4.223	3.458	3.444	3.128
3	7.856	6.143	4.383	3.584	3.554	3.203
4	7.873	6.155	4.398	3.593	3.577	3.238

Table 7. Eigenvalues of first-order buckling mode of JB5.

Working Condition	I	II	III	IV	V	VI
1	6.967	5.406	3.911	3.181	3.168	2.876
2	7.423	5.768	4.204	3.427	3.413	3.099
3	7.776	6.037	4.363	3.552	3.524	3.175
4	7.793	6.048	4.377	3.562	3.546	3.209

Tables 6 and 7 reveal that the first-order buckling mode eigenvalues of ZT1 and JB5 increase with the increment of the working condition number. For a given condition, the critical buckling load of the light steel frame materials decreases as the construction stage progresses. Specifically, the critical load factor for Stage I is approximately 2.5 times that of Stage VI, indicating that the light steel frame materials are more susceptible to eigenvalue buckling as the number of layers increases. As a result, working condition 1 is deemed the most unfavorable for the structure, and subsequent defect analyses were performed using loads of working condition 1.

2.4. Structural Model under the Action of Random Defects

The Monte Carlo method combines the defect-free structure with the most unfavorable working conditions to simulate random defects in the prefabricated light steel frame materials. The defect distribution pattern and amplitude are controlled to add overall and component unit random defects to the finite element model. The eigenvalue buckling analysis is then combined with ZT1, and the first ten orders of buckling modes are obtained when the overall random defects are imposed on the ideal materials. The random defects of the materials are simulated based on Equation (1) using the Monte Carlo method. Here, m represents the number of loading conditions, n represents the number of selected modes, c_{ij} is the modal participation coefficient, assumed to be normally distributed, and φ_{ij} is the jth-order normalized buckling mode of the structure for the ith condition.

$$\Delta X' = \sum_{i=1}^{m}\sum_{j=1}^{n} c_{ij}\varphi_{ij} \quad . \tag{1}$$

During the actual calculation process, an amplitude adjustment step aims to make the defect form more similar to the actual materials. The amplitude modulation coefficient, α, is defined as the maximum ratio between the value of the structural node's defective form and the defect's permissible value. Equation (1) can be transformed into Equation (2) by incorporating the amplitude modulation coefficient. The overall defects of the structure expressed in Equation (2) are brought into the perfected light steel frame materials to impose the initial defects.

$$\Delta X = \alpha \Delta X' \tag{2}$$

The initial defects in the component units are mainly initial buckling and residual stresses in the beams and columns. The simulation of random defects in component units is primarily achieved through the initial bending of beams and columns. Furthermore, the distribution forms of defects in different components are independent and randomized. For the initial defects of a single component, the distribution form can be modeled by a sinusoidal half-wave. The initial bending of beams and columns can be uniformly represented by Equation (3), where δ_{max} represents the defect amplitude, l represents the length of the component, and x represents the positional coordinate of the point on the component.

$$\delta = \delta_{\max} \sin\left(\frac{\pi x}{l}\right) \tag{3}$$

Due to the infinite number of possible modes of random encounters between defects in each component, the Monte Carlo method is used to simulate initial defects in components. Assuming that the initial defects of the components follow a normal distribution probability model, the initial bending amplitude of columns and beams is set at 1.05 times the Standard for Design of Steel Structure [42], meeting the requirements with a 95% probability. The Monte Carlo simulation is conducted using Matlab to generate random numbers for the defect amplitude of the distribution model. The sampling values are then randomly arranged to complete the random sampling process. As shown in Equation (4), where $\sigma_1 = l/980$.

$$P(\delta) = \frac{1}{\sqrt{2\pi\sigma_1}} e^{\frac{-(\delta)^2}{2\sigma_1}} \tag{4}$$

The process of establishing a random defect finite element model is shown in Figure 5.

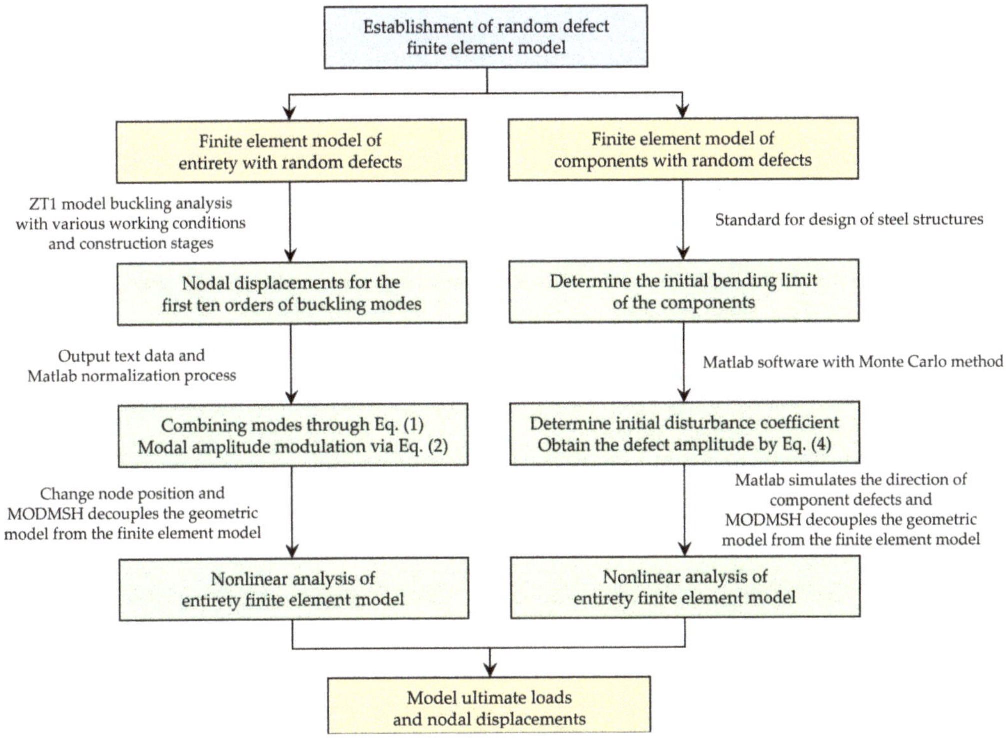

Figure 5. Flowchart for establishing the random defect finite element model.

3. Analysis of Buckling Characteristics under the Effect of Overall Random Defects

3.1. Ultimate Load Factor of the Light Steel Frame Materials

The simulation of the overall random defects of the light steel frame materials is based on the eigenvalue buckling analysis of the structure. Nonlinear buckling critical loads are solved using the arc length method. Due to the large number of models, the finite element model is solved in batches using the Ansys Batch program and the RESUME command. Based on the central limit theorem, thirty is determined to be the basic sample capacity of random defects. Thirty simulations with overall defects are performed on the light steel frame materials of ZT1 in each of the six construction stages, and the ultimate load factor is calculated. The results are presented in Figure 6. The structural ultimate load factor is the smallest in construction stage III and the largest in stage II. The ultimate load is not sensitive to the overall defect changes and is relatively small in construction stages I and III, with average values of 1.773 and 1.292, respectively. It can be found in Figure 4 that the structure of the two construction stages has a certain layer of the unfinished stage. The longitudinal stiffness of the light steel frame materials is not uniformly distributed during stages I and III, and the weaker floor of the stiffness controls the limit state. When the structure is in stages II, IV, V, and VI, the data on the structural ultimate load are relatively discrete, and the overall random defect changes greatly affect the load. In these stages, all layers of construction are complete, and the structure does not have a weak position. The ultimate load is changed with the obvious changes in the flexural morphology of the structure due to the overall defect changes.

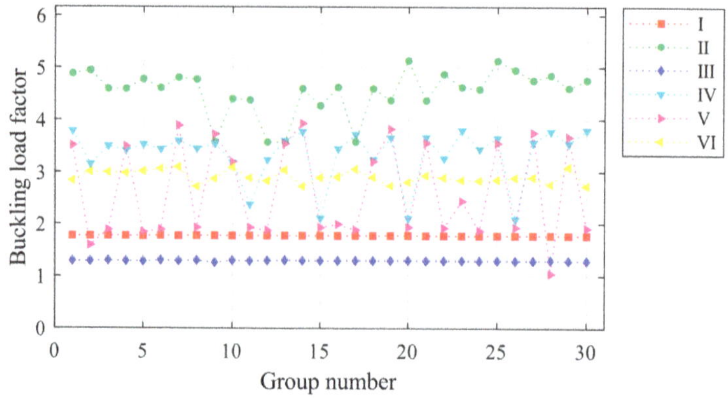

Figure 6. Ultimate load factors for 30 sets of overall random defect models for each construction stage.

3.2. Light Steel Frame Structural Deformation Properties

The structural deformation performance under varied global random defects during construction is examined by analyzing the ultimate displacements at eight control points of ZT1, subjected to thirty sets of defects across six construction stages. The load sub-steps where the ultimate displacements are located are the load sub-steps where the tangent stiffness matrix singularity is located, i.e., the critical state of the structure for the occurrence of instability. Nodes 1–4 represent the maximum displacement nodes in the first, second, and third floors—and the roof, respectively. Conversely, nodes 5–8 denote the maximum displacement nodes in the columns of the first, second, and third floors—and the roof, in that order. Furthermore, these maximum displacement nodes refer to the positions experiencing the greatest displacement within a column or on a floor at a given construction stage and defect set. Consequently, each model's specific node locations may vary but consistently symbolize the most hazardous displacements at their respective floors.

Based on the settings described above, the paper plots the ultimate displacements of control points for each construction stage, as shown in Figure 7, and the average values of limit displacements for control points are shown in Table 8. The ultimate displacements of the nodes increase as the construction stage progresses, with Stages I and III having significantly smaller ultimate displacements than the other stages. This indicates that as the structural completeness increases, the deformation performance of the light steel frame materials improves when it reaches the limit state. Even when the construction of a certain floor is completed, its ultimate displacement is significantly higher compared to the partially completed floor. For instance, Stage II has a higher ultimate displacement than Stage I, while Stage IV has a higher ultimate displacement than Stage III. The degree of variation of node ultimate displacements with defects in construction stages I and III is smaller than in other construction stages. This suggests that stages I and III are less affected by the overall random defects, while the other construction stages are more affected by the variation of defects. Generally, the floor ultimate displacements are larger than the column ultimate displacements for the control nodes of columns and floors on the same floor. This indicates that the overall defects did not affect the trend of the structure to increase the displacements along the height direction. However, construction stage VI is a special case (node 7 compared to node 3). This is because node 7 is located in the layer without floor panels, and compared with construction stage V, there is the role of roof loads. At this stage, node 7 is significantly impacted by overall defects, leading to a significant increase in ultimate displacement compared to node 3.

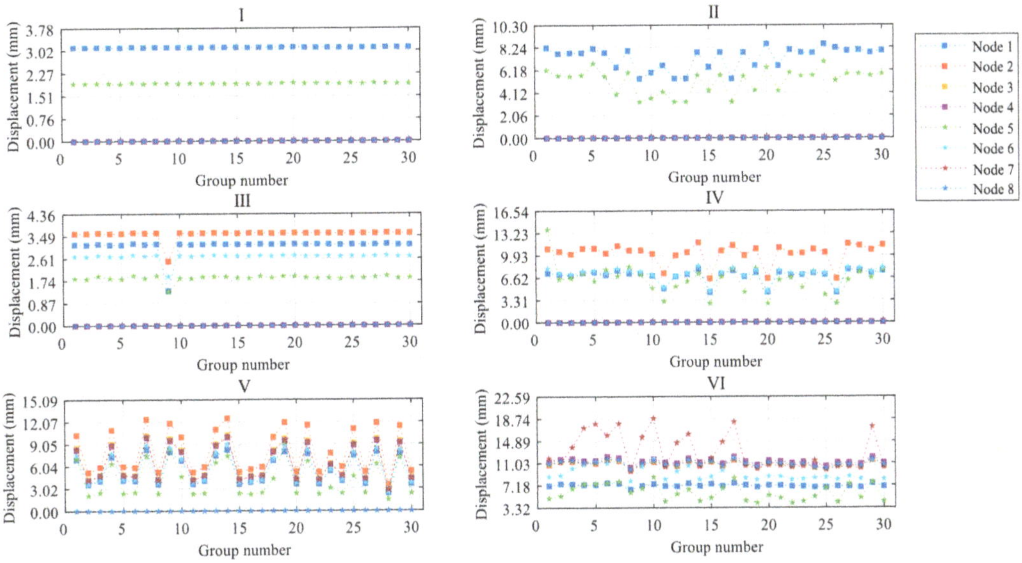

Figure 7. Limit displacement diagrams for control points of ZT1 at each construction stage.

Table 8. Average value of limit displacements for control points at each construction stage.

Node	I	II	III	IV	V	VI
1	3.136	7.324	3.107	6.945	5.491	7.197
2	0	0	3.575	10.226	8.084	10.875
3	0	0	0	0	6.640	11.026
4	0	0	0	0	6.466	11.354
5	1.925	5.188	1.859	6.355	4.213	5.965
6	0	0	2.690	7.218	5.808	9.433
7	0	0	0	0	6.514	12.982
8	0	0	0	0	0	10.987

3.3. *Two-Factor Analysis of ZT1 Model Calculation Results*

The previous calculations indicate that the ultimate load factor may be influenced by two factors: the construction stage and the overall random defects. A two-factor ANOVA without interaction was conducted on the calculated results to investigate the effect of these factors on the ultimate load factor. This analysis assumes that a controlling factor has no significant effect on the results and examines the degree of significance of the effect of each factor on the ultimate load factor by examining the magnitude of the contribution of the variance of the different factors to the total variance.

3.3.1. Two-Factor Analysis of Ultimate Loads

Table 9 shows the two-factor data source analysis results for the ultimate load factor. In the table, SS represents the sum of squares of the data source, df represents the degrees of freedom of the data source, MS represents the mean square of the data source, F represents the ratio of MS, and p represents the judgment factor. The resulting p-values are 0.273 and 5.05×10^{-62}. The first p-value is for the column factor (defective group number), and since it is greater than 0.05, the original hypothesis is accepted, i.e., there is no significant effect of defective group number on the ultimate load factor. The second p-value is for the row factor (construction stage), and since it is much less than 0.05, the original hypothesis is rejected, i.e., there is a very significant effect of the construction stage on the ultimate load factor. This indicates that the construction stage has a much greater effect on the ultimate load than the overall random defects. Still, the overall random defects may have a greater effect on a particular construction stage alone.

Table 9. Calculation table for two-factor analysis of ultimate loads.

Source	SS	df	MS	F	*P*
Defective group number	7.055	29	0.243	1.166	0.273
Construction stage	199.911	5	39.982	191.579	5.05×10^{-62}
Error	30.261	145	0.209	/	/
Total	237.227	179	/	/	/

To better observe the bias and tail weight of the calculated limit loads and compare the shapes of the data sets, box plots of the limit load factors for the two influences are presented in Figure 8. The box's upper and lower blue lines represent the 25% and 75% quantiles of the sample, respectively. The red line in the center indicates the sample's median, and the end of the dashed line indicates the outer limit of the sample. The shapes of the lines in the box plots in the later section have the same meaning. In Figure 8a, although the distribution of each group of data box plots fluctuates, the overall distribution of samples remains relatively unchanged. This is consistent with the results of the two-factor analysis, indicating that the impact of the overall random defects on the ultimate structural capacity is limited, and its impact is more reflected in the fluctuation of the load factor of a particular construction stage. In Figure 8b, the distribution of each group of data box plots has a significantly different distribution, indicating that the different construction stages of the ultimate load factor have a significant impact. Moreover, the limit load factors are closely distributed in construction stages I and III and more dispersed in other construction stages.

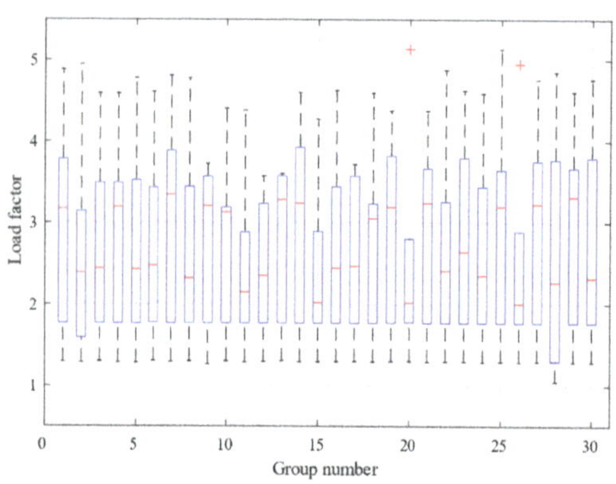

(**a**) defective group number

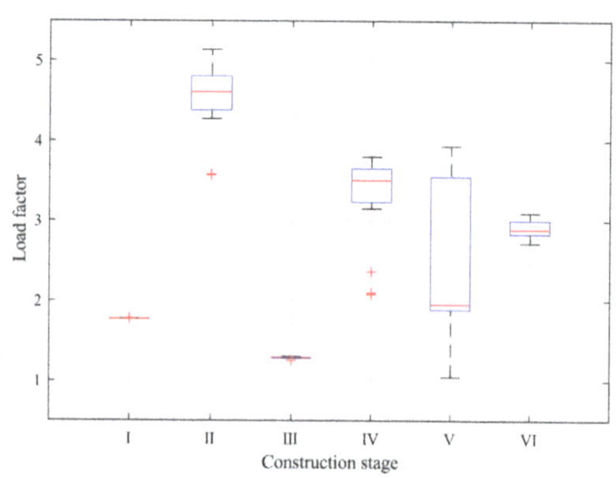

(**b**) construction stage

Figure 8. Box plots of the limit load factors for the two influences.

3.3.2. Two-Factor Analysis of Ultimate Displacement of Control Nodes

A two-factor ANOVA without interaction was conducted for the overall random defects and the construction stage, and the results are presented in Figure 9. The *p*-value calculated for the construction stage is much less than 0.05 for any node, indicating that the construction phase significantly affects the ultimate displacement of the node, similar to the ultimate load factor. The *p*-value calculated for overall random defects is only 0.04 at node 6 and much greater than 0.05 for other nodes, indicating that the ultimate displacement at node 6 is more affected by the overall random defects than other nodes.

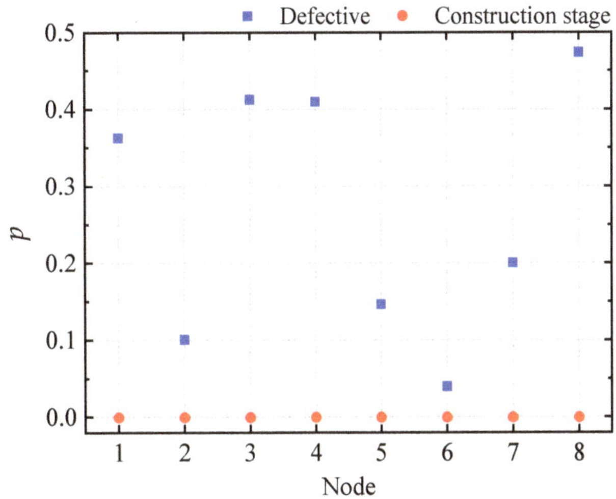

Figure 9. Two-factor analysis plot of the ultimate displacement of the control node for ZT1.

To better observe the distribution of node limit displacements, the paper plotted box plots of limit displacements of eight nodes under two different factors separately in Figures 10 and 11. According to the two-factor analysis calculations, node 6 is more affected by the overall defects. Although the distribution of the box plots is not the most obvious change at node 6 in Figure 10, the change of the median point of the sample distributions at node 6 is very noticeable. In contrast, nodes 3, 4, and 7 have a median point of the sample distribution of zero. Combined with the results of the two-factor analysis, it is reasonable to conclude that node 6 is more affected by the overall random defects. Additionally, since node 8 has non-zero displacement only in construction phase VI, the displacement distribution consists of discrete points without box plots. Figure 11 shows that the box plots of node ultimate displacements in the construction stage are significantly dispersed with respect to the overall random defects displacement box plots. This indicates that the construction stage changes significantly affect the node's ultimate displacements, which is the same as the significant effect of the construction stage on the ultimate load factors. Although the overall random defects do not have a significant effect on the overall node ultimate displacements, the specific node changes in a specific construction stage are still evident.

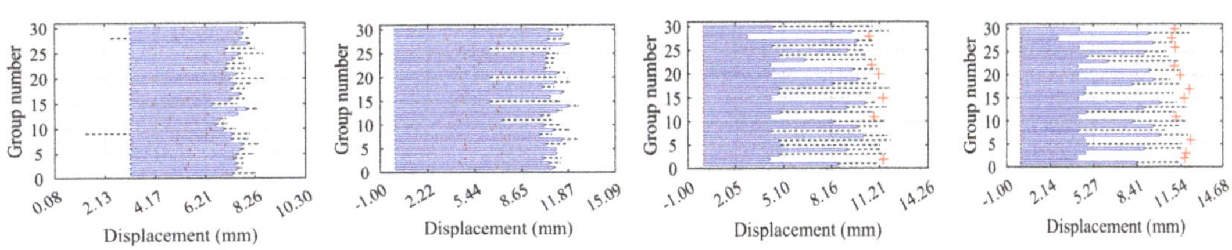

(**a**) node 1 (**b**) node2 (**c**) node 3 (**d**) node 4

Figure 10. *Cont.*

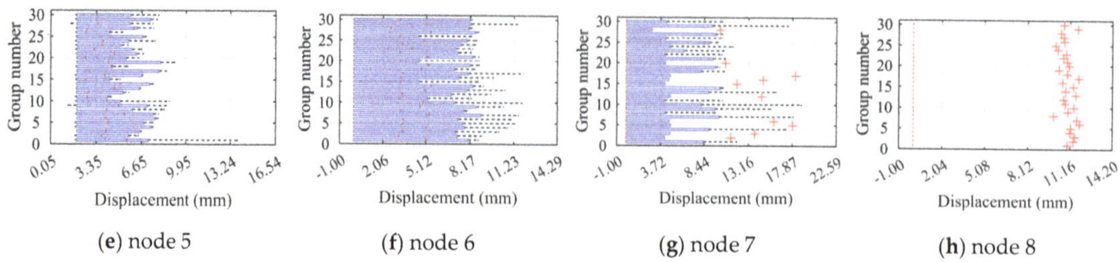

Figure 10. Box plots of ultimate displacements at the control nodes of ZT1 under the overall defects.

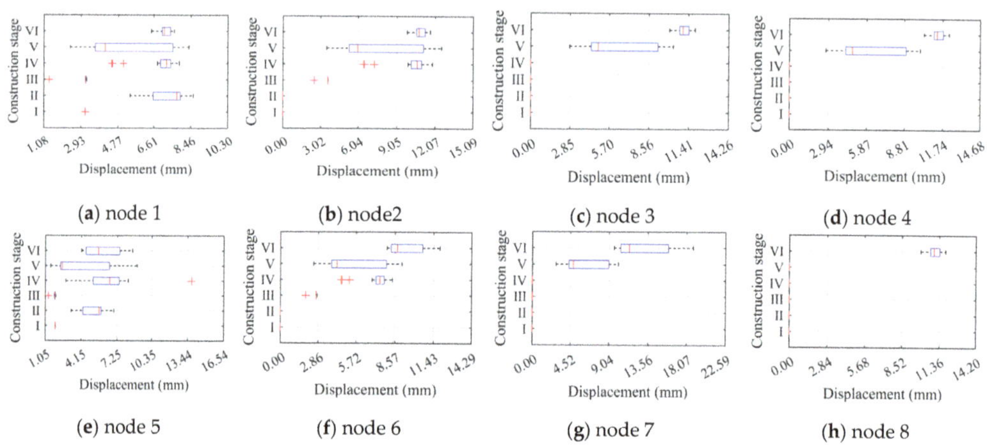

Figure 11. Box plots of ultimate displacements at the control nodes of ZT1 under each construction stage.

4. Analysis of Buckling Characteristics under the Effect of Component Random Defects

4.1. Ultimate Load Factor of the Light Steel Frame Materials

The component unit random defect simulation method was used to simulate 30 sets of component unit random defects for JB5 during construction stages I–VI. The ultimate load factors for each set of defects were calculated, and the results are presented in Figure 12. The ultimate load factor fluctuation is minimal during construction stages I and VI, indicating that changes in component unit defects have minimal influence on the ultimate load factor during these stages. The minimum ultimate load factor of the structure occurred during stage III at 0.9891, while the maximum ultimate load factor occurred during stage IV at 3.951. The maximum ultimate load factor varies significantly across construction stages. However, there is no significant difference between the ultimate load factor under the effect of overall random defects and component unit random defects.

4.2. Light Steel Frame Structural Deformation Properties

The ultimate displacements of eight control nodes of the JB5 were extracted under the effect of 30 sets of component unit defects during six construction stages to analyze the light steel frame materials' deformation performance. The significance of the load sub-steps and each control node for extracted displacements is the same as previously. The ultimate displacement diagrams for each node are presented in Figure 13. The ultimate displacements of the nodes in construction stages II, IV, and VI are significantly improved compared to their respective previous stages, indicating that the deformation performance of the structure improves with the perfection of the structural system. With the exception of construction stages I and III, the column displacement of the same level is larger than the layer displacement, demonstrating that component unit defects significantly affect the ultimate displacement of the column members. The phenomenon of overall defects only occurs during construction stage VI.

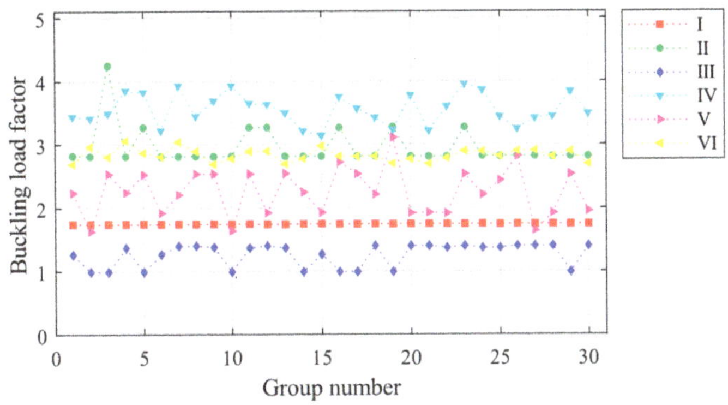

Figure 12. Ultimate load factors for 30 sets of component unit random defect models for each construction stage.

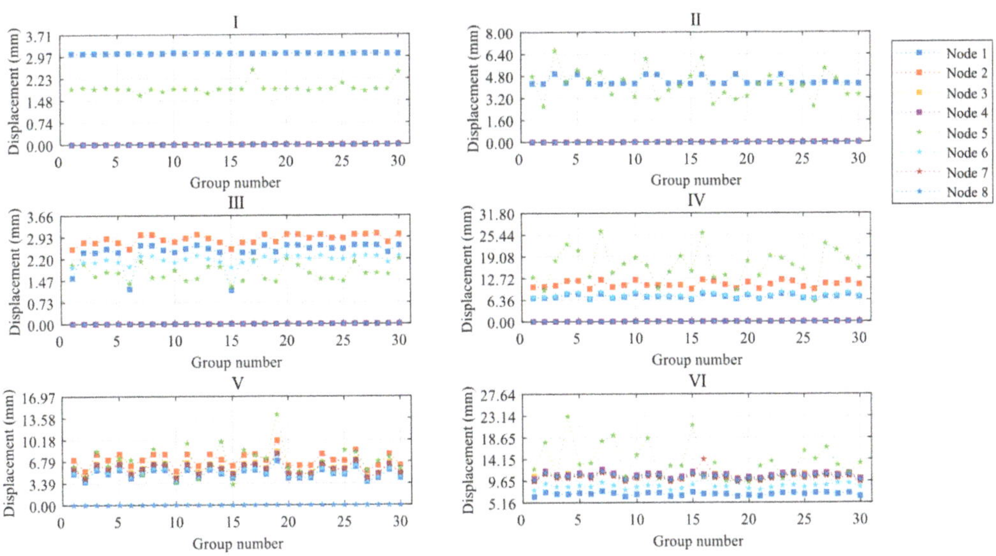

Figure 13. Limit displacement diagrams for control points of JB5 at each construction stage.

4.3. *Two-Factor Analysis of JB5 Model Calculation Results*

To investigate the effects of component unit random defects and construction stages on the ultimate loads of the light steel frame materials and the ultimate displacements of each control node, and to determine the degree of these effects, a two-factor analysis was performed on the results of JB5 calculations.

4.3.1. Two-Factor Analysis of Ultimate Loads

The two-factor method without interaction was used to analyze the effect of construction stages and component unit random defects on the ultimate load. Table 10 presents the calculation results for the two-factor analysis of the ultimate load. The column factor (defect group number) and row factor (construction stage) have judgment factors of 0.260

and 1.17 × 10^{-82}, respectively. The first p-value is greater than 0.05, indicating that different component unit random defects have no significant effect on the ultimate load factor. The second p-value is much less than 0.05, indicating that construction stages significantly affect the ultimate load factor. The effect of construction stages on the ultimate load is much greater than that of component unit random defects. However, component unit random defects may have a greater effect on a single construction stage. This conclusion is consistent with the findings for overall defects, where the effect of construction stages on the ultimate load factor as a whole is more significant than the effect of defects, but random defects may have more influence on the distribution of ultimate loads for a single construction stage.

Table 10. Calculation table for two-factor analysis of ultimate loads.

Source	SS	df	MS	F	p
Defective group number	1.851	29	0.064	1.179	0.260
Construction stage	107.319	5	21.464	396.395	1.17 × 10^{-82}
Error	7.851	145	0.054	/	/
Total	117.022	179	/	/	/

Figure 14 shows the box plots of the ultimate load factor for component unit random defects and construction stages. In Figure 14a, the ultimate load factor fluctuates somewhat with component unit random defects, but the distribution of ultimate loads throughout the construction stages does not show a significant difference. Therefore, the effect of component unit random defects on the distribution of ultimate load factors throughout the construction stages is not significant. The box plot is tighter than that in Figure 8a, and the difference is relatively obvious because the p-value obtained from the two-factor analysis of component unit random defects is smaller. In Figure 14b, the construction stages I and II samples are tightly distributed. The construction stage II discrete points are identified as sample anomalies, and the difference in sample distribution between construction stages is significant. Therefore, the ultimate load factor of the light steel frame materials is significantly affected by the construction stage. Compared to Figure 8b, the tightly distributed stages of the ultimate load factor of the structure have shifted from construction stages I and III to I and II, indicating that the structural factor affecting the ultimate load factor has changed from the stiffness difference between the floors to the number of structural floors.

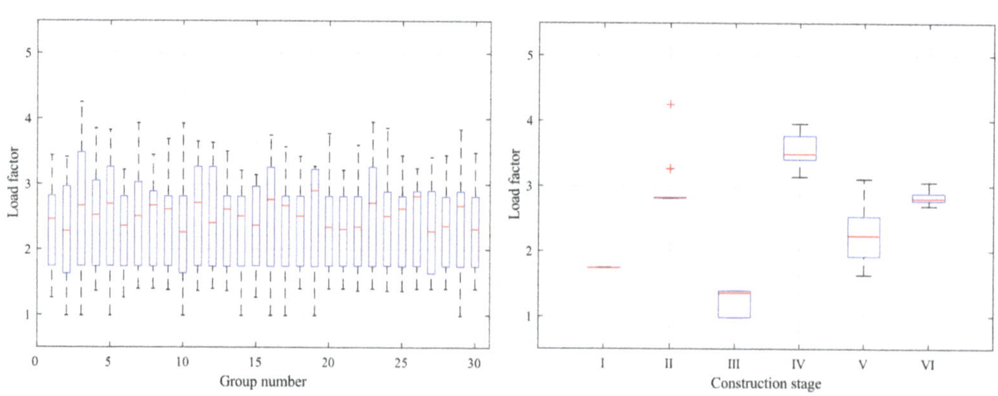

(**a**) defective group number (**b**) construction stage

Figure 14. Box plots of the limit load factors for the two influences.

4.3.2. Two-Factor Analysis of Ultimate Displacement of Control Nodes

A two-way ANOVA without interaction was conducted on the two factors of component unit random defects and construction stages, and the results are presented in Figure 15. The p-value calculated for construction stages at any node was much less than 0.05, indicating a significant effect on the ultimate displacements of the nodes. At node 1, the p-value calculated for component unit random defects was also less than 0.05, indicating a significant effect on ultimate displacement only at that node. Combined with Figure 13, we can conclude that the buckling location of the light steel frame materials occurs at the column with the largest displacement on the first floor, indicating that unit random defects significantly affect the critical buckling state of the structure.

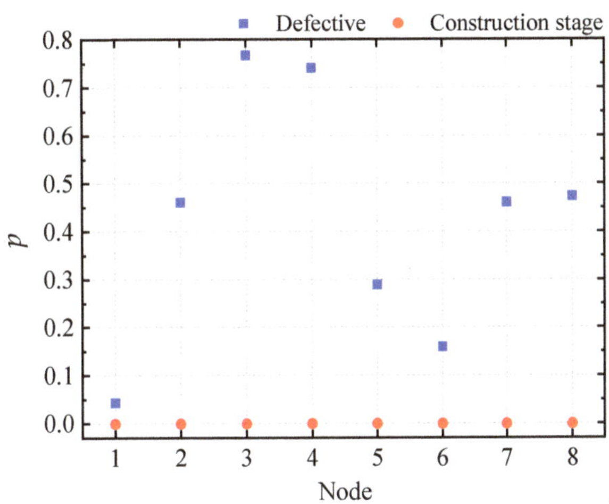

Figure 15. Two-factor analysis plot of the ultimate displacement of the control node for ZT1.

To better observe the distribution of node limit displacements, we plotted the box plots of the limit displacements of eight nodes separately for two different factors, as shown in Figures 16 and 17. In Figure 16, the change in the locus of the sample distribution of node 1 is more apparent than for the other nodes, indicating that node 1 is more affected by component unit random defects. Compared to Figure 10, the lower limit of the box plot for node 1 with defects changes significantly, and component unit defects widen the distribution of node limit displacements. Similar conclusions are found for node 6. Moreover, since node 8 has a non-zero displacement only at construction stage VI, its displacement distribution consists of discrete points with no box plot. Figure 17 shows that the construction stage change significantly affects the node ultimate displacements, and the length of the box plots is smaller under the pre-construction stage, which is consistent with the significant effect of the construction stages on the ultimate load factor. Compared to Figure 11, the box plots of each node displacement are tighter under component unit random defects, indicating that these defects have less effect on node ultimate displacement.

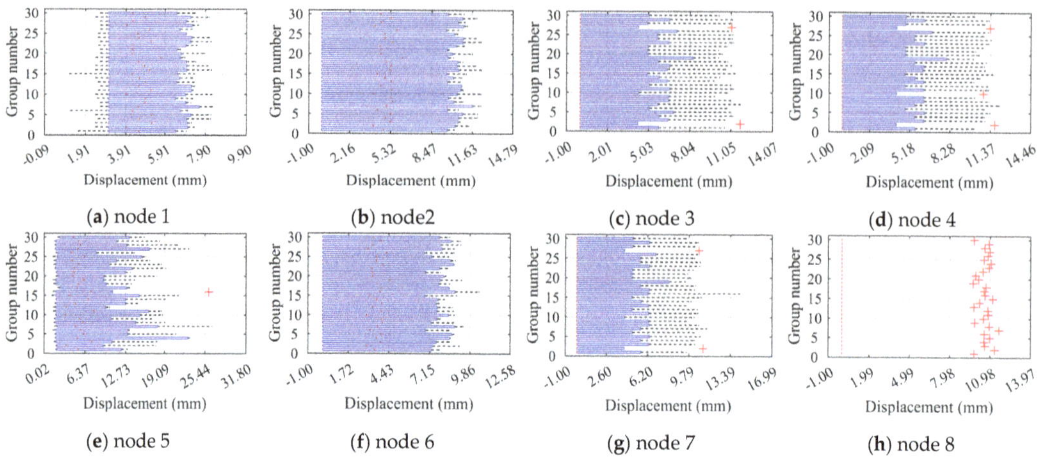

Figure 16. Box plots of ultimate displacements at the control nodes of JB5 under the unit defects.

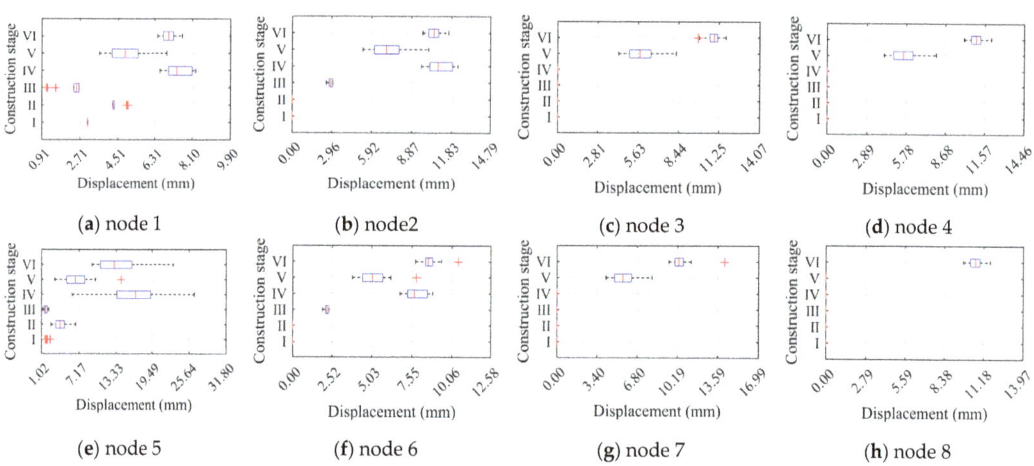

Figure 17. Box plots of ultimate displacements at the control nodes of JB5 under each construction stage.

5. Conclusions

This research presents a defect simulation approach suitable for light steel materials during construction. This method is based on the prevailing design specifications for initial defects in light steel framing materials and an analysis of random defects. The Ansys finite element software is employed to evaluate the ultimate load factor and structural deformation of the structure caused by overall random defects and component unit random defects during construction. The findings of the investigation are presented as follows:

(1) Compared to random defects, the construction stage is identified as the primary factor affecting the ultimate load factor (nodal ultimate displacement). When considering random defects as the influencing factor, the distribution of ultimate load factors (nodal ultimate displacements) remains relatively unchanged throughout the construction stage. However, the ultimate load factors of individual construction stages may fluctuate significantly. On the other hand, when the construction stage is considered the influencing factor, the distribution of ultimate load factors (nodal ultimate displacements) varies significantly.

(2) The response of the light steel frame structural ultimate load factor to construction stages and random defects varies depending on the stiffness difference between the layers and the number of layers. When the stiffness difference between the layers is smaller, the ultimate load factor is more affected by overall random defects. Conversely, when the number of layers is lower, the ultimate load factor is more affected by component unit random defects.
(3) When the light steel frame structural integrity is better, the ultimate displacements of both random defects increase significantly, and the deformation performance of the light steel frame materials improves correspondingly. The fluctuation of nodal ultimate displacements is more significantly affected by component unit random defects than overall random defects.
(4) The correlation between the ultimate load factor and the ultimate displacement of the light steel frame varies for different types of defects. There is no correlation between ultimate load factors and ultimate displacements for component unit random defects, whereas a correlation exists for overall random defects. Therefore, it is necessary to differentiate between component unit random defects and overall random defects.
(5) The displacements of the light steel frame tend to develop more rapidly under the action of component unit random defects than under overall random defects. However, the buckling critical loads of the light steel frame are not significantly different between the two types of defects.

Our research aims to offer references and assistance for the construction process of light steel structures. We hope to promote the standardized production of light steel materials, thereby minimizing defects. Additionally, we advocate for implementing temporary bracing during construction, strict adherence to welding procedures, and including specialized analyses to ensure structural stability. In the future, we suggest further simplifying the model, expanding the calculation scale, considering the structural dynamic response, and increasing the structural layers. These measures would enable finite element analysis results to better approximate the most unfavorable limit state of the structure in actual conditions.

Author Contributions: Conceptualization, G.Y.; methodology, G.Y., Y.C. and Y.Y.; software, Y.C. and W.M.; validation, Y.C. and Y.Y; formal analysis, G.Y.; investigation, G.Y. and Y.C.; resources, Y.Y.; data curation, Y.C. and X.M.; writing—original draft preparation, Y.C.; writing—review and editing, G.Y. and Y.Y.; visualization, Y.C. and X.M.; supervision, G.Y. and Y.Y.; project administration, W.M.; funding acquisition, Y.Y. All authors have read and agreed to the published version of the manuscript.

Funding: This research was funded by the Fundamental Research Funds for the Central Universities, grant number 2023CDJXY-031.

Institutional Review Board Statement: Not applicable.

Informed Consent Statement: Not applicable.

Data Availability Statement: The data presented in this study are available on request from the corresponding author.

Conflicts of Interest: The authors declare no conflict of interest.

References

1. Zhang, J.F.; Zhao, J.J.; Yang, D.Y.; Deng, E.F.; Wang, H.; Pang, S.Y.; Cai, L.M.; Gao, S.C. Mechanical-property tests on assembled-type light steel modular house. *J. Constr. Steel Res.* **2020**, *168*, 12. [CrossRef]
2. Zhang, X.C.; Zhang, X.Q. Comparison and sensitivity analysis of embodied carbon emissions and costs associated with rural house construction in China to identify sustainable structural forms. *J. Clean Prod.* **2021**, *293*, 12. [CrossRef]
3. Zhou, T.; Sun, K.Z.; Chen, Z.H.; Yang, Z.X.; Liu, H.B. Automated Optimum Design of Light Steel Frame Structures in Chinese Rural Areas Using Building Information Modeling and Simulated Annealing Algorithm. *Sustainability* **2023**, *15*, 9000. [CrossRef]
4. Egbon, B.; Tomlinson, D. Experimental investigation of longitudinal shear transfer in insulated concrete wall panels with notched insulation. *J. Build. Eng.* **2021**, *43*, 13. [CrossRef]

5. Han, W.L.; Zhao, Z.Z.; Qian, J.R.; Zhang, Y.B.; Ma, T. Experimental seismic behavior of squat shear walls with precast concrete hollow moulds. *Earthq. Eng. Eng. Vib.* **2019**, *18*, 871–886. [CrossRef]
6. Zhao, B.; Wu, D.; Zhu, H.Q. New modular precast composite shear wall structural system and experimental study on its seismic performance. *Eng. Struct.* **2022**, *264*, 23. [CrossRef]
7. Usefi, N.; Sharafi, P.; Mortazavi, M.; Ronagh, H.; Samali, B. Structural performance and sustainability assessment of hybrid-cold formed modular steel frame. *J. Build. Eng.* **2021**, *34*, 12. [CrossRef]
8. Feng, S.K.; Liu, Z.X.; Zhou, X.J.; Guo, Z.X. Seismic performance of curved haunched connections in modularized prefabricated steel structures. *J. Constr. Steel Res.* **2020**, *172*, 14. [CrossRef]
9. Skiadopoulos, A.; de Castro e Sousa, A.; Lignos, D.G. Experiments and proposed model for residual stresses in hot-rolled wide flange shapes. *J. Constr. Steel Res.* **2023**, *210*, 108069. [CrossRef]
10. Horvath, M.; Stoschka, M.; Fladischer, S. Fatigue strength study based on geometric shape of bulk defects in cast steel. *Int. J. Fatigue* **2022**, *163*, 9. [CrossRef]
11. Sun, J.M.; Dilger, K. Influence of initial residual stresses on welding residual stresses in ultra-high strength steel S960. *J. Manuf. Process.* **2023**, *101*, 259–268. [CrossRef]
12. Storheim, M.; Alsos, H.S.; Hopperstad, O.S.; Amdahl, J. A damage-based failure model for coarsely meshed shell structures. *Int. J. Impact Eng.* **2015**, *83*, 59–75. [CrossRef]
13. Yang, Y.; Yang, L.; Yao, G. Post-Processing of High Formwork Monitoring Data Based on the Back Propagation Neural Networks Model and the Autoregressive-Moving-Average Model. *Symmetry* **2021**, *13*, 1543. [CrossRef]
14. Kepple, J.; Herath, M.; Pearce, G.; Prusty, G.; Thomson, R.; Degenhardt, R. Improved stochastic methods for modelling imperfections for buckling analysis of composite cylindrical shells. *Eng. Struct.* **2015**, *100*, 385–398. [CrossRef]
15. Deng, D.; Murakawa, H. Numerical simulation of temperature field and residual stress in multi-pass welds in stainless steel pipe and comparison with experimental measurements. *Comput. Mater. Sci.* **2006**, *37*, 269–277. [CrossRef]
16. Kani, I.M.; McConnel, R.E. COLLAPSE OF SHALLOW LATTICE DOMES. *J. Struct. Eng. ASCE* **1987**, *113*, 1806–1819. [CrossRef]
17. Bielewicz, E.; Gorski, J. Shells with random geometric imperfections simulation—Based approach. *Int. J. Non-Linear Mech.* **2002**, *37*, 777–784. [CrossRef]
18. Lauterbach, S.; Fina, M.; Wagner, W. Influence of stochastic geometric imperfections on the load-carrying behaviour of thin-walled structures using constrained random fields. *Comput. Mech.* **2018**, *62*, 1107–1125. [CrossRef]
19. Kala, Z.; Vales, J. Imperfection sensitivity analysis of steel columns at ultimate limit state. *Arch. Civ. Mech. Eng.* **2018**, *18*, 1207–1218. [CrossRef]
20. Roy, K.; Lau, H.H.; Ting, T.C.H.; Chen, B.S.; Lim, J.B.P. Flexural behaviour of back-to-back built-up cold-formed steel channel beams: Experiments and finite element modelling. *Structures* **2021**, *29*, 235–253. [CrossRef]
21. Ananthi, G.B.G.; Roy, K.; Lim, J.B.P. Experimental and numerical investigations on axial strength of back-to-back built-up cold-formed steel angle columns. *Steel Compos. Struct.* **2019**, *31*, 601–615. [CrossRef]
22. Arrayago, I.; Rasmussen, K.J.R.; Real, E. Statistical analysis of the material, geometrical and imperfection characteristics of structural stainless steels and members. *J. Constr. Steel Res.* **2020**, *175*, 17. [CrossRef]
23. Mirzaie, F.; Myers, A.T.; Jay, A.; Mahmoud, A.; Smith, E.; Schafer, B.W. Analysis of geometric imperfections of spirally welded slender steel tubes. *Thin-Walled Struct.* **2020**, *146*, 16. [CrossRef]
24. Fina, M.; Weber, P.; Wagner, W. Polymorphic uncertainty modeling for the simulation of geometric imperfections in probabilistic design of cylindrical shells. *Struct. Saf.* **2020**, *82*, 20. [CrossRef]
25. Wang, X.X.; Chen, H.F.; Xuan, F.Z. Direct method-based probabilistic shakedown analysis for the structure under multiple uncertain design conditions. *Ocean Eng.* **2023**, *280*, 18. [CrossRef]
26. Yao, S.J.; Ma, Y.X.; Zhao, N.; Wang, Z.F.; Zhang, D.; Lu, F.Y. Equivalent scaling method on the dynamic response of box-shaped structures under internal blast. *Int. J. Impact Eng.* **2022**, *160*, 11. [CrossRef]
27. Chen, W.F. Structural stability: From theory to practice. *Eng. Struct.* **2000**, *22*, 116–122. [CrossRef]
28. Susmel, L. Notches, nominal stresses, fatigue strength reduction factors and constant/variable amplitude multiaxial fatigue loading. *Int. J. Fatigue* **2022**, *162*, 19. [CrossRef]
29. Prakash, P.R.; Srivastava, G. Distributed plasticity model for analysis of steel structures subjected to fire using the direct stiffness method. *Fire Saf. J.* **2019**, *105*, 169–187. [CrossRef]
30. Vu, Q.A.; Cao, T.S.; Nguyen, T.T.T.; Nguyen, H.H.; Truong, V.H.; Ha, M.H. An efficient differential evolution-based method for optimization of steel frame structures using direct analysis. *Structures* **2023**, *51*, 67–78. [CrossRef]
31. Li, L.X.; Chen, G.H.; Fang, M.X.; Yang, D.X. Reliability analysis of structures with multimodal distributions based on direct probability integral method. *Reliab. Eng. Syst. Saf.* **2021**, *215*, 14. [CrossRef]
32. Ding, T.; Li, G.Y.; Liu, Z. Cross-Scale Edge Purification Network for salient object detection of steel defect images. *Measurement* **2022**, *199*, 11. [CrossRef]
33. Banon, F.; Sambruno, A.; Batista, M.; Simonet, B.; Salguero, J. Evaluation of geometrical defects in AWJM process of a hybrid CFRTP/Steel structure. *Int. J. Mech. Sci.* **2021**, *210*, 16. [CrossRef]
34. Mordia, R.; Verma, A.K. Visual techniques for defects detection in steel products: A comparative study. *Eng. Fail. Anal.* **2022**, *134*, 12. [CrossRef]

35. Shi, T.; Li, H.; Ding, Q.X.; Wang, M.Q.; Zheng, Z.; Peng, C.; Mei, Q.L. Global variance reduction method based on multi-group Monte Carlo adjoint calculation. *Ann. Nucl. Energy* **2021**, *151*, 10. [CrossRef]
36. Zhao, K.Y.; Guo, R.; Liu, G.Y.; Li, Y.H. Monte Carlo simulation for exploring the mechanical properties of particle-reinforced composites based on the scale boundary finite element method. *Compos. Struct.* **2022**, *297*, 20. [CrossRef]
37. *GB 50009-2012*; Load Code for the Design of Building Structures. China Building Industry Press: Beijing, China, 2012.
38. *GB 50755-2012*; Code for Construction of Steel Structures. China Building Industry Press: Beijing, China, 2012.
39. Chen, Z.H.; Li, H.X.; Chen, A.Y.; Yu, Y.J.; Wang, H. Research on pretensioned modular frame test and simulations. *Eng. Struct.* **2017**, *151*, 774–787. [CrossRef]
40. Li, L.Y.; Fafard, M.; Boissonnade, N. Local and global instabilities of rolled T-section columns under axial compression. *Thin-Walled Struct.* **2022**, *178*, 17. [CrossRef]
41. Suman, S.; Biswas, P.; Baranwal, S.; Mekala, V. Finite Element Modelling of side heating for mitigation of residual stress and distortion in SAW welded P91 steel weld. In Proceedings of the 2nd International Conference on Advances in Mechanical Engineering and Nanotechnology (ICAMEN), Pink City, India, 28–29 February 2020; pp. 2511–2521. [CrossRef]
42. *GB 50017-2017*; Standard for Design of Steel Structure. China Building Industry Press: Beijing, China, 2017.

materials

MDPI

Article

Anti-Condensation Performance of a New Superhydrophobic Coating for Pavements

Kaijian Huang [1,*], Ruiyu Sun [1], Jiaqing Wang [1], Xijun Shi [2] and Hechang Lei [3]

1 College of Civil Engineering, Nanjing Forestry University, Nanjing 210037, China; sunruiyuyu@163.com (R.S.); jiaqingw@njfu.edu.cn (J.W.)
2 Ingram School of Engineering, Texas State University, San Marcos, TX 78666, USA; xijun.shi@txstate.edu
3 Beijing Key Laboratory of Opto-Electronic Functional Materials & Micro-Nano Devices, Department of Physics, Renmin University of China, Beijing 100872, China; hlei@ruc.edu.cn
* Correspondence: huangkaijian@njfu.edu.cn; Tel.: +86-138-13866459; Fax: +86-025-84498336

Abstract: Superhydrophobic coating ice suppression is an advanced and durable technology that shows great potential for application on pavements. Although many researchers have conducted experimental and theoretical validations to confirm the effectiveness of superhydrophobic surfaces in actively suppressing ice formation, there are still some who remain skeptical. They argue that the roughness of the surface may increase ice adhesion due to the mechanical interlocking effect of condensation droplets in low-temperature and high-humidity environments. In this study, we present a comprehensive investigation of a novel superhydrophobic coating specifically designed for pavement surfaces, aiming to address the question of its active anti-icing/ice-sparing capabilities in a condensing environment. The changes in contact angle before and after condensation for four material surfaces with varying wettability were investigated, as well as the morphology and ice adhesion of liquid water after it freezes on the material surface. The findings reveal that the proposed superhydrophobic coating for pavements effectively prevents condensate droplets from infiltrating the surface structure, resulting in delaying the surface icing time and reducing the attachment strength of the ice.

Keywords: superhydrophobic coating; contact angle; pavement; anti-condensation

Citation: Huang, K.; Sun, R.; Wang, J.; Shi, X.; Lei, H. Anti-Condensation Performance of a New Superhydrophobic Coating for Pavements. *Materials* **2023**, *16*, 5793. https://doi.org/10.3390/ma16175793

Academic Editor: Simon Hesp

Received: 19 July 2023
Revised: 18 August 2023
Accepted: 22 August 2023
Published: 24 August 2023

1. Introduction

The accumulation of snow and ice on the pavement surface in winter leads to a significant reduction in the skid resistance of the pavement surface, seriously weakening road capacity and posing a grave threat to vehicle safety, potentially leading to severe traffic accidents [1,2]. This threat is particularly prominent in southern China, where freezing rain and black ice form in high-altitude, cold, and wet areas [3,4], further jeopardizing road safety. Therefore, addressing the issue of snow and ice on pavements during winter is crucial to ensure traffic safety.

Currently, two main types of anti-icing technologies are employed for pavements, namely active de-icing technology and passive de-icing technology. Passive de-icing technology primarily relies on physical or chemical methods such as manual mechanical removal and the application of snow-melting agents. Although this approach is relatively efficient, it still has the potential to damage the pavement surface and bring about pollution to the surrounding ecological environment. What is more, the costs associated with passive de-icing methods are higher [5,6]. In contrast, active de-icing methods are more efficient and less damaging to pavement structures. Examples include self-heating pavements, pavements containing anti-freeze fillers, phase change energy storage pavements, carbon fiber conductive pavements, and emerging superhydrophobic materials [7–11]. However, the currently available active de-icing methods in the market are often not affordable.

Notably, the utilization of superhydrophobic materials, serving as a potential active de-icing solution, has garnered increasing attention due to its simple nature of preparation, cost-effectiveness, and, more importantly, free from environmental contamination [12,13].

There are several studies [14,15] that have investigated ice adhesion mechanisms on superhydrophobic surfaces. In 2003, Pilotek [16] first proposed that superhydrophobic coatings might exhibit low ice adhesion. It promotes the application of superhydrophobic surfaces in the field of anti-icing coverage. Some researchers [17,18] have compared ice adhesion for different surface structures and concluded that nanostructured surfaces with relatively lower roughness exhibit the lowest ice adhesion. Subramanyam's group at MIT, USA [19] fabricated four different surface structures, including smooth, micron, nano, and micro/nanocomposite surfaces, modified with surface energy substances. Their analysis revealed that the nanostructured and micro/nanocomposite surfaces exhibited excellent resistance to frost formation, with the nanostructured samples displaying the lowest ice adhesion. This finding has provided significant inspiration. Huang's non-uniform nucleation ice crystal growth model experimentally verifies the ice inhibition mechanism of superhydrophobic materials. The results show that the prepared superhydrophobic coating material exhibited excellent properties, including a high contact angle (>150°), good anti-slip nature, and the ability to retard water droplet crystallization and maintain the droplet shape even after freezing [20]. These findings highlight that superhydrophobic surfaces achieve extremely low surface energy due to the presence of micro-nano rough structures. Also, the grooves on the substrate surface are filled with air, resulting in a Cassie state for droplets, where only about 10% of the total contact area is occupied by the droplet and substrate contact [21]. This, combined with the surface's exceptionally low rolling angle, hinders the infiltration of droplets into pavement material upon impact, effectively reducing the icy surface area of the pavement. This, in turn, reduces the area of ice in contact with the pavement. The adhesion between the ice and the pavement material is reduced.

The surface affects the wettability of superhydrophobic coatings in terms of ice adhesion remains controversial. Some researchers argue [22–25] that ice adhesion is linked to surface wettability and studied the relationship between surface wettability and ice adhesion force. Their findings indicate that ice adhesion force is proportional to smaller hysteresis angles, resulting in lower adhesion forces. While some researchers hold the opposite opinion [26–30], for instance, Chen [31] prepared 13 kinds of silicon wafers with superhydrophilic or superhydrophobic wettability to study surface morphology and adhesion. It was found that ice adhesion on superhydrophilic and superhydrophobic surfaces was similar. Chen's research concluded that superhydrophobic coatings not only fail to reduce ice-cover bond strength but also exacerbate subsequent de-icing efforts due to their surface structure and mechanical interlocking effect between ice and the coating. Varanasi [32] investigated the application of superhydrophobic surfaces for de-icing and revealed that rough superhydrophobic surfaces in high humidity environments are susceptible to frost formation, therefore resulting in greater ice adhesion compared to smooth surfaces. These results raise questions about the effectiveness of superhydrophobic surfaces for de-icing applications.

Actually, superhydrophobic surfaces are frequently exposed to low-temperature and high-humidity environments, where gas–liquid phase change condensation or fog coalescence, commonly observed in nature, inevitably occurs on solid surfaces [32]. As the supercooling increases, water vapor condenses and nucleates within the micro- and nanostructures of the superhydrophobic surface and grows gradually. Failure to expel the condensed droplets from the surface irreversibly transforms the surface into a partially or completely wetted Wenzel state, adhering to the surface [33–35]. At this stage, when external droplets come into contact with the condensed surface, they fuse with the droplets present on the surface, causing a reduction in the apparent contact angle and altering the surface wettability. Consequently, the superhydrophobic ice-suppressing performance also diminishes. Therefore, to effectively utilize superhydrophobic ice-suppressing coatings on pavements, theoretical discussions and feasibility studies regarding the anti-condensation performance of superhydrophobic materials are valuable, ensuring the desired active anti-icing/ice-sparing effects.

This paper employs the "self-migration" movement of condensate droplets on superhydrophobic coatings to theoretically analyze factors influencing the anti-condensation performance of the coating. We characterize four material surfaces with varying wettability to investigate changes in the contact angle of different material surfaces before and after condensation and the morphology and ice adhesion of liquid water after freezing on the material surface to support the proposed theory. Condensed droplets do not affect the water contact angle of the superhydrophobic surface, and the superhydrophobic surface can delay the icing time of the droplets, as the droplets have a smaller contact area and adhesion force with the material surface after icing. The superhydrophobic coating has good anti-condensation properties and broad application prospects in active ice suppression on pavements.

2. Theoretical Analysis of Ice Suppression Performance of Superhydrophobic Surfaces due to Condensation

The findings from recent investigations [36–41] demonstrate that on certain nanostructured superhydrophobic surfaces, condensate droplets exhibit self-migration behavior or jumping behavior without gravitational influence. The mechanism behind the phenomenon is based on the release of surface energy into kinetic energy. Yet, the adhesion of the droplets to the superhydrophobic surface is a critical factor regulating the self-migration of droplets, specifically the adhesion between the droplets and the surface. Smaller interaction forces facilitate easier self-migration. The interaction force is minimized when the droplets are in the Cassie state, enabling self-migration. Even without self-migration, droplets in the Cassie state can easily roll off the surface under an external force, thus keeping the surface dry.

This "self-migration" of condensate droplets provides a new strategy for ice suppression: when the surface of a superhydrophobic material is in the Cassie state, the condensate droplets can jump off the surface without freezing in the microstructure of the material. The main processes include water vapor condensation nucleation, nucleation growth, merging and polymerization, self-jumping, and departing the surface. Therefore, when preparing a superhydrophobic ice-suppressing surface, as long as the Cassie state is maintained when condensate droplets are present on the material surface, these condensate droplets will not exist in the material microstructure, and with the occurrence of the "self-migration phenomenon", the condensate droplets remain unaffected, preserving the ice-suppressing effect.

Among these processes, water vapor condensation nucleation involves the aggregation of water vapor molecules and is typically induced by factors such as water vapor supersaturation and nucleation-inducing agents like dust and nanostructures (heterogeneous and homogeneous nucleation). The critical nuclear radius represents the minimum size at which water vapor molecules aggregate into stable droplets. It depends primarily on the supersaturation resulting from the dew point, subcooling temperature, and relative humidity. According to classical nuclear theory and previous research [42,43], the critical nuclear radius can be calculated using the following equation:

$$r_c = -\frac{2\gamma v}{\Delta G} \tag{1}$$

Based on the nucleation theory, the minimum radius at which a droplet can nucleate on a superhydrophobic surface can be calculated. The interfacial tension of water (0 °C) is $\gamma = 7.56 \times 10^{-2}$ J/m^2, the molar volume of water molecules is $v \approx 1.8 \times 10^{-5}$ m^3/mol, the Gibbs Free Energy is $\Delta G \approx -CpT[\ln(T/Td) + Td/T - 1]$, T = 273.15 K (0 °C), water dew point temperature vapor (70%RH, 25 °C) is Td = 291.15 K (15 °C), and the specific heat capacity of water vapor is C_p = 33.5 J/mol·K. Finally, the nucleation on the superhydrophobic surface can be obtained for about 145 nm.

The critical nuclear radius represents the smallest droplet size that forms a stable nucleus, and in most cases, water vapor undergoes heterogeneous nucleation upon contacting the surface. Consistent with the critical radius is the free energy barrier, which is another factor describing the ease of nucleation, especially for micro- and nano-structured surfaces. According to classical nucleation theory, the influence of surface structure on the heterogeneous free energy barrier of condensed water droplets can be analytically determined using the following equation:

$$\Delta G_c = \Delta G_c^{homo} f(m, x) \tag{2}$$

where ΔG_c^{homo} is the free energy barrier in the homogeneous nucleation of water droplets. $f(m, x)$ is the ratio of the nucleation-free energy barrier of a spherical water droplet relative to its volume. Since the critical radius of 145 nm is the definite value, $f(m, x)$ varies only with the radius (R_s) of the surface structure.

When water droplet nucleation occurs in nano-gaps that are less than 145 nm wide, the small nucleation barrier energy allows the nuclei to grow and merge into larger, micron-sized droplets. These droplets then undergo spontaneous upward movement to reach the top of the surface structure, forming the Cassie state. As time passes, the condensed droplets on the horizontally placed superhydrophobic surface begin to fuse and undergo self-migration, leading to the appearance of micron-sized water droplets on the surface. As such, water vapor tends to nucleate on the micron-sized large water droplets rather than in the gaps of the nanostructure because the gaps are nearly the same size or smaller than the critical nuclear radius of water droplets, bringing about high nuclear barrier energy. In this way, the condensate drops will not enter the nanometres of the superhydrophobic surface with a width less than the critical nuclear radius. This prevents the development of a mechanical interlocking effect, and the ice adhesion of the surface does not increase. Therefore, even in low temperature and high humidity environments, as long as the surface microform meets the requirements, the liquid water on the surface of the superhydrophobic coating will remain in a Cassie state and will not be affected by the condensation phenomenon. The superhydrophobic surface will maintain the droplet hemispherical shape after the liquid water freezes, and the ice adhesion will be smaller than that of the ordinary smooth surface.

3. Experiment

3.1. Sample Preparation

3.1.1. Testing Raw Materials

In this paper, four surfaces with different wettability were prepared for various experimental tests using the following raw materials (Table 1).

Table 1. Main test reagent specifications and manufacturers.

Reagent Name	Specification	Manufacturers
tetraethyl orthosilicate	98%	Sinopharm Chemical Reagent Co., Ltd. (Shanghai, China)
hydrochloric acid solution	0.01 mol/L	Shanghai Da Biotechnology Co., Ltd. (Shanghai, China)
silicon dioxide	7–40 nm	Sinopharm Chemical Reagent Co., Ltd.
stearic acid	AR	Sinopharm Chemical Reagent Co., Ltd.
anatase-structured titanium dioxide	25 nm	Sinopharm Chemical Reagent Co., Ltd.
rutile-structured titanium dioxide	60 nm	Sinopharm Chemical Reagent Co., Ltd.
anhydrous ethanol	AR	Sinopharm Chemical Reagent Co., Ltd.
deionized water	hyperpure	Laboratory homemade

3.1.2. Preparation Steps

In this paper, hydrophilic surfaces were prepared using SiO_2. Si(OH) sols were synthesized by mixing tetraethyl orthosilicate and hydrochloric acid solution at a mass ratio of 2.5:1. The pH was adjusted to 2.5, and the mixture was magnetically stirred for

2 h at room temperature. The specification of the magnetic mixer is DF-101S. The Si(OH) sol was then combined with a silica suspension, and the resulting mixture was stirred magnetically for 30 min. Subsequently, it was dispersed using an ultrasonic disperser for another 30 min to obtain the coating solution (Figure 1).

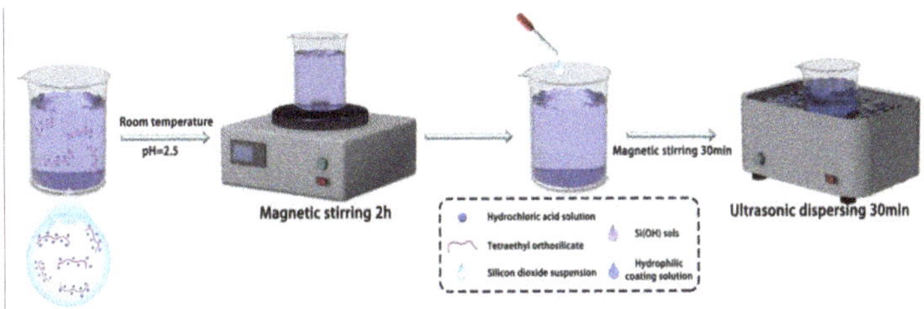

Figure 1. Preparation of hydrophilic coating.

Furthermore, based on the principles of hydrophobicity, surface modification of nanomaterials was conducted using low surface energy substances to prepare a hydrophobic surface with nanosilica. The silica particles and stearic acid were dispersed in anhydrous ethanol, sonicated, and magnetically stirred at room temperature. The heating reaction of stearic acid with anhydrous ethanol produces ethyl stearate with good lubricity and thermal stability, and the low surface energy substance produced by the reaction of stearic acid with ethanol constructs the rough structure of superhydrophobic materials. The hydrophobic modification effect of stearic acid on titanium dioxide nanoparticles played a decisive role in the hydrophobicity of the samples. The mixed solution was centrifuged to stabilize it and then dried to obtain hydrophobic nanomaterials (Figure 2). The specification of the centrifuge is TGL-16A, provided by Hunan Xiangli Equipment Co., Ltd. (Changsha, China).

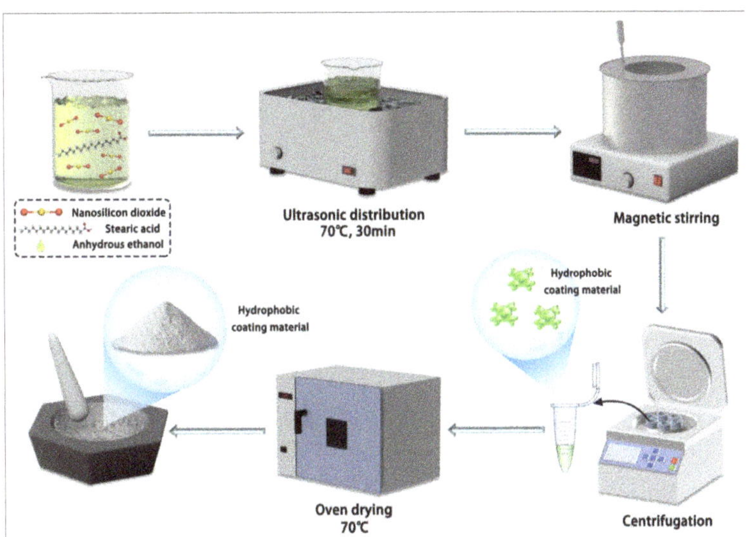

Figure 2. Preparation of hydrophobic coating.

The specific steps for the preparation of superhydrophobic materials are as follows: to prepare the TiO_2-SiO_2 composites, titanium dioxide nanopowder was initially dispersed into anhydrous ethanol through sonication. Subsequently, silicon dioxide and stearic acid were added to the initial solution. The mixture was then subjected to sonication and stirring at 70 °C for 6 to 12 h using a magnetic stirrer. Following this, the mixture was centrifugally dried and ground to obtain a superhydrophobic material powder; the process is shown in Figure 3. Finally, the hydrophilic, hydrophobic, and superhydrophobic materials were sprayed onto the surface of the material using a PQ-2 spray gun, provided of Shanghai Xuling Information Technology Co., Ltd. (Shanghai, China)

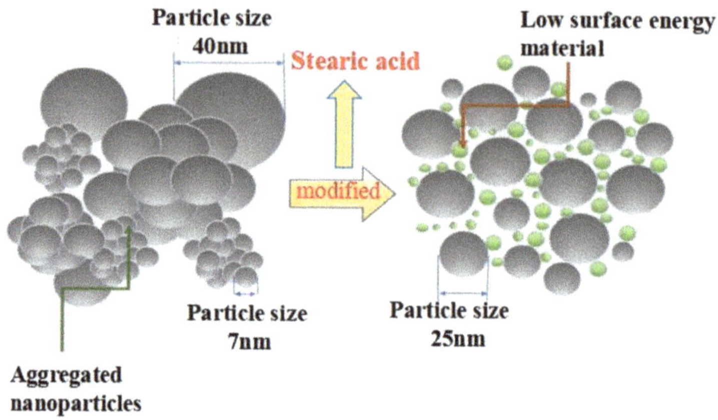

Figure 3. Preparation of superhydrophobic coatings.

3.2. *Test Methods*

3.2.1. Apparent Morphology

The morphology and particle distribution of the samples were observed using a Regulus 8100 field emission scanning electron display microscope, provided by Shanghai Na Scientific Instrument Co., Ltd. (Shanghai, China). The sample coated with superhydrophobic material was firmly secured to the sample stage with a conductive adhesive, ensuring that the viewing surface was facing upwards. Then, after the conductive adhesive was air-dried, a thin layer of gold was sprayed on the surface of the sample to ensure good electrical conductivity [44]. The magnification of the sample morphology could be adjusted accordingly.

3.2.2. Water Contact Angle Test

Four different surfaces with varying wettability were designed for this paper: hydrophilic surface, hydrophobic surface, superhydrophobic surface 1 (made from modified anatase titanium dioxide), and superhydrophobic surface 2 (made from modified rutile titanium dioxide). The contact angles produced by condensate (at 0 °C, humidity RH = 70%) were measured in a room temperature environment (25 °C, humidity RH = 30%) both before and after condensation (with a temperature difference of over ten degrees Celsius). As the temperature decreased and humidity increased, the contact angles of the characterized surfaces changed. They were carefully deposited onto the upper sample surface using a syringe. The droplet mass was 5 mL. A contact angle tester (Data Physics OCA40Micro, provided of Beijing Oriental Defei Instruments Co., Ltd., Beijing, China) was used to measure the static contact angle of the drop on the sample surface. Five different positions on the surface of the material were selected for contact angle measurement, and their average values were calculated.

3.2.3. Freezing Experiments

To simulate the icing morphology of water droplets under low temperature and high humidity conditions, drip icing tests were conducted on superhydrophobic coatings at temperatures of −4 °C, −8 °C, and −10 °C with a relative humidity of 70%. The specific implementation steps were as follows:

a. Three specimens with identical dimensions were prepared for ordinary uncoated specimens, hydrophobic coated specimens, and superhydrophobic coated specimens 1 and 2, respectively.
b. The sample was placed in a temperature control box, and 5 mL of water droplets were carefully deposited onto the surface of each specimen.
c. The morphology of the frozen water droplets was observed. To minimize operational errors, five sets of experiments were conducted for each condition.

3.2.4. Ice Adhesion Strength Test

The adhesion of ice on pavement surfaces plays a critical role in road traffic safety, as the presence of snow and ice can significantly increase the bond size between the ice and the pavement, making de-icing efforts challenging. Since spray coating on the pavement surface can alter the original characteristics of the pavement, it is essential to evaluate the ice adhesion of novel superhydrophobic coatings after icing. The experimental schematic diagram is presented below, along with specific test steps (Figure 4).

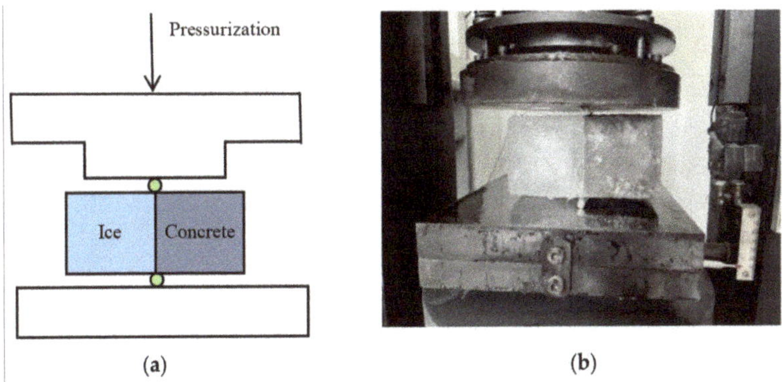

Figure 4. Ice adhesion strength test. (**a**) Schematic diagram of test principle. (**b**) Iced concrete specimen with pressure splitting diagram.

a. Prepare concrete specimens measuring 100 mm × 100 mm × 100 mm in advance and place them into a mold measuring 400 mm × 100 mm × 100 mm. Separate the two specimens with a plastic film, leaving a 100 mm gap at each end.
b. Fill the empty slots at both ends with water, ensuring that the water comes into contact with the side of the specimen that will be used for the test surface. Transfer the mold to a freezer set at a temperature of 10 °C. After 4 h, remove the mold from the freezer. At this point, the specimens and water will have frozen into two rectangular iced concrete specimens measuring 200 mm × 100 mm × 100 mm.
c. Demold the frozen concrete specimens and place them onto a press. Position two arc-shaped pads at the interface between the specimens and the ice on both ends. Prior to use, ensure that the arc-shaped pads are subjected to the same freezing conditions as the specimens, ensuring that pads with different temperatures do not impact the bond between the ice and the concrete specimens. Apply pressure to the specimen using the HTC-1068 pressure tester (provided by Beijing Fuhaida Technology Co., Ltd., Beijing, China) until splitting occurs.
d. Record the results and compile the data.

4. Results and Discussion

4.1. Microscopy of Superhydrophobic Materials

The micromorphology of the superhydrophobic material was examined using FESEM, and the results are presented in Figure 5. During the hydrothermal reaction, SiO_2 particles were observed to encapsulate the surface of TiO_2 particles. The nano-TiO_2 reacted with stearic acid, leading to a rough morphology of the nanostructures and the formation of a large number of papillae. The lotus leaf is a natural superhydrophobic surface due to its special micro-nanostructure and self-cleaning characteristics. It can be observed in the electron micrograph that the superhydrophobic surface has the same micronano papillary structure as the lotus leaf surface. Thus, the field emission scanning electron microscopy (FESEM, provided of Shanghai Sinu Optical Technology Co., Ltd., Shanghai, China) analysis confirmed the presence of superhydrophobic materials with low surface energy.

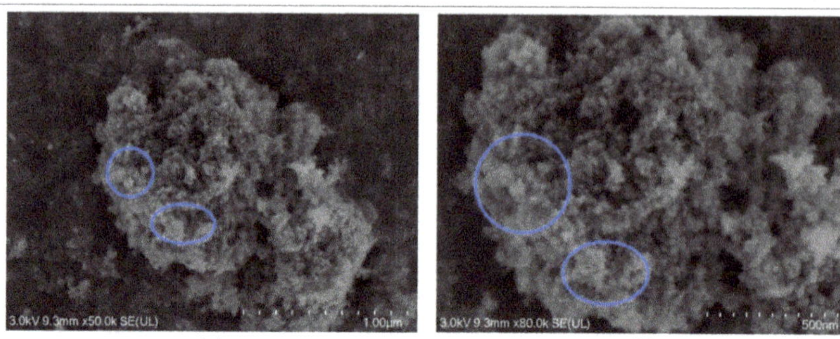

(a) (b)

Figure 5. Field emission scanning electron microscopy images of the superhydrophobic surface modified with rutile titanium dioxide nanoparticles. (**a**,**b**) Electron microscopy images at different magnifications: images (**a**,**b**) are under ×50,000 and ×80,000 magnification scales, respectively.

4.2. Effect of Condensation on the Water Contact Angle of Materials

The contact angle test was performed to evaluate the wettability of water droplets on the surface of the superhydrophobic material. Multiple measurements were taken at different locations on the coated sample. A surface material can be classified as superhydrophobic when the contact angle of water droplets exceeds 150°. The contact angle results are summarized in Table 2.

Table 2. Static contact angles of representative surfaces at room temperature and after condensation.

Sample	Water Contact Angle (25 °C, RH = 30%)	Water Contact Angle (0 °C, RH = 70%)
Hydrophilic surface	22.1°	13.9°
Hydrophobic surface	137.3°	82.0°
Superhydrophobic surface 1 (Modified by anatase type nano titanium dioxide)	151.9°	147.8°
Superhydrophobic surface 2 (Modified by rutile type nano titanium dioxide)	154.0°	151.9°

The results show that the water drop contact angle on the prepared superhydrophobic coating 2 was 154.0° (>150°) at room temperature and 151.9° (>150°) after the condensation phenomenon. These findings demonstrate that the coating material still has a large water contact angle at low temperatures and high humidity, conforming to the characteristics of wettability as superhydrophobicity, thereby confirming its robustness in maintaining superhydrophobicity.

4.3. Water Droplet Icing on Different Wettability Surfaces

Following the freezing test procedure described in Section 3.2.3, the samples were placed in a temperature control box set to −4 °C. After 20 min, the droplets on the surface of the normal sample were observed to freeze, while those on the hydrophobic coating exhibited a mixed state of ice and water. Remarkably, the droplets on the superhydrophobic coating remained in a liquid state. As time progressed, after 35 min, the droplets on the superhydrophobic coating eventually froze. This indicates that at a temperature of −4 °C, the solidification time of droplets on superhydrophobic surfaces was significantly longer compared to the ordinary specimen surface, thus demonstrating the ability of the superhydrophobic coating to extend the solidification time of droplets. Next, the temperature of the high-temperature control box was changed to −8 °C, and another set of comparison specimens was placed inside. It was observed that the droplets on the surface of the normal specimen started to freeze after about 8 min, while the water droplets on the hydrophobic coating surface began freezing after about 11 min. In contrast, the water droplets on the superhydrophobic coating surface exhibited freezing after approximately 14 min. Finally, a final group of specimens was placed in the high and low-temperature control box set to −10 °C to observe the time required for water droplets to freeze. The results indicated that the droplets on the regular specimen surface started freezing after about 3 min, the droplets on the hydrophobic coating surface started freezing after approximately 6 min, and the droplets on the superhydrophobic coating surface started icing after approximately 8 min. Figure 6 shows the photos of water droplet states on the regular specimen surface and the superhydrophobic-coated specimen at the same moment. Among them, Figure 6A–C represent the photos of droplets freezing on the surface of the regular specimen, Figure 6D–F represent the photos of drops of water on the surface of the hydrophobic coating specimen, and Figure 6G–I represent the photos of droplets on the surface of the superhydrophobic coating specimen.

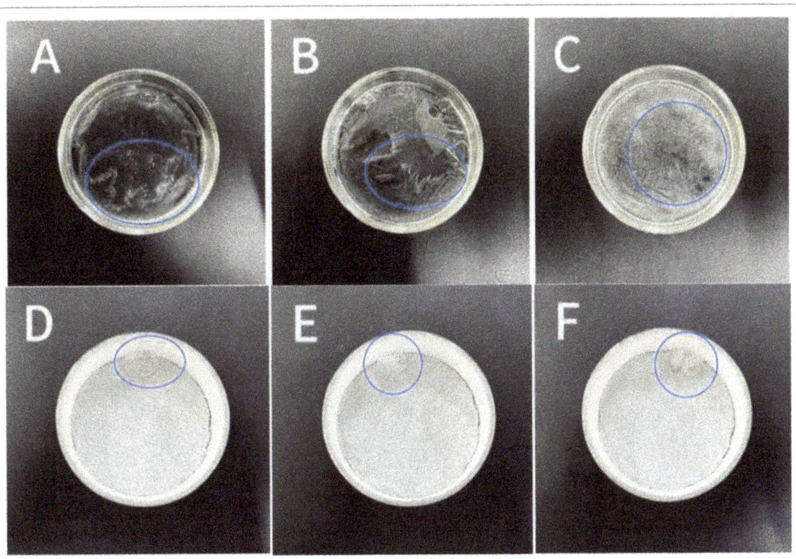

Figure 6. *Cont.*

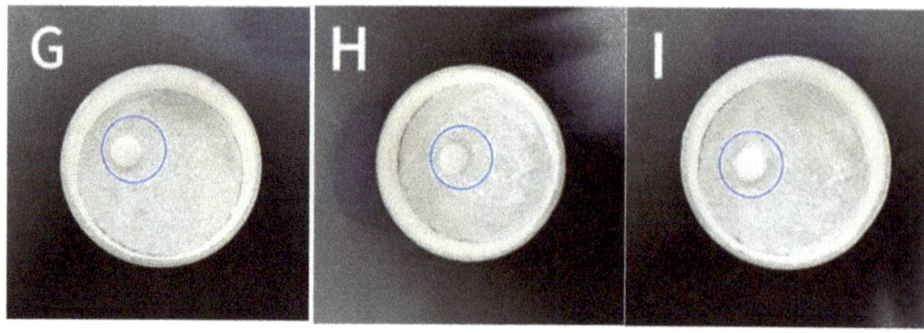

Figure 6. After the water droplets freeze on the sample surface. (**A–C**) Photos of droplets freezing on the surface of the regular specimen; (**D–F**) photos of drops of water on the surface of the hydrophobic coating specimen; (**G–I**) photos of droplets on the surface of the superhydrophobic coating specimen.

Analyzing the figures, under −4 °C conditions, Figure 6A shows the state when the droplets on the regular specimen surface start to freeze, with small scattered ice crystals already present. Figure 6D represents the hydrophobic coating specimen at the same moment, while Figure 6G depicts the superhydrophobic coating specimen, where the surface water droplets remained in a liquid state. Under −8 °C and −10 °C conditions, Figure 6B,E,H represent the state after freezing at the same moment, while Figure 6C,F,I represent the state after icing at the same moment. It can be observed that the icing time of Figure 6B,C is shorter, while the icing time of Figure 6E,F,H,I is longer. Additionally, the ice droplet contact area with the wall of the dish after icing was larger in both regular specimens and hydrophobic coating specimens compared to the superhydrophobic coating. The water droplets in Figure 6G–I gradually increased in size as the temperature decreased, suggesting nucleation of water vapor condensing on the large droplets in a low-temperature and high-humidity environment.

4.4. The Law of Ice Adhesion Cover on Different Surfaces

To evaluate the ice cover adhesion of the new pavement superhydrophobic coating after icing, specimens of hydrophilic coating, hydrophobic coating, superhydrophobic coating modified with anatase titanium dioxide nanoparticles, and superhydrophobic coating modified with rutile titanium dioxide nanoparticles were prepared, corresponding to No. 1, No. 2, No. 3, and No. 4, as shown in Figure 7a. These specimens were placed in molds, as depicted in Figure 7b, inside high- and low-temperature control boxes set to −10 °C, allowing the water to condense into rectangular iced concrete specimens of dimensions 200 mm × 100 mm × 100 mm.

(**a**) (**b**)

Figure 7. Sample preparation: (**a**) experimentally prepared hydrophilic coating specimen, hydrophobic coating specimen, superhydrophobic coating specimen; (**b**) ice adhesion strength test sample making mold diagram.

The iced concrete specimens were positioned on a press, and two circular mat strips were inserted at both ends of the interface between the specimens and the ice. Pressure was applied until splitting occurred, and the experimental results are presented in Figure 8, illustrating the ice adhesion to the four specimens. The flat hydrophilic surface exhibited the highest ice adhesion, measuring 947.75 N. This is due to the smoothness of the hydrophilic surface, which results in a larger actual contact area between the ice surface and the specimen. The ice adhesion of the hydrophobic surface and the superhydrophobic surface modified with rutile titanium dioxide was below that of the smooth hydrophilic surface, measuring 252.75 N and 237.75 N, respectively. Remarkably, the ice adhesion of superhydrophobic surfaces modified with rutile titanium dioxide was as low as 214 N. This can be attributed to the enhanced stability and superior weather resistance of rutile titanium dioxide compared to anatase titanium dioxide nanoparticles. Consequently, the impact of environmental changes on the superhydrophobic surface was minimized.

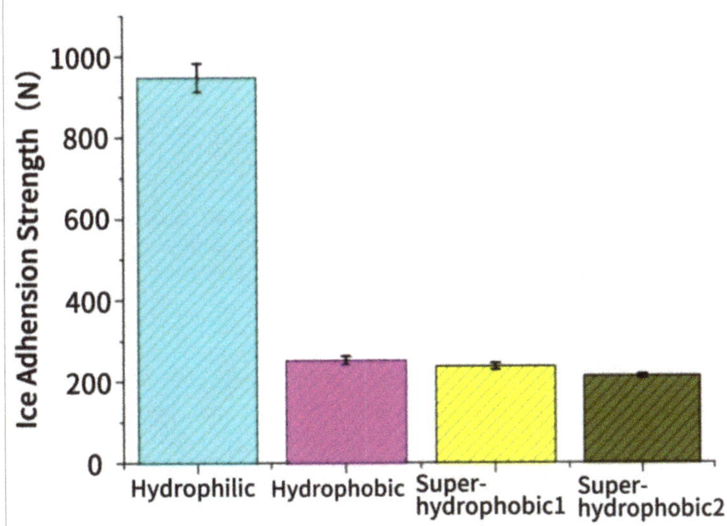

Figure 8. Adhesion strength of ice cover.

For superhydrophobic surfaces, at room temperature, liquid water resided above the surface texture, known as the Cassie state, while the air trapped in the surface structure beneath the liquid water remained in thermodynamic equilibrium with it. As the temperature decreased, the surface liquid water gradually began to freeze, and the condensed droplets of air did not freeze within the superhydrophobic surface texture where the nano-gap fell below the critical nuclear radius. Instead, they condensed and nucleated on the surface of larger droplets. At this stage, the liquid water existed in a state of coexistence between Wenzel and Cassie states, as shown in Figure 9. As a result, the actual contact area between the ice surface and the sample decreased, resulting in a reduction in the ice adhesion.

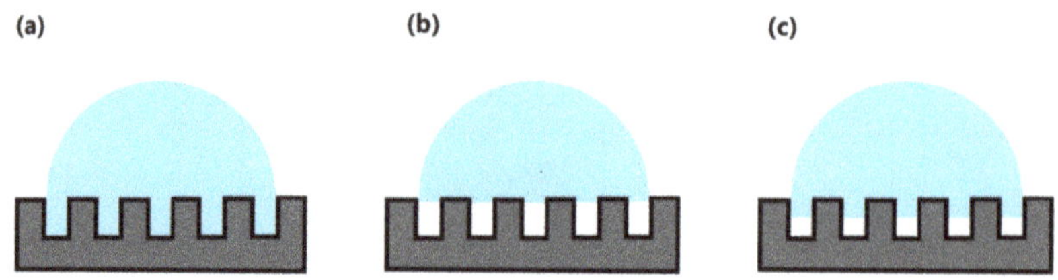

Figure 9. Mechanism of ice suppression on superhydrophobic surfaces: (**a**) Wenzel model; (**b**) Cassie model; (**c**) Wenzel and Cassie coexistence model [45].

5. Conclusions

Applications of superhydrophobic coatings as anti-icing materials for pavements can effectively delay the icing of static water before icing and reduce the adhesion of ice to the pavement surface after icing due to their unique micro- and nano-structures and excellent anti-condensation properties. Through both theory and experimentation, the study yielded important conclusions about the mechanisms behind these coatings:

(1) Through theoretical discussions, we have uncovered the ice suppression capabilities of the superhydrophobic surface in relation to the condensate. It can be concluded that when the nano-gap of the superhydrophobic surface texture is lower than the critical nuclear radius of 145 nm under low humidity and high moisture environment, it prevents condensate droplets from entering the surface texture, allowing liquid water to remain in the Cassie state on the superhydrophobic surface. This inhibits the formation of mechanical interlocking effects, which can increase ice adhesion on the surface.

(2) The surface of the superhydrophobic coating is composed of micro- and nano-papillae. Even in conditions of low temperatures and high humidity, condensation on the superhydrophobic surface does not affect its water contact angle, which is 151.9° with little change compared to the pre-condensation angle of 154.0°, ensuring continued strong superhydrophobicity. The large contact angle conceals the air in the gap of the coarse structure, which delays the icing time of the droplets and reduces the contact surface of the droplet with the superhydrophobic coating.

(3) After investigating the adhesion of ice on a variety of wettability surfaces, our findings revealed that the adhesion force of hydrophilic surfaces with ice is the most robust, up to 947.75 N. In contrast, it can be observed that the presence of air within the microstructure of superhydrophobic surfaces significantly reduces the ice adhesion, with an adhesion of 214 N, without any observable mechanical interlocking effect.

Author Contributions: Conceptualization, K.H. and R.S.; methodology, K.H.; software, J.W.; validation, J.W., X.S. and H.L.; formal analysis, K.H.; investigation, X.S.; resources, H.L.; data curation, J.W.; writing—original draft preparation, K.H.; writing—review and editing, R.S.; visualization, R.S.; supervision, J.W.; project administration, K.H.; funding acquisition, K.H. All authors have read and agreed to the published version of the manuscript.

Funding: This research was funded by he Natural Science Foundation of Jiangsu Province (Grant number: BK20210617) and the National Natural Science Foundation of China (Grant number: 52108408, 51608273, 51778300).

Institutional Review Board Statement: Not applicable.

Informed Consent Statement: Not applicable.

Data Availability Statement: The data can be provided if needed by the corresponding author.

Acknowledgments: The corresponding author and the third author would like to appreciate the support from the Natural Science Foundation of Jiangsu Province (Grant number: BK20210617) and the National Natural Science Foundation of China (Grant number: 52108408, 51608273, 51778300). The fourth author did not receive any support from these grants. The results and conclusions of this manuscript only represented the author's opinion, but not the funding programs.

Conflicts of Interest: The authors declare no conflict of interest.

References

1. Wenqi, Z. Analysis of skid resistance of icy pavement and its impact on traffic safety. *J. Wuhan Univ. Technol.* **2014**, *36*, 58–62.
2. Qiu, X.; Xu, J.-X.; Tao, J.-Q.; Yang, Q. Asphalt Pavement Icing Condition Criterion and SVM-Based Prediction Analysis. *J. Highw. Transp. Res. Dev.* **2018**, *12*, 1–9. [CrossRef]
3. Lina, Z.; Qingyun, M.; Guiming, Y. Analysis of the impact of low temperature rain and snow freeze on key industries and the causes of disaster in China in early 2008. *Clim. Environ. Res.* **2008**, *13*, 11.
4. Bingchen, Z.; Shumin, L. Exploration of snow removal on roads in winter. *Shandong Traffic Sci. Technol.* **2004**, *2*, 4.
5. Yiqiu, T.; Huining, X.; Chi, Z. Old methods and new technologies of snow removal. *China Highw.* **2018**, *511*, 53–57.
6. Fay, L.; Shi, X. Environmental Impacts of Chemicals for Snow and Ice Control: State of the Knowledge. *Water Air Soil Pollut.* **2012**, *223*, 2751–2770. [CrossRef]
7. Xu, X.; Xiong, F.; Meng, J.; Wang, X.; Niu, C.; An, Q.; Mai, L. Vanadium-Based Nanomaterials: A Promising Family for Emerging Metal-Ion Batteries. *Adv. Funct. Mater.* **2020**, *30*, 1904398. [CrossRef]
8. Liu, Z.; Sha, A.; Xing, M.; Li, Z. Low temperature property and salt releasing characteristics of antifreeze asphalt concrete under static and dynamic conditions. *Cold Reg. Sci. Technol.* **2015**, *114*, 9–14. [CrossRef]
9. Datta, K.K.R. Exploring the Self-Cleaning Facets of Fluorinated Graphene Nanoarchitectonics: Progress and Perspectives. *ChemNanoMat* **2023**, *9*, e202300135. [CrossRef]
10. Bhushan, B.; Multanen, V. Designing liquid repellent, icephobic and self-cleaning surfaces with high mechanical and chemical durability. *Philos. Trans. R. Soc. A Math. Phys. Eng. Sci.* **2019**, *377*, 20180270. [CrossRef]
11. Archer, R.J.; Becher-Nienhaus, B.; Dunderdale, G.J.; Hozumi, A. Recent Progress and Future Directions of Multifunctional (Super)Wetting Smooth/Structured Surfaces and Coatings. *Adv. Funct. Mater.* **2020**, *30*, 1907772. [CrossRef]
12. Arabzadeh, A.; Ceylan, H.; Kim, S.; Gopalakrishnan, K.; Sassani, A. Superhydrophobic Coatings on Asphalt Concrete Surfaces: Toward Smart Solutions for Winter Pavement Maintenance. *Transp. Res. Board* **2016**, *2551*, 10–17. [CrossRef]
13. Antonini, C.; Innocenti, M.; Horn, T.; Marengo, M.; Amirfazli, A. Understanding the effect of superhydrophobic coatings on energy reduction in anti-icing systems. *Cold Reg. Sci. Technol.* **2011**, *67*, 58–67. [CrossRef]
14. Jiajia, L.; Kaijian, H.; Jiaqing, W. Progress of ice suppression research on superhydrophobic materials for pavements. *New Chem. Mater.* **2023**, *51*, 223–228.
15. Wang, J.; Dai, Q.; Si, R. Experimental and Numerical Investigation of Fracture Behaviors of Steel Fiber-Reinforced Rubber Self-Compacting Concrete. *J. Mater. Civ. Eng.* **2022**, *34*, 04021379. [CrossRef]
16. Pilotek, S.; Schmidt, H.K. Wettability of Microstructured Hydrophobic Sol-Gel Coatings. *J. Sol-Gel Sci. Technol.* **2003**, *26*, 789–792. [CrossRef]
17. Farhadi, S.; Farzaneh, M.; Kulinich, S.A. Anti-icing performance of superhydrophobic surfaces. *Appl. Surf. Sci.* **2011**, *257*, 6264–6269. [CrossRef]
18. Chaudhuri, X.S. Design of anti-icing coatings using supercooled droplets as nano-to-microscale probes. *Langmuir* **2012**, *28*, 4434–4446.
19. Subramanyam, S.B.; Kondrashov, V.; Rühe, J.; Varanasi, K.K. Low Ice Adhesion on Nano-Textured Superhydrophobic Surfaces under Supersaturated Conditions. *ACS Appl. Mater. Interfaces* **2016**, *8*, 12583–12587. [CrossRef]
20. Shi, X. Characterization and Mechanism of a New Superhydrophobic Deicing Coating Used for Road Pavement. *Crystals* **2021**, *11*, 1304.
21. Jisr, R.M.; HRmaile, H.; Schlenoff, J.B. Hydrophobic and Ultrahydrophobic Multilayer Thin Films from Perfluorinated Polyelectrolytes. *Angew. Chem. Int. Ed.* **2004**, *44*, 782–785. [CrossRef]
22. Sarkar, D.K.; Farzaneh, M. Superhydrophobic Coatings with Reduced Ice Adhesion. *J. Adhes. Sci. Technol.* **2009**, *23*, 1215–1237. [CrossRef]
23. Dotan, A.; Dodiuk, H.; Laforte, C.; Kenig, S. The Relationship between Water Wetting and Ice Adhesion. *J. Adhes. Sci. Technol.* **2009**, *23*, 1907–1915. [CrossRef]
24. Petrenko, V.F.; Peng, S. Reduction of ice adhesion to metal by using self-assembling monolayers (SAMs). *Can. J. Phys.* **2003**, *81*, 387. [CrossRef]
25. Petrenko, V.F.; Whitworth, R.W. Elastic, Thermal, and Lattice Dynamical Properties. In *Physics of Ice*; Oxford University Press: Oxford, UK, 2002.
26. Bharathidasan, T.; Kumar, S.V.; Bobji, M.; Chakradhar, R.; Basu, B.J. Effect of wettability and surface roughness on ice-adhesion strength of hydrophilic, hydrophobic and superhydrophobic surfaces. *Appl. Surf. Sci.* **2014**, *314*, 241–250. [CrossRef]

27. Kulinich, S.A.; Farzaneh, M. How wetting hysteresis influences ice adhesion strength on superhydrophobic surfaces. *Langmuir ACS J. Surf. Colloids* **2009**, *25*, 8854. [CrossRef]
28. Murase, H.; Nanishi, K.; Kogure, H.; Fujibayashi, T.; Tamura, K.; Haruta, N. Interactions between heterogeneous surfaces of polymers and water. *J. Appl. Polym. Sci.* **1994**, *54*, 2051–2062. [CrossRef]
29. Wang, J.; Luo, X.; Huang, X.; Ye, Y.; Ruan, S. Analysis on Effects of Joint Spacing on the Performance of Jointed Plain Concrete Pavements Based on Long-Term Pavement Performance Database. *Materials* **2022**, *15*, 8132. [CrossRef]
30. Su, P.; Li, M.; Dai, Q.; Wang, J. Mechanical and durability performance of concrete with recycled tire steel fibers. *Constr. Build. Mater.* **2023**, *394*, 132287. [CrossRef]
31. Chen, J.; Liu, J.; He, M.; Li, K.; Cui, D.; Zhang, Q.; Zeng, X.; Zhang, Y.; Wang, J.; Song, Y. Superhydrophobic surfaces cannot reduce ice adhesion. *Appl. Phys. Lett.* **2012**, *101*, 41–932. [CrossRef]
32. Varanasi, K.K.; Deng, T.; Smith, J.D.; Hsu, M.; Bhate, N. Frost formation and ice adhesion on superhydrophobic surfaces. *Appl. Phys. Lett.* **2010**, *97*, 234102. [CrossRef]
33. Liu, T.; Sun, W.; Li, X.; Sun, X.; Ai, H. Growth Modes of Condensates on Nanotextured Surfaces and Mechanism of Partially Wetted Droplet Formation. *Acta Phys. Chim. Sin.* **2013**, *9*, 9807–9815.
34. Narhe, R.D.; Beysens, D.A. Growth Dynamics of Water Drops on a Square-Pattern Rough Hydrophobic Surface. *Langmuir ACS J. Surf. Colloids* **2007**, *23*, 6486. [CrossRef]
35. Wilke, K.L.; Preston, D.J.; Lu, Z.; Wang, E.N. Toward Condensation-Resistant Omniphobic Surfaces. *ACS Nano* **2018**, *12*, 11013–11021. [CrossRef]
36. Boreyko, J.B.; Chen, C.-H. Self-propelled dropwise condensate on superhydrophobic surfaces. *Phys. Rev. Lett.* **2009**, *103*, 184501. [CrossRef]
37. Dorrer, C.; Rühe, J. Wetting of silicon nanograss: From superhydrophilic to superhydrophobic surfaces. *Adv. Mater.* **2008**, *20*, 159–163. [CrossRef]
38. Chen, C.H.; Cai, Q.; Tsai, C.; Chen, C.L.; Xiong, G.; Yu, Y.; Ren, Z. Dropwise condensation on superhydrophobic surfaces with two-tier roughness. *Appl. Phys. Lett.* **2007**, *90*, 173108. [CrossRef]
39. Feng, J.; Pang, Y.; Qin, Z.; Ma, R.; Yao, S. Why Condensate Drops Can Spontaneously Move Away on Some Superhydrophobic Surfaces but Not on Others. *ACS Appl. Mater. Interfaces* **2012**, *4*, 6618–6625. [CrossRef]
40. Feng, J.; Qin, Z.; Yao, S. Factors Affecting the Spontaneous Motion of Condensate Drops on Superhydrophobic Copper Surfaces. *Langmuir* **2012**, *28*, 6067–6075. [CrossRef]
41. Chen, X.; Wu, J.; Ma, R.; Hua, M.; Koratkar, N.; Yao, S.; Wang, Z. Nanograssed Micropyramidal Architectures for Continuous Dropwise Condensation. *Adv. Funct. Mater.* **2011**, *21*, 4617–4623. [CrossRef]
42. Liu, X.Y. A new kinetic model for three-dimensional heterogeneous nucleation. *J. Chem. Phys.* **1999**, *111*, 1628–1635. [CrossRef]
43. Cao, L.; Jones, A.K.; Sikka, V.K.; Wu, J.; Gao, D. Anti-Icing Superhydrophobic Coatings. *Langmuir ACS J. Surf. Colloids* **2009**, *25*, 12444–12448. [CrossRef] [PubMed]
44. Ren, S.; Liu, X.; Li, M.; Fan, W.; Xu, J.; Erkens, S. Experimental characterization of viscoelastic behaviors, microstructure and thermal stability of CR/SBS modified asphalt with TOR. *Constr. Build. Mater.* **2020**, *261*, 120524. [CrossRef]
45. Wang, Z.; Han, C.; Yu, Y.; Deng, H.; Li, C. Preparation and properties of superhydrophobic PVDF microporous composite membrane. *Chem. Ind. Eng. Prog.* **2018**, *37*, 673–680.

materials

MDPI

Article

Experimental Study on the Impact of Using FRP Sheets on the Axial Compressive Performance of Short-Circular Composite Columns

Jie Liu [1,*], Deliang Ma [1], Feifei Dong [2] and Zhongxiang Liu [3]

1 College of Civil Engineering, Nanjing Forestry University, Nanjing 210037, China; deliang_ma@outlook.com
2 National Engineering Research Center of Biomaterials, Nanjing Forestry University, Nanjing 210037, China; feifei_dong@outlook.com
3 School of Transportation, Southeast University, Nanjing 210096, China; zhongxiang@seu.edu.cn
* Correspondence: jieliu@outlook.com

Abstract: This paper conducts an experimental study on the axial compressive performance of FRP-steel-concrete composite columns. Nine short columns were produced and evaluated in the study, comprising of three concrete-filled steel tube reference columns and six FRP-steel-concrete composite columns, respectively denoted as "reference columns" and "composite columns". Two categories of failure modes, including shear failure and waist drum, were observed from the experiments. The failure mode may trend toward waist drum from shear failure as more FRP layers were used. The number of FRP layers had a direct effect on the level of compressive strength attained, with a greater number of layers resulting in a greater increase in compressive strength. Moreover, a greater tensile strength and higher elastic modulus of CFRP sheets are more effective at improving the compressive stiffness of the columns. Finally, a four-stage confinement mechanism for FRP-wrapped steel tube concrete composite columns is proposed and discussed, through which the damage mechanisms of the composite structures are more rationally characterized.

Keywords: FRP; composite column; axial compress; bearing capacity; refinement

Citation: Liu, J.; Ma, D.; Dong, F.; Liu, Z. Experimental Study on the Impact of Using FRP Sheets on the Axial Compressive Performance of Short-Circular Composite Columns. *Materials* **2023**, *16*, 6373. https://doi.org/10.3390/ma16196373

Academic Editor: Yuri Ribakov

Received: 17 August 2023
Revised: 15 September 2023
Accepted: 15 September 2023
Published: 24 September 2023

1. Introduction

Concrete-filled steel tube (CFST) columns are widely utilized in civil engineering structures due to their significant enhancement of the performance of inner concrete facilitated by steel tube confinement [1–5]. However, the outer steel tubes of these columns are susceptible to experiencing local buckling and erosion damage in service [6–10].

In order to improve the pressure-bearing performance of CFSTs and their ability to resist environmental erosion, high-performance or functional materials have been adopted to wrap and reinforce CFSTs. The most typical method is the augmentation method of wrapping FRP layers [11–13] and using shape memory alloy (SMA) [14,15]. These reinforcement techniques all make use of the superior mechanical properties of external materials, such as the light weight and high strength of FRP materials [16–20], as well as the high ductility and good energy dissipation capacity of SMA [15].

One pioneer study on the axial bearing performance of FRP-wrapped CFSTs was written by Tao et al. [21]. In their study, both rectangular and circular composite specimens retrofitted with CFRP were tested. The experimental findings unequivocally demonstrate that the load-bearing strength of retrofitted circular column specimens was significantly improved by CFRP composites, whereas only limited improvement was observed in the case of rectangular columns. After that, Liu et al. [22] carried out an experimental investigation of 11 short column specimens under axial compressing loads and proposed corresponding formulas for calculating the bearing strength of the FRP-CFST columns. Hu et al. [23] carried out an experimental study to investigate the behavior of FRP-confined

CFSTs under axial compression. The findings of the study revealed that the application of FRP confinement can result in a noteworthy enhancement of both the carrying strength and ductility of CFSTs. Na et al. [24] subsequently conducted an inquiry into the slenderness effects of CFTs confined with CFRP. Their study demonstrated that the composite wraps provided effective confinement to the inner CFST, thus enhancing the performance of the CFST columns. However, it was observed that the reinforcing effect of the CFRP wrapping layers decreased with an increasing slenderness ratio. More recently, Zhang et al. [25] proposed a stress–strain analysis model of FRP-confined CFSTs under axial compression, based on an experimental investigation. Wang et al. [7] explored the axial compressive performance and confinement mechanisms of circular FRP-steel tube concrete stub columns and determined that the capacity of the CFST specimens was augmented by reducing the steel tube's diameter-to-thickness ratio. In 2022, Onyelowe et al. [26,27] investigated the compressive strength of CFRP-wrapped concrete columns using AI (artificial intelligence) techniques and concluded that the strength of wrapped concrete columns depends on the FRP thickness and tensile strength.

Although previous studies have addressed the bearing strength of FRP-wrapped CFST columns, few studies have investigated the complex load-bearing effects of composite columns—particularly the failure mechanism of composite columns under axial forces, and the effect of different FRP materials (and thicknesses) on load-bearing capacity. This paper examines the characteristics and mechanism of compression failure in CFST columns strengthened with wraps composed of two different FRP materials with varying thicknesses. The failure mechanism of composite columns will be investigated to gain a deeper understanding of the factors affecting such mechanisms. Finally, based on the experimental results, a four-stage failure mechanism of an FRP-steel-concrete column under axial compress is proposed. The research methodology of this study is illustrated in Figure 1.

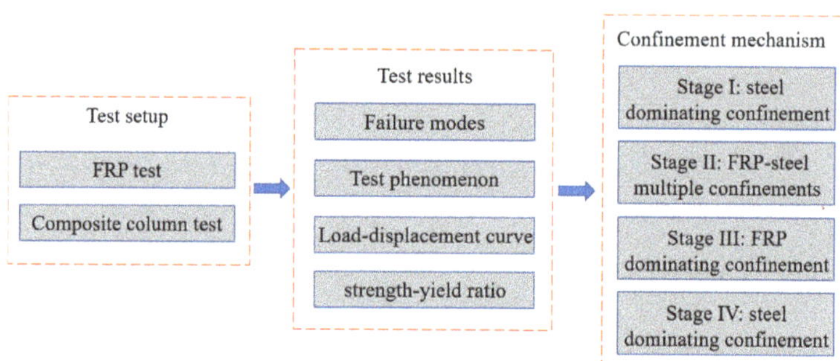

Figure 1. Research methodology of this study.

2. Experimental Program

2.1. Specimen Configurations and Material Properties

In order to study the effect of FRP wraps on the compressive performance of CFST, 9 short columns were fabricated, including 3 CFST columns (referred to as "reference columns") and 6 FRP-steel-concrete composite columns (FRP sheets wrapped outside the reference column, referred to as "composite columns") were manufactured. Detailed information about the columns is listed in Table 1. The CFST specimens were all poured into hollow steel tubes, and the outer diameter D and thickness of the steel tubes were 100 mm and 2 mm, respectively. The composite column and the test setup are plotted in Figure 2. The thickness of a single layer of FRP was 0.167 mm. For each composite column, FRP was affixed onto the outer surface of the steel tube using steel glue, and the FRP interface of adjacent layers was also bonded together via steel glue. The tensile strength of the steel glue material was 44.2 MPa, which was provided by the manufacturer.

Table 1. Information about the columns.

Category	Specimen No.	FRP Category	Number of FRP Layer	The Total Thickness of FRP Layer/mm
Reference column	N1	N/A	0	0
	N2	N/A	0	0
	N3	N/A	0	0
composite column	CC1	CFRP	1	0.167
	CC2	CFRP	2	0.334
	CC3	CFRP	3	0.501
	CB1	BFRP	1	0.167
	CB2	BFRP	2	0.334
	CB3	BFRP	3	0.501

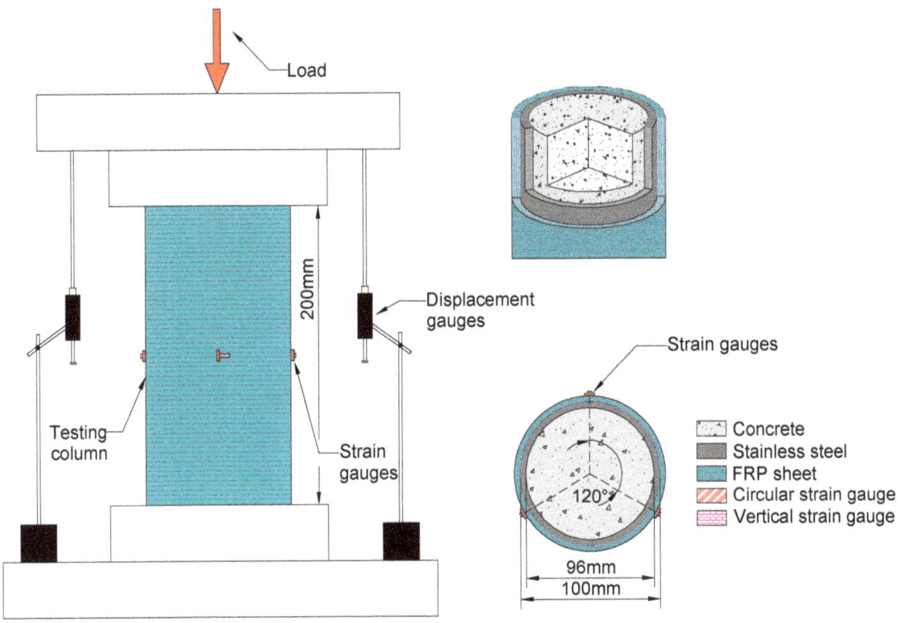

(**a**) Loading schematic. (**b**) 3-D drawing of specimen.

Figure 2. Diagram of the composite column (unit: mm).

For the convenience of expression, each specimen was named separately. As in Table 1, the 3 reference columns are donated as N1, N2, and N3. The 6 composite column specimens are donated as CC1, CC2, CC3, CB1, CB2, and CB3. The adhesive was epoxy resin-type steel glue with a tensile strength of 44.2 MPa.

The FRP wrapping utilized in this study was sourced from Jiangsu Green Material Valley New Material Technology Development Co., Ltd. Both CFRP (Carbon Fiber Reinforced Polymer) and BFRP (Basalt Fiber Reinforced Polymer) composites were employed in the sample specimens. The Young's moduli of CFRP and BFRP were 1.67×10^5 and 9.24×10^4 MPa respectively, as provided by the manufacturer. The steel tubes were made of 06Cr19Ni10 stainless steel (i.e., SUS 304 stainless steel in the US and Japan), and the tensile strength of the steel material was 550 MPa. The average yield strength and Young's modulus were measured as 207 MPa and 198 GPa, respectively, by the manufacturer. The concrete inner core was C60 grade. The compressive strength of the concrete cube was measured via a cube test block with a side length of 150 mm, according to the "Concrete

Strength Inspection and Evaluation Standard" (GB/T50107-2019). The average compressive strength of the cubes was measured to be 61.4 MPa. To precisely determine the material properties of the FRP sheets, a tensile test of six FRP specimens (including three CFRP and three BFRP) was fabricated. The sizes of the FRP specimens were designed as 100 mm (length) × 10 mm (width) × 0.167 mm (thickness), respectively, according to the recommendation of "Carbon fiber sheets for structural reinforcement and repair" (JG/T 167-2016). The loading and failure diagram of the test for FRP sheets is depicted in Figure 3. The average tensile strengths of the CFRP and BFRP sheets were measured as 4101.0 and 2005.1 MPa, respectively. The material properties of FRP and steel are listed in Table 2.

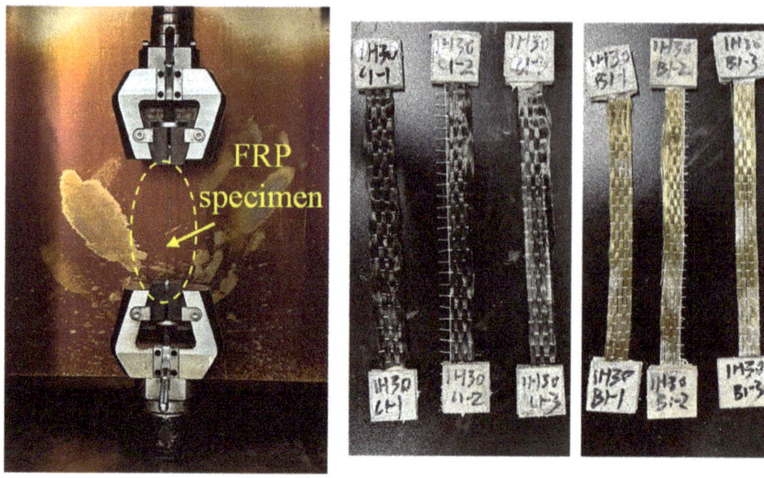

Figure 3. Loading and failure diagrams of FRP specimens.

Table 2. Material properties of FRP and steel.

	Young's Moduli/MPa	Tensile Strength/MPa
CFRP	1.67×10^5	4101.0 (on average)
BFRP	9.24×10^4	2005.1 (on average)
Steel	1.98×10^5	550

2.2. Test Setup

The axial compression loading test was conducted utilizing the Bonwe 3000 kN high-rigidity geotechnical concrete testing machine, and the test setup is depicted in Figure 4. The loading mechanism was executed in the following manner:

Preloading: The column specimen's axis was positioned at the central axis of the testing machine to ensure that the specimen bore the force on the axis and remained within the range of the elastic loading stage of the specimen. The loading process was initiated by preloading the specimen up to approximately 10% of the estimated ultimate load at a loading rate of 2 mm/min, and the specimen was subjected to preloading three times. This ensured that the instrument worked correctly and the influence of the gap between the test specimen and the tool was eliminated.

Formal loading: Each specimen underwent monotonic axial compression testing, with the complete loading process utilizing a program for automatic control of axial displacement. The load was applied with the displacement-control mode at a rate of 2 mm/min. The specimen's load increased until it was damaged (FRP fracture, steel pipe yield) or the loading was stopped when the load declined to 80% of the peak load. The specimen was unloaded afterward.

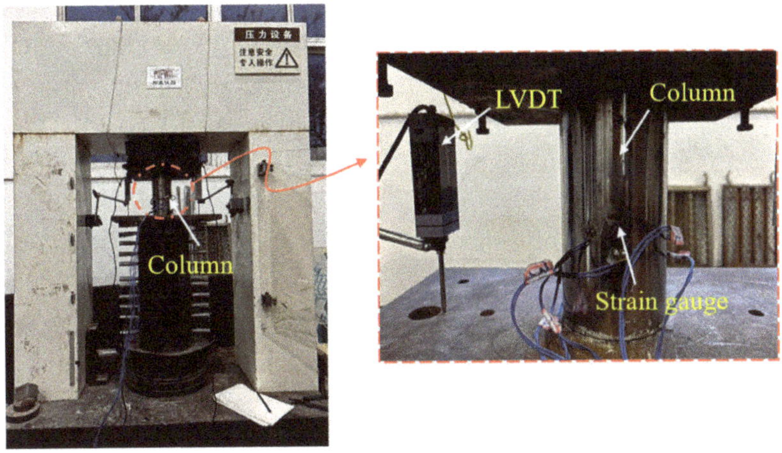

Figure 4. Test setup and instrumentation of the composite columns.

The measurement items for the columns' monotonic axial compression test primarily entailed information on the vertical load, vertical compression, and horizontal and vertical strain in the middle plane of the specimen. The vertical load was evaluated using the built-in dynamometer of the press, and the vertical compression was calculated utilizing the vertical relative displacement of the loading plate at both ends of the specimen during the loading process as measured by the displacement gauges, as demonstrated in Figures 1 and 3. The horizontal and vertical strain gauges in the middle of the specimen measured the horizontal and vertical strain, respectively. The strain gauges were arranged at the middle height of the specimen in the trisecting circle, with three pairs of vertically arranged strain gauges employed to record the lateral strain gauge distribution. The strain gauge model BX120-15AA, with a grid length of 15 mm × 3 mm and a resistance value of 120 ± 0.19 Ω, was adopted. During the loading process, the load and displacement were collected by the data acquisition system TDS630 automatically at a sampling frequency of 50 Hz.

2.3. Mechanical Conditions of Composite Columns

Considering that the FRP-steel-concrete composite short column is a composite structural system, the load-bearing state of the concrete core is the key factor affecting the bearing strength of the composite columns. The mechanical behavior of the concrete inside is then analyzed. The inner concrete of the columns is in a typical triaxial stress condition. Based on the discussion of Liu et al. [22], the axial compressive strength of the inner concrete is in a linear relationship with the lateral confining pressure as follows:

$$f_{cc} = f_c + kp, \quad (1)$$

where f_{cc} and f_c are the axial compressive strength of the inner concrete with and without confinement, respectively, p denotes the magnitude of the lateral confining pressure, and k denotes the strength-increasing coefficient of the inner concrete. From this formula, it can be seen that the stronger the confinement effect of the outer wrapped FRP and the steel pipe, the higher the bearing strength of the inner core concrete column.

Figure 5 plots the force of the inner concrete and FRP-steel layer. σ_{c1} is the axial press of the concrete. σ_{c2} and σ_{c3} are the principal stresses in the other two directions; they have the following relation:

$$\sigma_{c1} = f_{cc} = f_c + kp, \quad (2)$$

$$\sigma_{c2} = \sigma_{c2} = p, \quad (3)$$

Based on the force diagram of the FRP-steel layer in Figure 4, the lateral confining pressure p can be adopted as follows:

$$p = (\sigma_s \times 2t_s + \sigma_{FRP} \times 2t_{FRP})/D, \quad (4)$$

where σ_s and σ_{FRP} denote the transverse stresses of the steel pipe and FRP sheets, respectively; t_s and t_{FRP} denote the thicknesses of the steel pipe and FRP sheets, respectively; and D denotes the diameter of the inner concrete column. By combining Equations (1) and (4), the final expression can be obtained:

$$f_{cc} = f_c + k(\sigma_s \times 2t_s + \sigma_{FRP} \times 2t_{FRP})/D, \quad (5)$$

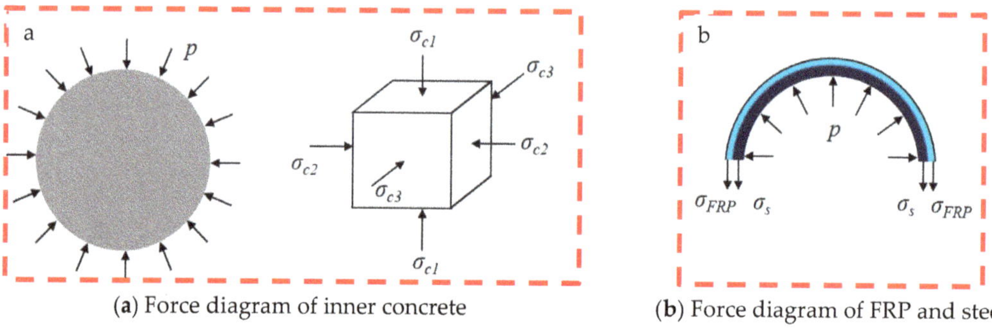

Figure 5. Force diagram of inner concrete and FRP-steel layer.

Equation (5) shows that for steel-concrete columns wrapped in FRP, if the circumferential confinement stress or the thickness of the confining circle of the outer steel pipe and FRP layer increases, the compressive strength of the composite column will be effectively improved accordingly.

For the CFST column, the confinement effect of the steel on the inner concrete can be expressed as the following:

$$\xi_{ss} = \frac{A_{ss}\sigma_{0.2}}{A_c \times 0.833 \times f_{cu}} \quad (6)$$

where A_{ss} is the cross-sectional area of the steel tube; $\sigma_{0.2}$ is the 0.2% proof stress of the stainless steel; A_c is the cross-sectional area of the concrete; and f_{cu} is the strength of the concrete cube.

Han et al. [28] proposed the equation for bearing capacities of circular CFST specimens, which is shown as the following:

$$N_u = (1 + b\xi_{ss})f_c' A_c \quad (7)$$

where N_u is the bearing capacity of the CFST column, f_c' is the strength of a standard concrete cylinder, and b is the influence parameter of steel refinement.

3. Experimental Results

3.1. Failure Modes

After testing the specimens, the damaged specimens are shown in Figure 6. After continuous axial compression, the FRP layer on the surface of the specimen was fractured and degummed. As shown in Figure 7, the FRP constraining layer of the composite columns was peeled off to study the failure mode of the composite columns. Based on the axial compression tests, two failure modes of waist drum and shear failure were observed. Typical shear failure was found in specimens CC1, CC2, CB1, and CB2, whereas waist drum failure was found in CC3 and CB3, indicating that as the number of FRP layers increased, the failure mode of the composite columns may trend to be waist drum from shear failure. The destruction mechanism can be described as follows: for the reference columns, as the concrete compressed, due to the limited restraint capacity of the outer wrapping of the steel tube, the shear failure of the inner core concrete column occurred at a lower load. However, for composite columns, the expansion of internal concrete is restrained by FRP-steel composite confinement, leading to a significant enhancement of the compressive performance of internal concrete. Due to the strong external constraints of the double layers of FRP and steel pipe, the inner core concrete column was always in a state of obvious triaxial compression. The failure of the concrete column is more inclined to material destruction, resulting in the waist drum.

Figure 6. Pictures of the composite columns after destruction.

(**a**) CC1, CC2, and CC3 (**b**) CB1, CB2, andCB3

Figure 7. Two failure modes of composite specimens.

3.2. Test Phenomenon

Figure 8 plots the load–displacement relationship of specimen N1, within which "a" represents the loading moment, "b" indicates the initiation of the nonlinear stage, "c" denotes the bearing capacity (peak), and "d" signifies the residual load after bearing capacity (peak). Based on the experimentally observed phenomena, during the initial loading stage, the load–displacement curve of the specimen exhibited a nearly linear increase, with no apparent surface damages observed, as shown in Figure 9a. Upon reaching the yield load point of the steel tube, a slight bulging was noticed on its surface, as depicted in Figure 9b. Eventually, as the load approached the ultimate bearing capacity, clear instances of buckling deformation occurred on the surface of the steel tube, as illustrated in Figure 9c. Subsequently, the bearing capacity of the specimens gradually decreased. Upon the load reaching its lowest point, the original steel tube's buckling deformation heightened, with novel buckling deformation emerging, as shown in Figure 9d. As the specimen endured continuous compression, its lateral deformation rose persistently, culminating in the formation of several distinct local plastic deformations, as depicted in Figure 9e.

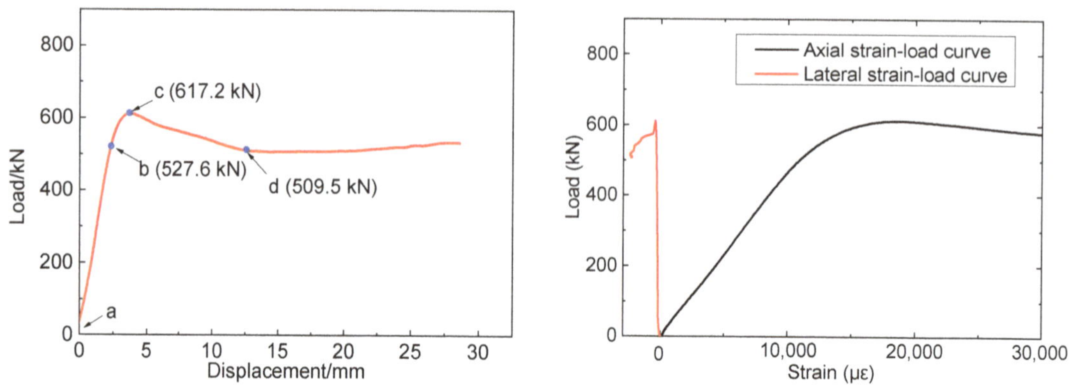

(**a**) Load–displacement. (**b**) Strain–load curve.

Figure 8. Figure of the load–displacement and strain–load curve of N1.

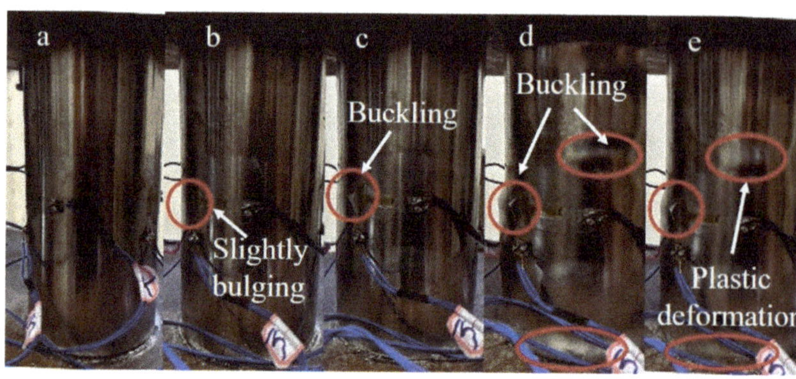

Figure 9. Figure of failure process for N1 under axial compression. (**a**) The initial loading stage; (**b**) The stage of yielding of steel tube; (**c**) The stage when the axial compression load reaches its maximum value; (**d**) The stage when the axial compressive load reaches the lowest value after the peak value; (**e**) Continuous compression stage after buckling.

Similar testing phenomena were observed in the six composite columns. Take CC3 as an example. The load–displacement curve is plotted in Figure 10, where *"a"*~*"b"* have identical meanings to those of Figure 8. Similar to the reference columns, during the initial loading stage (Figure 10a), the load–displacement curve of the composite columns increased approximately linearly, with no discernable damage observed on the specimen's surface. As the loads increased, a slight glue cracking sound was heard, with no obvious change observed, as depicted in Figure 11b. When the loads reached the specimen's ultimate load (Figure 11c), evident glue cracking sounds were observed, with localized cracks appearing on the FRP surface, indicating that the cemented and reinforced outer layer had begun to disintegrate and fail in large quantities. Subsequently, the FRP cracks expanded rapidly, with the surface of the steel pipe bulging slightly and the bearing load beginning to gradually decline. The load continued to decline to the lowest point; at this point, there were hardly any new cracking positions on the FRP surface, as shown in Figure 11d. Ultimately, several distinct local plastic deformations were formed at the FRP fracture, as shown in Figure 11e. The load–displacement curve of specimen CB3 was plotted in Figure 10b, and a similar phenomenon to CC3 was observed.

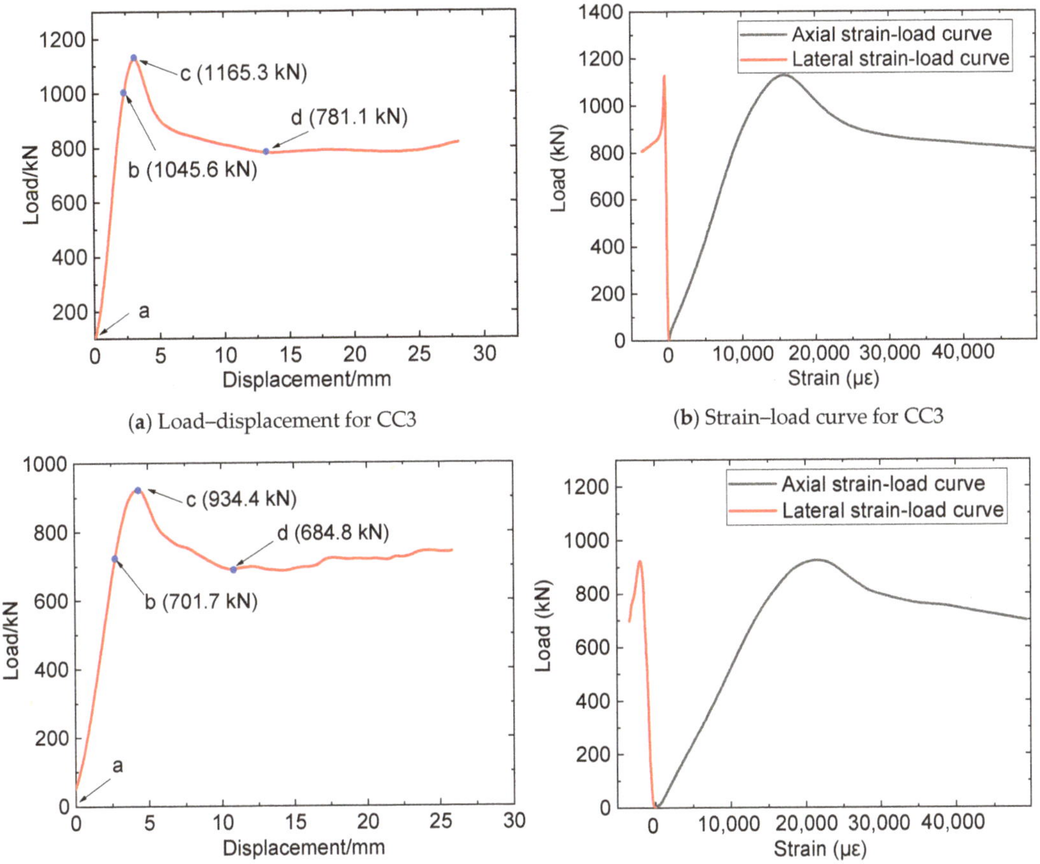

(a) Load–displacement for CC3

(b) Strain–load curve for CC3

(c) Load–displacement for CB3

(d) Strain–load curve for CB3

Figure 10. Figures of the load–displacement and strain–load curves for CC3 and CB3.

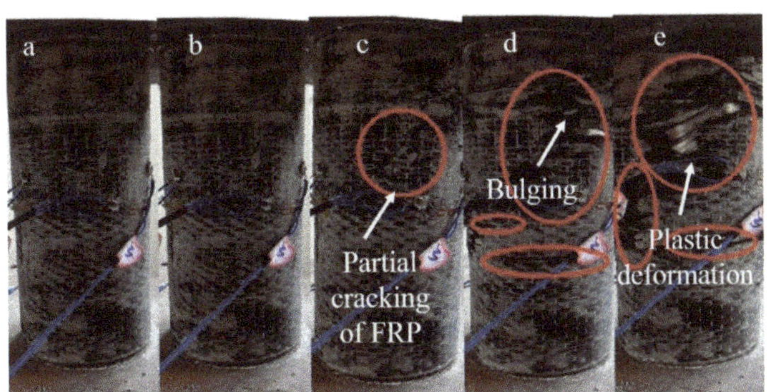

Figure 11. Failure process of the CC3 specimen under axial compression. (**a**) The initial loading stage; (**b**) The stage of slight glue cracking; (**c**) The stage when the axial compression load reaches its maximum value; (**d**) The stage when the axial compressive load reaches the lowest value after the peak value; (**e**) Continuous compression stage after buckling.

3.2.1. Load–Displacement Relationship

The experimental findings revealed that the load–displacement curves of both the reference and composite specimens can be discerned into four distinct stages: linear elastic, elastic–plastic, descending, and steady. Figure 12 presents a comparative analysis of the load–displacement relationships for the N1, CC1, CC2, and CC3 specimens.

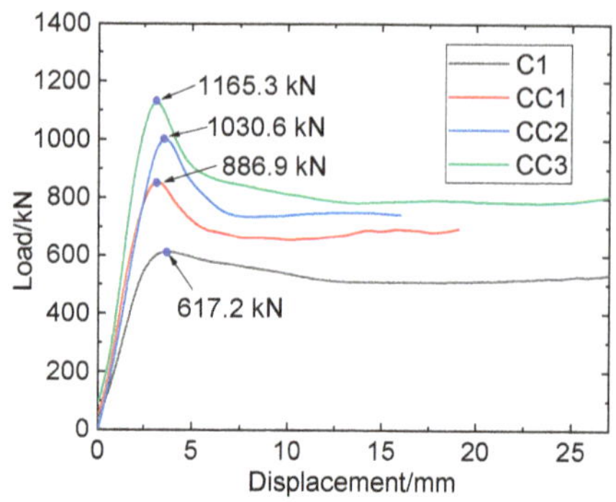

(**a**) N1, CC1, CC2, CC3

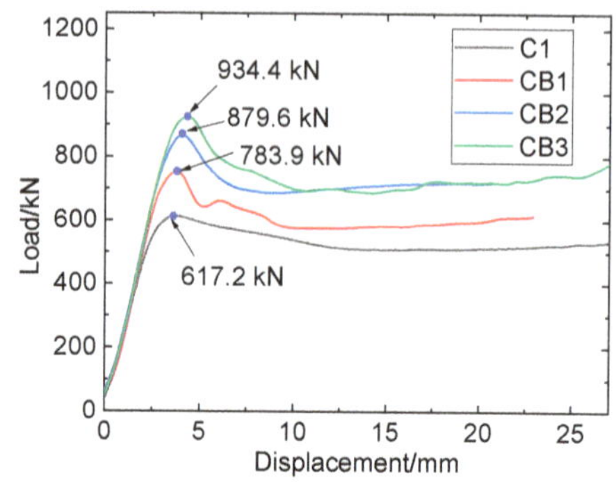

(**b**) N1, CB1, CB2, CB3

Figure 12. Load–displacement relationship.

From Figure 12, it can be seen that, at the initial stage, as the compression displacement increases, the load of the N1 specimen also linearly increases, indicating that the specimen is in the elastic stage. With the further increase of compression displacement (segment b–c), the load–displacement curve shows a quadratic feature, indicating that the specimen is in the elastic–plastic stage. After a short elastic–plastic stage, the load of the specimen reaches its peak at point B, where the bearing capacity is 617.2 kN. Thereafter, as the displacement continues to increase, steel tube buckling deformation increases steeply, and the load begins to decrease. With the continuous expansion of concrete, steel material continues to deform and enters the hardening stage, resulting in a slight increase in confining pressure on the concrete and entering the steady stage.

Compared with the N1 specimen, CC1, CC2, and CC3 show longer elastic–plastic stages and larger peak loads due to the constraint effect of CFRP on the steel-concrete column, with the peak loads (bearing capacities) of 886.9 kN, 1030.6 kN, and 1165.3 kN, respectively. When the compressing load reaches the peak, FRP begins to rupture, and the load–displacement curve enters the descending stage. When the load reaches the lowest point, FRP fractures extensively, and thereafter, the specimen continues to be compressed and the steel material remains in the yield stage. When entering the steady state, the load of the specimen is in the micro-fluctuation stage. Comparing the curves of the CC1, CC2, and CC3 specimens, it can be concluded that, with the increase in the number of CFRP layers, the slope of the load–displacement curve in the elastic stage gradually increases, and the starting point of the elastic–plastic stage increases while the corresponding displacement decreases.

The load–displacement curves of the CB1, CB2, and CB3 specimens are illustrated in Figure 11b. They show similar characteristics to the CC1, CC2, and CC3 specimens. It is noteworthy that, compared with specimens reinforced by CFRP, the peak loads of specimens reinforced by BFRP are relatively small, with the compressing capacities of 783.9 kN (CB1), 879.6 kN (CB2), and 934.4 kN (CB3).

In summary, during the linear elastic stage, a similar phenomenon of the load–displacement curve of CFRP- or BFRP-constrained specimens could be observed. This may be attributed to the small expansion deformation of the concrete at this stage, resulting in limited compression on the steel pipe. Consequently, the steel tube remains at the linear elastic stage with the FRP and the steel tube deforming in coordination, resulting in small forces on the FRP. As the specimen was continuously compressed and entered the elastic–plastic stage, the concrete underwent significant expansion deformation, the steel tube endured large stress and produced plastic deformation, and the restraining effect of the FRP became increasingly significant. The ultimate bearing capacity of the steel tube enhanced by CFRP is more significant than that by BFRP, which can be attributed to the bigger elastic modulus and higher tensile strength of CFRP. In addition, more FRP wrapping layers cause an increase in the bearing capacity of the composite columns, indicating that wrapping FRP sheet is an effective method to improve the bearing performance of CFST columns.

3.2.2. Mechanical Characteristics Analysis

Based on the results of Figure 12, the bearing capacities of N1, CC1, CC2, and CC3 were 617.2 kN, 886.9 kN, 1030.6 kN, and 1165.3 kN, respectively. Compared with the N1 specimen, the increasing ratio of the bearing capacities of CC1, CC2, and CC3 was calculated as 43.7%, 66.8%, and 88.6%, respectively. Those of CB1, CB2, and CB3 were 26.9%, 42.3%, and 51.1% compared to the N1 specimen. The strength-yield ratio is defined as the ratio of the ultimate bearing strength to the yield strength of the tested specimens, as Equation (8) shows.

$$R_{sy} = \frac{f_y}{f_c}, \quad (8)$$

where R_{sy} is the strength-yield ratio, and f_y and f_c are the yield strength and ultimate strength of the columns, respectively.

The strength-yield ratios of all nine specimens were calculated as 1.17 (N1), 1.17 (N2), 1.17 (N3), 1.11 (CC1), 1.23 (CC2), 1.20 (CC3), 1.13 (CB1), 1.17 (CB2), and 1.19 (CB3), as presented in Figure 13a and Table 2. The strength-yield ratios of all nine specimens ranged between 1.11 and 1.20, indicating that the wrapped FRP sheets have a negligible effect on the strength-yield ratio of the composite specimens. Further study was conducted to analyze the influence of the FRP layer on the compressive stiffness of the composite column. This was performed by investigating the initial stiffness, which is defined as the linear stiffness of the specimen during the initial linear elastic compression process (i.e., section a~b in the load–displacement curve). The calculated outcomes are presented in Figure 13b and Table 2. The results indicate that the initial stiffness of the composite specimens ranges between 100 kN/mm and 180 kN/mm, reaching a maximum of 178.2 kN/mm for CC2 and a minimum of 106.9 kN/mm for CB1. The average initial stiffness for the N1, N2, and N3 specimens was recorded at 91.2 kN/mm. These findings show that the outer FRP sheets can notably enhance the initial stiffness of the CFST specimens. Notably, the initial stiffness values for the CFRP-reinforced specimens are significantly higher than those of the BFRP-reinforced specimens. This supports the inference that the higher elastic modulus of the CFRP sheets is more effective in improving the compressive stiffness of the column specimens. Table 3 illustrates a summary of the characteristic parameters for all the tested specimens.

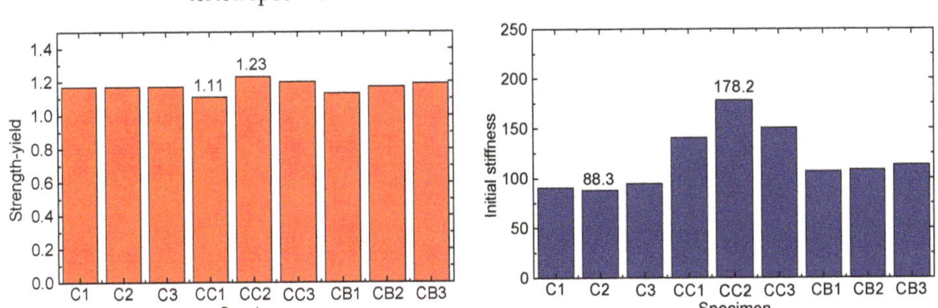

(a) Strength-yield ratios (b) Initial stiffness

Figure 13. Strength-yield ratios and initial stiffnesses of the nine columns.

Table 3. Tested results of the columns.

Specimen No.	Extreme Load/kN	Yield Load/kN	Strength-Yield, μ	Initial Stiffness/(kN/mm)
N1	617.2	528.8	1.17	90.9
N2	617.8	529.3	1.17	88.3
N3	617.1	528.2	1.17	94.8
CC1	886.9	800.1	1.11	140.6
CC2	1030.6	839.6	1.23	178.2
CC3	1165.3	974.7	1.20	150.6
CB1	783.9	691.8	1.13	106.9
CB2	879.6	748.9	1.17	108.6
CB3	934.4	788.3	1.19	113.4

3.3. Confinement Mechanism of Composite Columns

The aforementioned loading test results of composite columns demonstrate that the outer coating of FRP will change the load-bearing failure mechanism of the short column and enhance its bearing capacity. The change in this load-bearing mechanism is embodied in the confinement mechanism of the FRP layer, which is discussed in this section, as shown in Figure 14 (red curve). The confinement mechanism being proposed can be classified into four distinct stages.

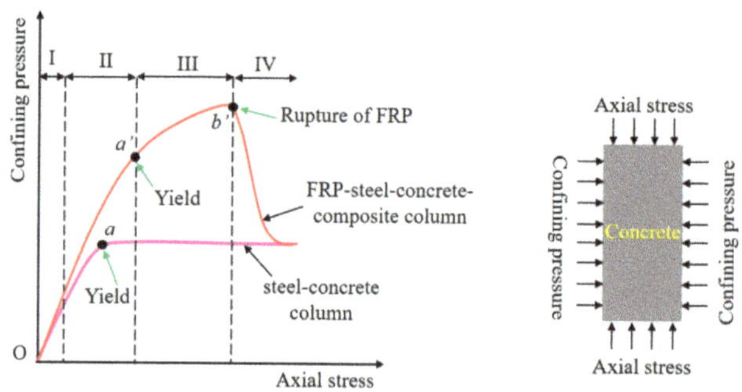

Figure 14. Confinement mechanism of the composite column and reference column.

(1) Stage I: steel dominating confinement

In the initial loading stage, the axial compression effect is minimal, and the lateral deformation of the inner concrete results primarily from Poisson's effect and the emergence of microcracks. As activation is insufficient at this stage, only a slight rise in total confining pressure is observed. The bulk of the confinement occurs as a result of the steel pipe, with the FRP being inconsequential due to inadequate activation. Throughout this stage, the axial stress–strain reaction of the internal concrete registers on the first ascending branch.

(2) Stage II: FRP-steel multiple confinements

With an increase in axial strain, gradual radial dilation is observed in the internal concrete, accompanied by the initiation and propagation of minor cracks. Consequently, the steel and FRP confinement forces become fully activated, inducing a much faster acceleration in their increase rate compared to that observed earlier. This surge in forces leads to the swift development of total confining pressure. To accurately determine the boundary of this phase, this deviation must be captured. Additionally, the steel tube exhibits yielding at the end of the second stage, which can be defined as a critical boundary line. The duration of this stage hinges on the yield strength and yield strain of the steel tube, with higher values of these parameters extending the duration of the second stage proportionately.

(3) Stage III: FRP dominating confinement.

As the bearing pressure continues to increase, the FRP confining pressure increases gradually owing to the progressive radial expansion of interstitial concrete, which are accompanied by the appearance of numerous macroscopic cracks. Conversely, the steel confining pressure stays mostly stable after yielding. As a consequence, the total confining pressure increases at a slower pace. 'Stage III' culminates when the internal concrete reaches its ultimate axial strain (ε_u), leading to the rupture of the FRP hoop (represented by point b' in Figure 13). The axial stress–strain response of the internal concrete during this stage aligns roughly with the second ascending branch. Hence, the mechanical characteristics of the FRP layers play a critical role in the confinement effect during 'Stages II and III,' with stronger FRP layers being most effective in enhancing the ultimate load-bearing capacity of the specimen.

(4) Stage IV: steel dominating confinement

Following the third stage, the FRP hoop layers are pulled off while the steel pipe remains in a plastic stress state following yielding. During the fourth stage, the confinement pressure is primarily borne by the steel pipe, resulting in the external confinement force being returned to the confining effect of the steel pipe in the concrete inner core. The confinement mechanism during this stage was similar to that observed in a pure steel tube concrete column (i.e., the reference column). As the compression continued, the final failure of the specimen is reflected in the crushing of the concrete core.

4. Conclusions

Drawing on experimental research, this study investigated the bearing mechanism and damage process of short columns, including pure steel tube concrete columns and FRP-wrapped steel tube concrete composite columns. The failure process, failure mode, load–displacement relationship, strength-yield ratio, and initial stiffness were discussed. The following conclusions can be drawn:

(1) Two failure modes, shear failure and waist drum, were observed during testing. For FRP-wrapped steel tube concrete composite columns, the failure mode may trend toward waist drum from shear failure as more FRP layers are used. Due to the strong external combining constraints of the FRP and steel tube, the inner core concrete column is always in a state of obvious triaxial compression. Consequently, the failure of the concrete column is more inclined to material destruction, resulting in the waist drum failure mode.

(2) When compared with the pure steel tube concrete column, the CFRP- and BFRP-wrapped columns exhibit larger peak loads, indicating that the effective use of FRP wraps can significantly increase the compressive strength of steel-concrete columns. Furthermore, the number of FRP layers has a direct impact on the level of compressive strength attained, with more layers resulting in a greater increase in the compressive strength.

(3) The incorporation of wrapped FRP sheets has no direct effect on the strength-yield ratio of the composite specimens. However, the initial stiffnesses (the ratio of the load to compressive displacement during the initial stage) of CFRP-reinforced specimens are significantly greater than those reinforced with BFRP. This discrepancy indicates that the higher elastic modulus and greater tensile capacity of CFRP sheets are more effective at improving the compressive stiffness of columns.

(4) This research proposed a four-stage confinement mechanism for FRP-wrapped steel tube concrete composite columns, comprising the following stages: Stage I—steel dominating confinement, Stage II—multiple confinements by both FRP and steel, Stage III—FRP dominating confinement, and Stage IV—steel dominating confinement. The confinement effect during stages II and III is primarily influenced by the mechanical properties of the FRP layer, with higher-strength FRP layers being the most effective at improving the ultimate bearing strength of the specimens.

Author Contributions: Data curation, J.L. and D.M.; formal analysis, J.L.; conceptualization, J.L.; methodology, J.L. and D.M.; software, J.L. and D.M.; validation, J.L. and Z.L.; funding acquisition, J.L. and Z.L.; writing-original draft preparation, J.L. and F.D.; writing-review and editing, Z.L. All authors have read and agreed to the published version of the manuscript.

Funding: This research was funded by the National Natural Science Foundation of China (Grant Nos: 52108151 and 52108272).

Institutional Review Board Statement: Not applicable.

Informed Consent Statement: Not applicable.

Data Availability Statement: Data will be made available on request.

Acknowledgments: The authors gratefully acknowledge the financial support provided by the National Natural Science Foundation of China (Grant Nos: 52108151 and 52108272).

Conflicts of Interest: The authors declare no conflict of interest.

References

1. Lai, M.H.; Ho, J.C.M. A theoretical axial stress-strain model for circular concrete-filled-steel-tube columns. *Eng. Struct.* **2016**, *125*, 124–143. [CrossRef]
2. Dong, C.X.; Kwan, A.K.H.; Ho, J.C.M. Effects of external confinement on structural performance of concrete-filled steel tubes. *J. Constr. Steel Res.* **2017**, *132*, 72–82. [CrossRef]
3. Alatshan, F.; Osman, S.A.; Hamid, R.; Mashiri, F. Stiffened concrete-filled steel tubes: A systematic review. *Thin-Walled Struct.* **2020**, *148*, 106590. [CrossRef]
4. Lai, M.H.; Song, W.; Ou, X.L.; Chen, M.T.; Wang, Q.; Ho, J.C.M. A path dependent stress-strain model for concrete-filled-steel-tube column. *Eng. Struct.* **2020**, *211*, 110312. [CrossRef]
5. Reda, M.A.; Ebid, A.M.; Ibrahim, S.M.; El-Aghoury, M.A. Strength of composite columns consists of welded double CF sigma-sections filled with concrete—An experimental study. *Designs* **2022**, *6*, 82. [CrossRef]
6. Zhang, Y.R.; Wei, Y.; Bai, J.W.; Wu, G.; Dong, Z. A novel seawater and sea sand concrete filled FRP-carbon steel composite tube column: Concept and behaviour. *Compos. Struct.* **2020**, *246*, 112421. [CrossRef]
7. Wang, Y.L.; Chen, G.P.; Wan, B.L.; Han, B.; Ran, J. Axial compressive behavior and confinement mechanism of circular FRP-steel tubed concrete stub columns. *Compos. Struct.* **2021**, *256*, 113082. [CrossRef]
8. Wei, Y.; Zhu, C.; Miao, K.T.; Zheng, K.; Tang, Y. Compressive performance of concrete-filled steel tube columns with in-built seawater and sea sand concrete-filled FRP tubes. *Constr. Build. Mater.* **2022**, *317*, 125933. [CrossRef]
9. Guo, F.; Al-Saadi, S.; Raman, R.K.S.; Zhao, X.L. Durability of fiber reinforced polymer (FRP) in simulated seawater sea sand concrete (SWSSC) environment. *Corros. Sci.* **2018**, *141*, 1–13. [CrossRef]
10. Miao, K.T.; Wei, Y.; Dong, F.H.; Zheng, K.; Wang, J. Experimental study on concrete-filled steel tube columns with inner distributed seawater and sea sand concrete-filled fiber-reinforced polymer tubes under axial compression. *Compos. Struct.* **2023**, *320*, 117181. [CrossRef]
11. Cao, S.; Wu, C.Q.; Wang, W.Q. Behavior of FRP confined UHPFRC-filled steel tube columns under axial compressive loading. *J. Build. Eng.* **2020**, *32*, 101511. [CrossRef]
12. Liang, J.F.; Lin, S.Q.; Li, W.; Liu, D. Axial compressive behavior of recycled aggregate concrete-filled square steel tube stub columns strengthened by CFRP. *Structures* **2021**, *29*, 1874–1881. [CrossRef]
13. Zeng, J.J.; Liang, S.D.; Li, Y.L.; Guo, Y.C.; Shan, G.Y. Compressive behavior of FRP-confined elliptical concrete-filled high-strength steel tube columns. *Compos. Struct.* **2021**, *266*, 113808. [CrossRef]
14. Eilbeigi, S.; Tavakkolizadeh, M.; Masoodi, A.R. Nonlinear Regression Prediction of Mechanical Properties for SMA-Confined Concrete Cylindrical Specimens. *Buildings* **2022**, *13*, 112. [CrossRef]
15. Eilbeigi, S.; Tavakkolizadeh, M.; Masoodi, A.R. Enhancing Mechanical Behavior and Energy Dissipation in Fiber-Reinforced Polymers through Shape Memory Alloy Integration: A Numerical Study on SMA-FRP Composites under Cyclic Tensile Loading. *Materials* **2023**, *16*, 5695. [CrossRef]
16. Tavakol, M.; HajiKazemi, H.; Rajabzadeh-Safaei, N.; Masoodi, A.R. Influence of shape memory alloy on seismic behaviour of hollow-section concrete columns. *Proc. Inst. Civ. Eng. Struct. Build.* **2021**, 1–18. [CrossRef]
17. Wei, Y.; Bai, J.W.; Zhang, Y.R.; Miao, K.; Zheng, K. Compressive performance of high-strength seawater and sea sand concrete-filled circular FRP-steel composite tube columns. *Eng. Struct.* **2021**, *240*, 112357. [CrossRef]
18. Tang, H.Y.; Chen, J.L.; Fan, L.Y.; Sun, X.; Peng, C. Experimental investigation of FRP-confined concrete-filled stainless steel tube stub columns under axial compression. *Thin-Walled Struct.* **2020**, *146*, 106483. [CrossRef]
19. Li, Y.L.; Zhao, X.L.; Raman, R.K.S. Durability of seawater and sea sand concrete and seawater and sea sand concrete-filled fibre-reinforced polymer/stainless steel tubular stub columns. *Adv. Struct. Eng.* **2021**, *24*, 1074–1089. [CrossRef]
20. Aziz, Y.H.A.; Zaher, Y.A.; Wahab, M.A.; Khalaf, M. Predicting temperature rise in Jacketed concrete beams subjected to elevated temperatures. *Constr. Build. Mater.* **2019**, *227*, 116460. [CrossRef]

21. Tao, Z.; Han, L.H.; Zhuang, J.P. Axial loading behavior of CFRP strengthened concrete-filled steel tubular stub columns. *Adv. Struct. Eng.* **2007**, *10*, 37–46. [CrossRef]
22. Liu, L.; Lu, Y.Y. Axial Bearing Capacity of Short FRP Confined Concrete-filled Steel Tubular Columns. *J. Wuhan Univ. Technol. Mater. Sci. Ed.* **2010**, *25*, 454–458. [CrossRef]
23. Hu, Y.M.; Yu, T.; Teng, J.G. FRP-Confined Circular Concrete-Filled Thin Steel Tubes under Axial Compression. *J. Compos. Constr.* **2011**, *15*, 850–860. [CrossRef]
24. Na, L.; Yiyan, L.; Shan, L.; Lan, L. Slenderness effects on concrete-filled steel tube columns confined with CFRP. *J. Constr. Steel Res.* **2018**, *143*, 110–118. [CrossRef]
25. Zhang, Y.R.; Wei, Y.; Bai, J.W.; Zhang, Y. Stress-strain model of an FRP-confined concrete filled steel tube under axial compression. *Thin-Walled Struct.* **2019**, *142*, 149–159. [CrossRef]
26. Onyelowe, K.C.; Jayabalan, J.; Ebid, A.M.; Samui, P.; Singh, R.P.; Soleymani, A.; Jahangir, H. Evaluation of the Compressive Strength of CFRP-Wrapped Circular Concrete Columns Using Artificial Intelligence Techniques. *Designs* **2022**, *6*, 112. [CrossRef]
27. Onyelowe, K.C.; Ebid, A.M.; Mahdi, H.A.; Soleymani, A.; Jayabalan, J.; Jahangir, H.; Samui, P.; Singh, R.P. Modeling the confined compressive strength of CFRP-jacketed noncircular concrete columns using artificial intelligence techniques. *Cogent Eng.* **2022**, *9*, 2122156. [CrossRef]
28. Han, L.H.; Yao, G.H.; Zhao, X.L. Behavior and calculation on concrete-filled steel CHS (Circular Hollow Section) beam-columns. *Steel Compos. Struct. Int. J.* **2004**, *4*, 169–188. [CrossRef]

materials

MDPI

Article

Investigation of Self-Healing Performance of Asphalt Mastic—From the Perspective of Secondary Aging

Bo Li [1,2], Yu Wang [1], Peng Xiao [1,2,*], Aihong Kang [1,2], Yao Zhang [1,2] and Zhengguang Wu [1,2]

1 College of Civil Science and Engineering, Yangzhou University, Yangzhou 225127, China; libo@yzu.edu.cn (B.L.); yuwang4012@126.com (Y.W.); ahkang@yzu.edu.cn (A.K.); yaozhang@yzu.edu.cn (Y.Z.); zgwu@yzu.edu.cn (Z.W.)

2 Research Center for Basalt Fiber Composite Construction Materials, Yangzhou 225127, China

* Correspondence: pengxiao@yzu.edu.cn; Tel.: +86-0514-8797-9408

Abstract: Reclaimed asphalt pavement (RAP) has been widely utilized because it is an environmentally friendly and economical material. The performance of recycled asphalt mixtures will deteriorate gradually with the secondary aging process of asphalt, including the self-healing property. To further understand the self-healing characteristics of asphalt after secondary aging, taking 70# petroleum asphalt, SBS-modified asphalt, and extracted old asphalt mastics as objects, the fatigue self-healing test and fracture self-healing test were conducted to simulate the intermediate-and low-temperature healing behaviors of different asphalt mastics. The impact of healing time, healing temperature, and aging degree of mastics on the healing performance was systematically investigated. The results show that the original unaged asphalt mastics present excellent fatigue healing properties with an index of 0.796 and 0.888 for 70# petroleum and SBS-modified asphalt mastics, respectively. The secondary aging process causes significant impact on the healing properties, leading to a great drop in the corresponding index, which decreased to 47.5% and 72.5% of that of the unaged ones. The fracture healing ability of all mastics was much inferior to the fatigue healing. After secondary aging, the fracture healing index values of 70# petroleum asphalt, SBS-modified asphalt, and extracted old asphalt mastics were all as low as around 0.3, indicating similar performance can be found in the secondary aged SBS-modified asphalt mastics and 70# asphalt mastics. Overall, after secondary aging, the fatigue damage of SBS-modified asphalt mastics can be cured effectively by self-healing, but the fatigue and fracture self-healing properties of 70# asphalt mastics are difficult to recover. These results could provide an innovative view to understand the fatigue and fracture healing characteristics of recycled asphalt pavement after secondary aging.

Keywords: asphalt mastic; recycling; fatigue self-healing; fracture self-healing; secondary aging

Citation: Li, B.; Wang, Y.; Xiao, P.; Kang, A.; Zhang, Y.; Wu, Z. Investigation of Self-Healing Performance of Asphalt Mastic—From the Perspective of Secondary Aging. *Materials* **2023**, *16*, 7567. https://doi.org/10.3390/ma16247567

Academic Editor: Giovanni Polacco

Received: 4 November 2023
Revised: 29 November 2023
Accepted: 6 December 2023
Published: 8 December 2023

1. Introduction

Reclaimed asphalt pavement (RAP) has been gradually widely used because of its advantages of saving resources and reducing pollution. However, recycled asphalt mixtures containing a high amount of RAP normally present deteriorated performance, especially in the fatigue cracking and low temperature cracking resistance [1,2].

The self-healing performance of recycled asphalt is an important aspect to address due to its comprehensive performance. Self-healing refers to the ability of a material to repair itself when damaged by the environment or external force. As an elastic–plastic material, asphalt has a certain self-healing characteristic. During the rest period without vehicle load, asphalt will present obvious self-healing behavior and repair the micro cracks inside the pavement, so as to restore the pavement performance and prolong the service life [3]. Therefore, the study of the self-healing performance of recycled asphalt for the mixture design is of great significance. Since asphalt is a viscoelastic material with complex compositions, there is no unified theory for the study of asphalt self-healing mechanism.

At present, the mainstream self-healing mechanisms include molecular diffusion energy, surface energy theory, capillary flow theory [4–8], etc.

Based on these theories, many evaluation methods and indicators have been derived. Through fatigue-healing-fatigue tests, three-point bending tests, and linear amplitude sweep test, cumulative dissipation energy and (or) tensile fracture stress-based healing indexes were proposed [9,10]. In addition, many scholars have also studied the influencing factors of asphalt self-healing performance, mainly including healing time, healing temperature, asphalt type, and damage degree, etc. Through the above experiments, a general conclusion is drawn that after aging, the lightweight components decrease and the asphaltene and resin contents increase, making the aged asphalt harder and leading to deterioration in flowability and self-healing performance.

(1) Healing time

Some scholars introduced healing time into the study of asphalt fatigue characteristics. Norambuena-Contreras et al. [11] found that there is an optimal healing time for the self-healing of asphalt. When the healing time is too short, the cracks cannot be completely repaired. When the healing time is too long, it leads to the aging of asphalt and reduces its mechanical properties. Zhang et al. [12] have found that the fatigue resistance of asphalt mixtures is significantly improved by introducing a certain healing time after loading. Therefore, the growth in fatigue life can be used to evaluate its healing performance.

(2) Healing temperature

The healing temperature is of great significance to the self-healing performance of asphalt. Sun et al. [13] used the molecular dynamics model to explain the effect of temperature on the self-healing ability of asphalt. The optimal healing temperature range of asphalt is determined by differential scanning calorimetry (DSC), which is 40.3–48.7 °C. Grossegger et al. [14] used different healing temperatures to study the capillary flow of asphalt. The results show that the surface energy and contact angle of asphalt decrease with the increase of temperature, resulting in faster capillary flow of asphalt. Combined with the additional pressure caused by the increase in temperature, the self-healing performance of asphalt is greatly enhanced. Fan et al. [15] evaluated the thermal sensitivity of asphalt mixtures by microwave heating and analyzed the self-healing performance before and after microwave heating by semi-circular bending fracture healing test, the experimental results indicate that the self-healing performance of asphalt mixtures is directly proportional to temperature, but the healing rate gradually slows down.

(3) Asphalt type

Scholars hold different views on the influence of modifiers on the self-healing performance of asphalt. Ding et al. [16] found through their research that the bonding performance of rubber modified asphalt is significantly improved, which can effectively prevent the occurrence of cracks and repair fine cracks better than unmodified asphalt. Xu et al. [17] studied the influence of different modifiers on the self-healing performance of asphalt. The results show that no matter what modifier is used, the increase of modifier content will reduce the self-healing ability of modified asphalt, which proves that petroleum asphalt possesses superior self-healing ability. Zhou et al. [18] selected five representative modified asphalts and petroleum asphalts to compare their self-healing properties. The experimental results show that all modifiers have a negative effect on the self-healing capability of asphalt in the early stage, but with the increase of healing time, self-healing ability becomes comparable for the modified and petroleum asphalt. Fakhri et al. [19] developed a nano-silica modified asphalt, which shows better self-healing performance than neat asphalt when the amount of modifier is less than 8%. Teknologi and Arshad et al. [20,21] experimentally investigated the performance of nano-modified asphalt, and the results indicate that nano modification can be one of the effective ways to improve the self-healing and road performance of asphalt.

(4) Damage Degree

Damage degree is another important impact factor on the self-healing performance of asphalt. Li et al. [22] used DSR fatigue test to simulate different damage degrees of asphalt. The results show that the healing index increases with the decrease of damage degree, which also can be simulated by Ramberg–Osgood model. Lv et al. [23] developed a new calculation formula based on the calculation of conventional self-healing index. Based on the rest-damage superposition principle (RDSP), in which the linear amplitude sweep frequency test (LAS) was used to quantify the self-healing performance of asphalt binders with different degrees of damage and rest periods, the ratio of loading times required to reach the same modulus before and after healing was calculated. Finally, a self-healing performance prediction model was established, with the fitting degree of over 0.9. Zhao et al. [24] explored the multiple damage healing laws of laboratory-simulated recycled asphalt by dynamic shear rheometer (DSR), fourier transform infrared (FTIR), thermogravimetric analysis (TGA), and gel permeation chromatograph (GPC). The results show that the cycles of damage-healing times had a greater impact on the healing performance than the healing time. Wang et al. [25] used different loading times to simulate different damage degree, and the test results show that if the asphalt is in minor damage state, the short-term self-healing performance will be better than the long-term performance. However, when the asphalt is in serious damage state (such as enduring low temperature fracture), the asphalt almost lost the healing performance.

(5) Aging degree

With the deepening aging of asphalt, the content of macromolecular compounds such as asphaltene and resin increases, which hinders the flow and diffusion between asphalt molecules and thus affects the self-healing performance of asphalt. Xiang et al. [26] investigated the internal structure of asphalt before and after aging by different microstructure tests. The final test data show that the self-healing performance of asphalt decreases with the increase in aging times, and the presence of rejuvenating agent can restore partly the self-healing performance of aged asphalt. Sun et al. [27] explored the chemical composition, colloidal stability, and macroscopic phase transition of asphalt aged in a laboratory. The results show that the self-healing sensitivity of asphalt to temperature decreases with the deterioration of aging degree.

Asphalt mastic is an essential component in asphalt mixture, of which the self-healing property directly affects the recovery ability of pavement performance. In particular, recycled asphalt pavement undergoes secondary aging during service, and the self-healing properties of asphalt mastic subsequently change significantly. However, there is a lack of understanding of the self-healing performance of recycled asphalt mastic after and before aging process.

In this paper, various types of asphalt mastics under different aging states were fabricated to study the changing laws in self-healing properties by fatigue-healing-fatigue test (intermediate temperature cracking) and fracture-healing-fracture test (low temperature cracking). The main research goals include (1) exploring the self-healing properties of different types of asphalt after secondary aging process; (2) after secondary aging, observing whether the asphalt still has the potential to be healed; (3) recording the self-healing properties of recycled asphalt under different conditions. The research results of this paper can enrich the findings about the change laws in the self-healing performance of different types of asphalt mastics after secondary aging and promote the understanding of multiple recycling applications of RAP.

2. Raw Materials and Methodologies

2.1. Materials

2.1.1. Raw Materials

The new asphalts used in this study mainly include petroleum asphalt with the penetration grade of 70# and SBS-modified asphalt with the PG grade of 76-22. The performance indicators of the asphalts are shown in Tables 1 and 2.

Table 1. Performance indicators and test results of 70# asphalt.

Index	Requirements in JTG-F40 [28]	Results	Test Method in JTG-E20 [29]
Penetration value (25 °C, 0.1 mm)	60~80	68	T0604
Softening point (°C)	≥45	49	T0606
Ductility (5 cm/min, 15 °C, cm)	≥100	165.1	T0605
Flashpoint (°C)	≥260	321	T0661
Viscosity at 60 °C (Pa·s)	≥180	213	T0604

Table 2. Performance index and test results of SBS-modified asphalt.

Index	Requirements in JTG-F40 [28]	Results	Test Method in JTG-E20 [29]
Penetration value (25 °C, 0.1 mm)	30~60	56	T0604
Softening point (°C)	≥60	82	T0606
Ductility (5 cm/min, 5 °C, cm)	≥20	38	T0605
Separation (°C)	≤2.5	1.4	T0661
Elastic recovery (25 °C, %)	≥75	76	T0662

2.1.2. Extracted Old Asphalt from RAP

The RAP materials used in this study originated from the Suzhou section of the Nanjing-Shanghai expressway, which suffered 6 years of service life. According to the standard methods in JTG E20-2011 (T0722), the old asphalt was extracted from RAP, and the corresponding properties were tested and summarized in Table 3. It can be seen from Table 3 that the penetration of the extracted old asphalt still reaches 34 (0.1 mm), which meets the requirements for hot in-place recycling according to the JTG/T 5521-2019 [30].

Table 3. Performance index and test results of extracted old asphalt.

Index	New SBS-Modified Asphalt	Extracted Old Asphalt	Test Method in JTG-E20 [29]
Penetration value (25 °C, 0.1 mm)	56	34	T0604
Softening point (°C)	82	66	T0606
Ductility (5 cm/min, 5 °C, cm)	38	8.3	T0605
Viscosity at 135 °C (Pa·s)	2.35	3.46	T0613

2.1.3. New Fillers

Limestone mineral powder was selected as the filler and the specific technical indicators are shown in Table 4.

Table 4. Performance indexes and test results of mineral powder.

Index		Requirements in JTG-E42 [31]	Results	Test Method in JTG-E42 [31]
Water content (%)		≤1.0	0.3	Oven dry
Hydrophilic coefficient		<1.0	0.63	T0353
Plasticity index (%)		<4.0	2.5	T0354
Passing (%)	<0.6 mm	100	100	T0351
	<0.15 mm	92.6	90–100	
	<0.075 mm	92.2	75–100	

2.1.4. Rejuvenating Agent

The rejuvenating agent used in this study is Runqiang-RA102 high-performance asphalt rejuvenator produced by Jiangsu Subote Company (Nanjing, China). The specific technical indicators are shown in Table 5.

Table 5. Performance indexes of rejuvenating agent.

Index	Requirements in JTG/T 5521 [30]	RA-102	Test Method in JTG-E20 [29]
Viscosity at 90 °C (cP)	-	4000	T0619
Flashpoint (°C)	≥220	248	T0633
Saturated hydrocarbons content (%)	≤30	25.6	T0618
Aromatic content (%)	≥30	53	T0618
Mass loss after RTFOT (%)	≤4%	1.02	T0603

2.2. Preparation of Asphalt Mastic Samples

2.2.1. Asphalt Aging Process and Asphalt Mastic Fabrication

The asphalt mastic samples were prepared according to the following steps. The overall mastic preparation scheme is shown in Figure 1.

(1) Original mastic samples were prepared by adding mineral powders into the new 70# asphalt and SBS-modified asphalt.

(2) Primary aged mastic samples were prepared by adding mineral powders into the aged asphalts and the extracted old asphalt from RAP. The aged 70# asphalt and SBS-modified asphalts were artificially made by rolling thin-film oven test (RTFOT) and pressurized aging vessel (PAV) test.

Recycled mastic samples were prepared by adding the rejuvenating agent into the primary aged asphalts and extracted old asphalt, and then mixed with mineral powders.

Secondary aged mastics were prepared by adding mineral powders into the secondary aged asphalts by RTFOT short-term and PAV long-term aging.

Besides, the mixing procedures for the asphalt mastics were showing as the following steps. Firstly, asphalt and mineral powders were weighed and preheated. Then, the preheated asphalt was put on a heating furnace to maintain a temperature of 155–165 °C. Thirdly, mineral powders were added into the asphalt continuously, and the asphalt mastic was stirred at a speed of 2000 r/min for 20 min to make it homogeneous.

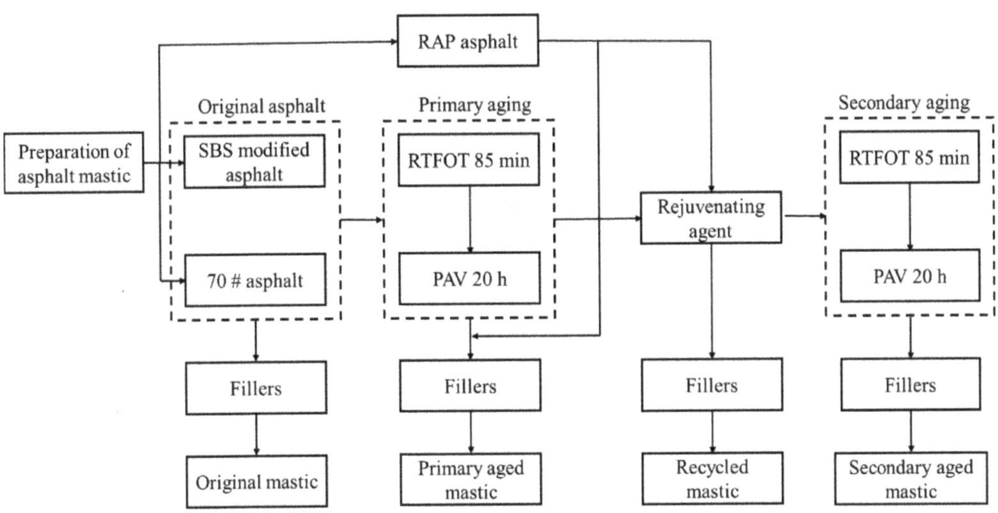

Figure 1. Preparation procedures of asphalt mastics.

2.2.2. Rejuvenating Agent Dosing

The recycled asphalt samples were prepared by mixing the aged asphalt with a rejuvenating agent. Four dosage levels of 2%, 3%, 4%, and 6% were set to conduct these tests, by weight of the aged asphalt. Then, the needle penetration, softening point, and viscosity were measured, and the results are listed in Tables 6–8. It can be observed from Tables 6–8 that the indexes of all three types of asphalt recovered to an acceptable level when 6% of rejuvenating agent was used. In order to avoid the negative impact caused by excessive rejuvenating agent, 6% of rejuvenating agent was selected for this study.

Table 6. Physical properties of recycled 70# asphalt.

Index	Original	Rejuvenating Agent Content (%)					Test Method in JTG-E20 [29]
		0	2	4	6	8	
Penetration value (25 °C, 0.1 mm)	68.2	30.7	43.8	52.4	61.2	68.1	T0604
Softening point (°C)	48.5	62.2	57.3	54.5	50.7	46.3	T0606
Viscosity at 135 °C (Pa·s)	0.54	1.01	0.94	0.83	0.66	0.46	T0613

Table 7. Physical properties of recycled SBS-modified asphalt.

Index	Original	Rejuvenating Agent Content (%)					Test Method in JTG-E20 [29]
		0	2	4	6	8	
Penetration value (25 °C, 0.1 mm)	58.0	31.0	38.4	45.2	51.4	57.8	T0604
Softening point (°C)	68.2	76.2	72.1	68.6	66.3	61.5	T0606
Viscosity at 135 °C (Pa·s)	2.89	4.75	4.33	3.51	2.88	2.54	T0613

Table 8. Physical properties of recycled RAP old asphalt.

Index	Rejuvenating Agent Content (%)					Test Method in JTG-E20 [29]
	0	2	4	6	8	
Penetration value (25 °C, 0.1 mm)	34.1	39.2	44.4	52.7	56.2	T0604
Softening point (°C)	66.3	63.7	61.6	59.1	54.0	T0606
Viscosity at 135 °C (Pa·s)	3.46	3.15	2.98	2.64	2.17	T0613

2.3. *Fatigue-Healing-Fatigue Test Method*

2.3.1. Experimental Parameters

The fatigue-healing-fatigue test was carried out by DSR time scanning. The diameter of the parallel plate of 8 mm and the spacing of 2 mm were set. The test temperature of 25 °C, loading frequency of 10 Hz, and 4% of the fatigue loading strain were applied. The healing temperatures for 70# asphalt were set to be 25 °C, 35 °C, and 45 °C, while the healing temperatures for SBS-modified asphalt and RAP extracted old asphalt were 25 °C, 45 °C, and 65 °C. The healing times for all the three types of mastics were 10 min, 20 min, 30 min, 40 min, 50 min, and 70 min. Three duplicates were used for each test condition. In addition, the timing when the selected complex modulus dropped to 50% of the initial value was chosen as the termination point to simulate 50% fatigue damage degree of the samples. The specific test parameters are summarized in Table 9.

Table 9. Basic parameters of fatigue-healing-fatigue test.

Asphalt Type	Fatigue Test Temperature/°C	Loading Frequency/Hz	Applied Strain/%	Damage Degree	Healing Temperature/°C	Healing Time/min
70#	25	10	4	50%	25, 35, 45	10,20,30,40,50,70
SBS, RAP					25, 45, 65	

2.3.2. Evaluating Indicator

In this study, the cumulative dissipated energy ratio of asphalt mastic before and after healing was used as the evaluation index of fatigue healing performance. Figure 2 is the schematic diagram of the test, and the calculation processes are shown in Equation (1) to Equation (3).

$$W_{\rm i} = \pi\sigma\varepsilon\sin(\delta) \tag{1}$$

$$W = \sum W_{\rm i} \tag{2}$$

$$HI_1 = Wa/Wb \tag{3}$$

where HI_1 is the healing index, %; W is dissipation energy, MJ/m^3; σ is the stress amplitude, MPa; ε is the applied strain, %; δ is the phase angle, °; W_b and W_a are the cumulative dissipated energy before and after the healing process of asphalt mastic, MJ/m^3.

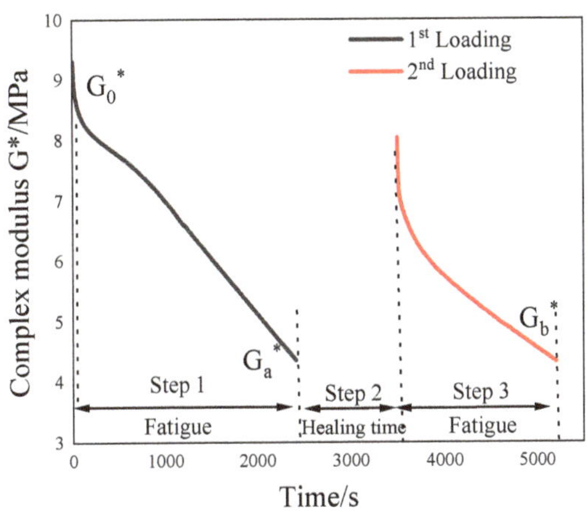

Figure 2. Schematic diagram of F-H-F test.

2.4. Fracture-Healing-Fracture Test

Based on direct tension test (DTT), the fracture-healing-fracture test was employed to simulate the brittle fracture process of asphalt materials under low temperature conditions and recovery process after a period of healing time at certain temperatures [32,33].

2.4.1. Specimen Preparation

The fracture-healing-fracture specimens were prepared according to the requirements of T0629 method in JTG E20, which is similar to the specimens for ductility test. In order to ensure that the specimens break at the middle position during the direct tensile process, a pre-cut seam with the width of 2 mm and depth of 1 mm was artificially made in the middle of the specimens, as shown in Figure 3.

2.4.2. Experimental Parameters and Evaluating Indicator

The test temperature was set to be −10 °C. The tensile rate was 100 mm/min, and the test was terminated after the specimen was broken. In order to ensure the full healing of asphalt mastic, the healing time was determined to be 4 h, 8 h, and 12 h, respectively. Three duplicates were used for each test condition. The healing temperatures were the same as those of fatigue-healing-fatigue test. The specific test parameters are shown in Table 10.

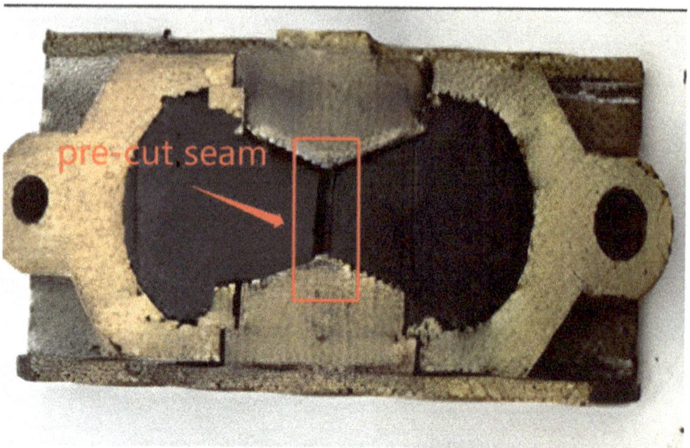

Figure 3. Illustration of an asphalt mastic specimen for fracture-healing-fracture test.

Table 10. Basic parameters of fracture-healing-fracture test.

Asphalt Type	Loading Rate/mm·min^{-1}	Healing Temperature/°C	Healing Time/h
70#	100	25, 35, 45	4,8,12
SBS, RAP		25, 45, 65	

According to the load-displacement curve, the peak stress σ_0 corresponding to the fracture of asphalt mastic can be used to evaluate its low temperature crack resistance. After the primary tensile fracture, the specimen was immediately placed in the module and kept warm at the targeted healing temperature. After a certain healing time, the tensile test was carried out again to obtain the peak stress σ_1 after healing. The ratio of the two stresses after and before the healing process was defined as the fracture healing index HI_2. The specific test processes are shown in Figure 4. The calculation formula is shown in Formula (4).

$$HI_2 = \sigma 1/\sigma 0 \tag{4}$$

In the formula: σ_0 is the peak value of the first tensile fracture stress, MPa; σ_1 is the peak value of the secondary tensile fracture stress, MPa.

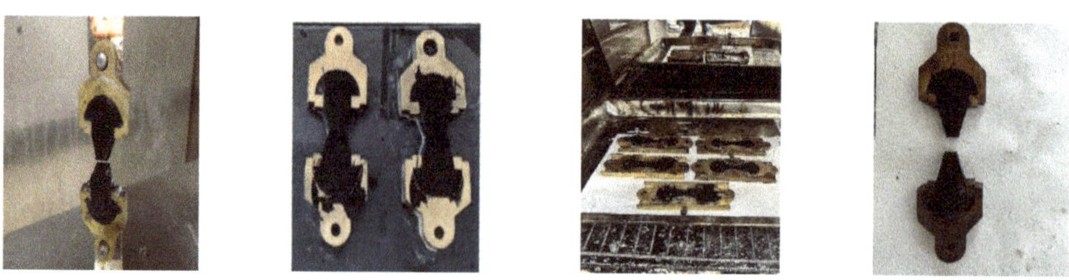

(a) (b) (c) (d)

Figure 4. Fracture-healing-fracture test procedures. (**a**) first tensile; (**b**) bonding; (**c**) healing; (**d**) secondary tensile.

3. Results and Discussions

3.1. Fatigue-Healing-Fatigue Test Results

The healing index values for fatigue-healing-fatigue tests of various asphalt mastics under different test conditions were calculated and recorded as HI_1. The specific results are illustrated in Figures 5–7.

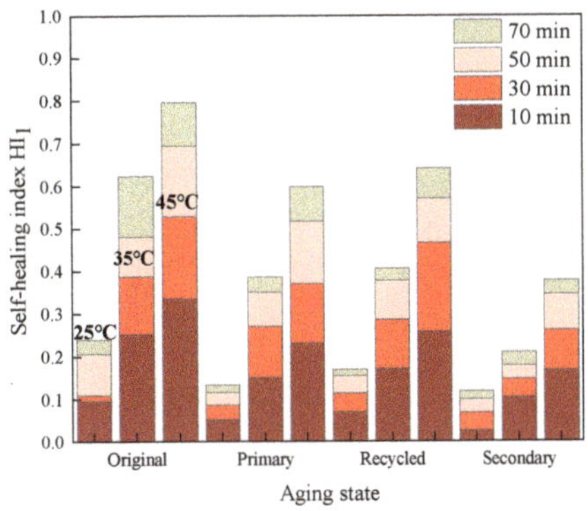

Figure 5. HI_1 of 70# asphalt mastics.

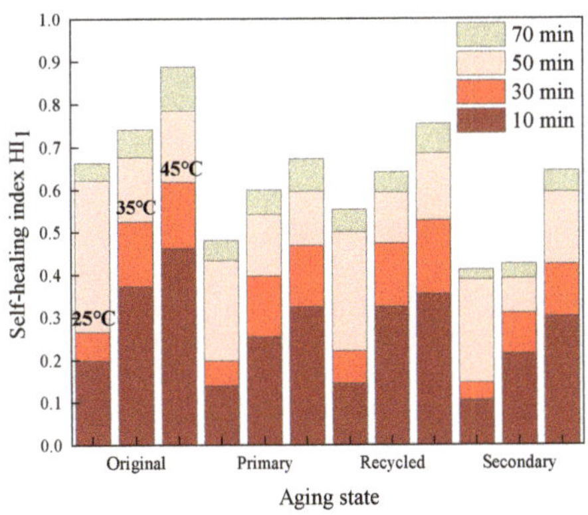

Figure 6. HI_1 of SBS-modified asphalt mastics.

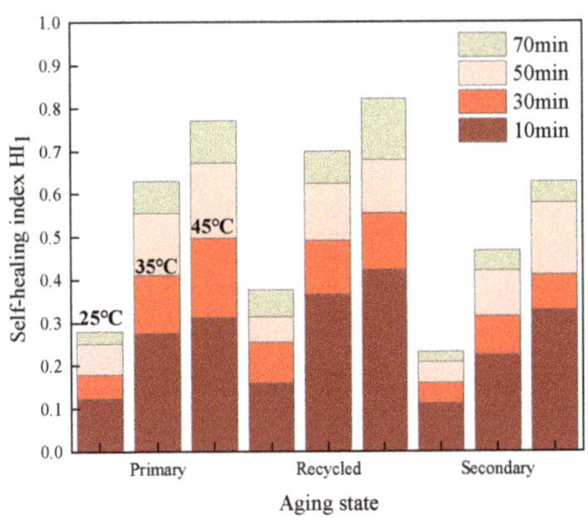

Figure 7. HI_1 of extracted old asphalt mastics.

Comparing the results of Figures 5–7, it becomes evident that within a certain range, a higher healing temperature, a longer healing time, and the addition of a rejuvenating agent can contribute to enhancing the fatigue healing properties of asphalt mastics. Furthermore, deepening the aging degree will lead to a deterioration in the healing performance to some extent.

3.1.1. Effect of Healing Time on Fatigue Self-Healing Properties

Taking the results from the test temperature of 45 °C as examples, the HI_1 results are shown in Figure 8.

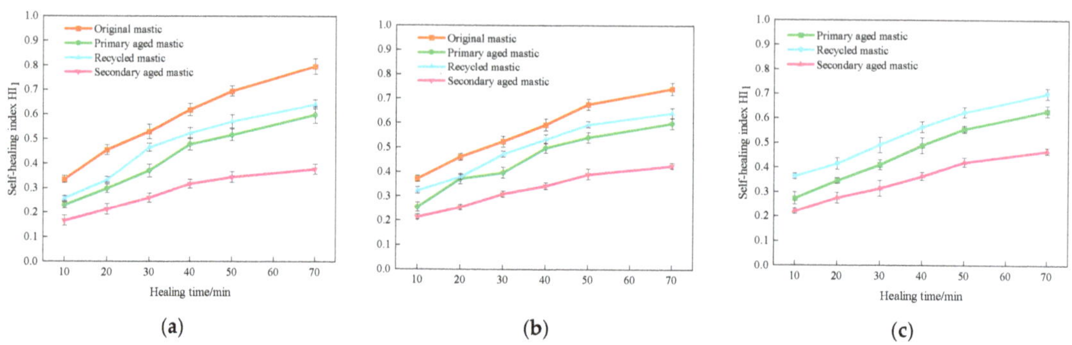

Figure 8. Results of self-healing index HI_1 of asphalt mastics at different healing times. (**a**) HI_1 of 70# asphalt mastics at 45 °C; (**b**) HI_1 of SBS-modified asphalt mastics at 45 °C; (**c**) HI_1 of extracted old asphalt mastics at 45 °C.

It can be seen from Figure 8 that the fatigue healing index HI_1 values of all kinds of asphalt mastics, despite the aging states, increase with an increasing healing time, indicating that the self-healing performance is gradually recovering. For the unaged original asphalt mastics, the fatigue healing index HI_1 of 70# asphalt mastic reaches 0.335, 0.529, 0.694, and 0.796, respectively, at the healing timing of 10 min, 30 min, 50 min, and 70 min. Meanwhile, the corresponding HI_1 value of SBS-modified asphalt mastic is 0.372, 0.524, 0.676, and 0.741, respectively. It infers that the unaged 70# and SBS-modified asphalt mastics present comparable healing ability after fatigue damage.

After primary aging, the HI_1 values of all kinds of asphalt mastics reduce significantly. The HI_1 values of primary aged 70# asphalt mastics decrease from 65.5% to 77.4% of the unaged original mastics, at healing timing from 10 min to 70 min. Meanwhile, the corresponding HI_1 values of primary aged SBS-modified asphalt mastics reduce from 68.5% to 84.1% of the original ones. It can be noticed that with healing time, the healing capability of primary aged mastics tends to approach the original ones. After adding rejuvenating agent, the HI_1 values of recycled mastics recover to some extent, but cannot reach the same level as the original mastics.

After secondary aging, the HI_1 values of all kinds of asphalt mastics further reduce. The HI_1 values of secondary aged 70# asphalt mastics are around 50% of the corresponding original unaged ones, while the HI_1 values of secondary aged SBS-modified asphalt mastics decreases to 65.3–75.7% of the original ones, at the healing timing of 10–70 min. In addition, the HI_1 values of extracted old asphalt mastics are close to those of SBS-modified asphalt mastics at the same test condition. Generally speaking, after secondary aging, 70# asphalt mastics show the worst healing capability, while SBS-modified asphalt and extracted old asphalt presenting the similar superior healing ability, indicating that 70# asphalt being more sensitive to secondary aging process [34].

3.1.2. Effect of Healing Temperature on Fatigue Self-Healing Properties

Taking the results from the healing time of 70 min as examples, the HI_1 results are illustrated in Figure 9.

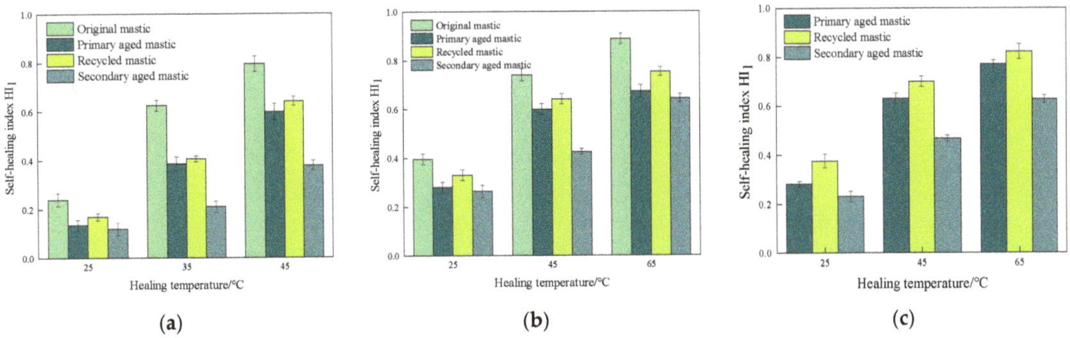

Figure 9. The results of HI_1 of asphalt mastics at different healing temperatures. (**a**) HI_1 of 70# asphalt mastic at 70 min; (**b**) HI_1 of SBS-modified asphalt mastic at 70 min; (**c**) HI_1 of extracted old asphalt mastics at 70 min.

It can be seen from Figure 9 that the healing index HI_1 is positively correlated with the temperature. As the healing temperature increases, the HI_1 values of the asphalt mastics increase to a great extent. In terms of original asphalt mastics, the HI_1 value of 70# asphalt mastic increases by 232% at 45 °C than that at 25 °C, while the HI_1value of SBS-modified asphalt increases by 87.1%. It infers that the self-healing characteristic of 70# asphalt mastics is more sensitive to temperatures than that of SBS-modified asphalt mastics [35].

After primary aging, the self-healing performance of all types of asphalt mastics deteriorates. At the healing temperatures of 25 °C, 35 °C, and 45 °C, the HI_1 values of primary aged 70# asphalt mastics decrease to 55.8%, 61.9%, and 75.1% of the original unaged ones, respectively. Meanwhile, at the healing temperatures of 25 °C, 45 °C, and 65 °C, the HI_1 values of primary aged SBS-modified asphalt mastics reduce to 71.0%, 80.8%, and 75.7% of the original ones, respectively. It reflects that the self-healing ability of primary aged asphalt mastics is approaching the original asphalt mastics with an increasing healing temperature but cannot reach the same level, indicating primary aging causes some damage to the fatigue healing characteristics of asphalt mastics [36].

After adding rejuvenating agent, the self-healing performance of recycled asphalt mastics was recovered to some certain extent. The fatigue healing index HI_1 values of recycled 70#, SBS, and old asphalt mastics increase to 107.2%, 107.0%, and 111.3% of that of corresponding primary aged mastics at 45 °C.

After secondary aging, the healing index HI_1values of all mastics still increase with an increasing healing temperature. However, some differences can be found in the HI_1 values of different mastics. At the healing temperatures of 25 °C, 35 °C, and 45 °C, the HI_1 values of secondary aged 70# asphalt mastics reduce to 69.8%, 51.7%, and 59.0% of the recycled mastics, respectively. Meanwhile, the HI_1 values of secondary aged SBS-modified asphalt mastics decrease to 79.8%, 66.5%, and 85.4% of the corresponding recycled ones, respectively, while the extracted old mastics presenting similar HI_1 values to secondary aged SBS asphalt mastics. This suggests that after secondary aging, the healing ability of 70# asphalt mastics can hardly recover to a satisfactory level compared to that of primary aged ones, even at a higher healing temperature. However, the healing ability of SBS-modified asphalt mastics still shows the comparable level to that of primary aged ones. These findings also can indicate that SBS-modified asphalt have better anti-aging capability than 70# asphalt, resulting in superior healing performance [35].

3.1.3. Effect of Aging State of Asphalt on Fatigue Self-Healing Properties

Taking the HI_1 results at the healing temperature of 45 °C and the healing time of 70 min as examples, the effect of aging state of asphalt on the of self-healing index HI_1 of 70# asphalt, SBS-modified asphalt, and extracted old asphalt mastics are shown in Figure 10.

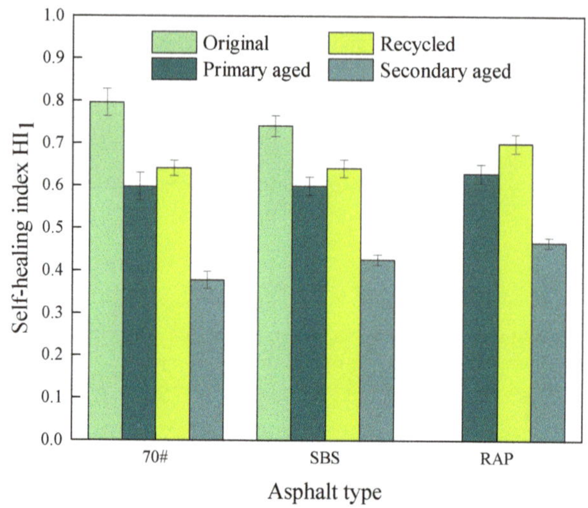

Figure 10. Results of HI_1 of asphalt mastics at different aging states.

It can be seen from Figure 10 that aging state has great impact on the healing properties of asphalt mastics. In terms of 70# asphalt mastics, the HI_1 values reduce to 75.1%, 80.5%, and 47.5% of the original unaged mastics after primary aging, recycling process, and secondary aging, respectively. As for SBS-modified asphalt mastics, the corresponding HI_1 values decrease to 80.8%, 86.5%, and 57.5%, respectively. This suggest that the deeper the aged degree achieves, the worse healing ability of the mastics will be, though the rejuvenating agent can recover the healing ability to some degree. In summary, as for the original asphalt mastics, the self-healing properties of 70# asphalt mastics are superior to those of SBS-modified asphalt mastic. After primary aging and secondary aging, the healing ability of SBS-modified asphalt mastics becomes better than that of 70# asphalt mastics. This is mainly because the SBS modifier absorbs some light oils of the matrix asphalt and swells to form a spatial three-dimensional network structure, which improves the viscoelastic properties of the asphalt after the aging process [37]. After a period of healing time, its complex modulus can restore to a large value, thus presenting good healing ability [38].

3.2. *Fracture-Healing-Fracture Test Results*

The fracture healing index HI_2 values of various asphalt mastics under different test conditions were calculated, and are illustrated in Figures 11–13.

It can be seen from the experimental results that the change laws of fracture healing index are similar to that of fatigue healing ones. The healing time, healing temperature, and rejuvenating agent are positively correlated with healing index HI_2, with aging degree also leading to a negative correlation with HI_2.

3.2.1. Effect of Healing Time on Fracture Self-Healing Properties

The results of the test at temperature of 45 °C are used and illustrated in Figure 14.

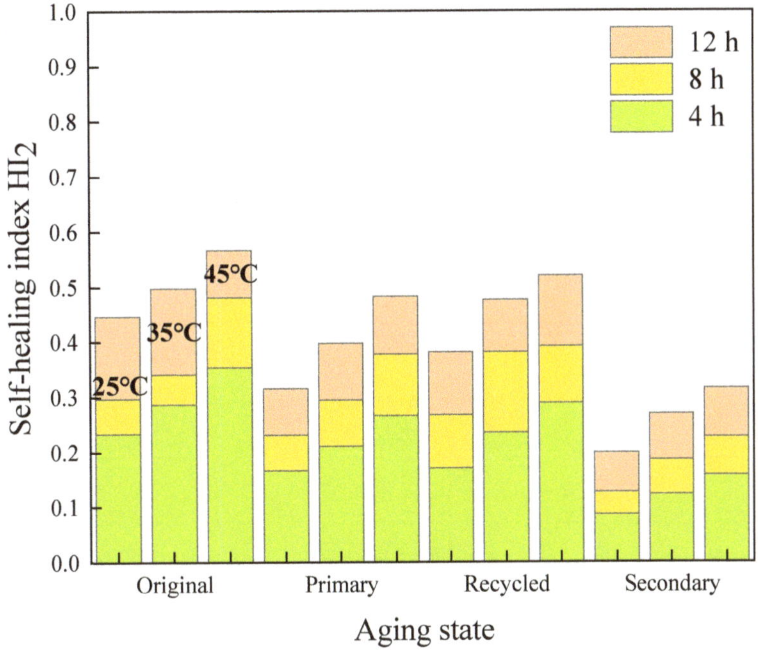

Figure 11. HI_2 of 70# asphalt mastics.

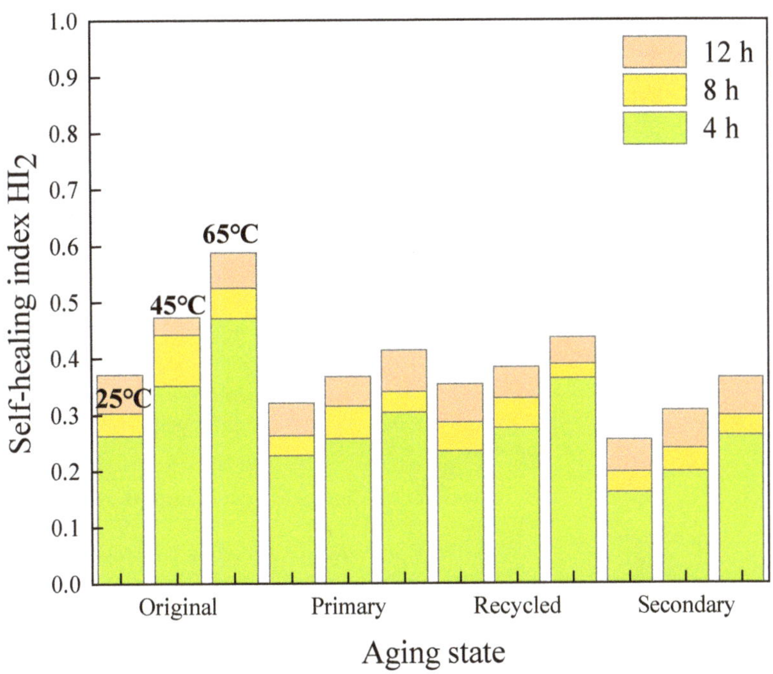

Figure 12. HI_2 of SBS-modified asphalt mastics.

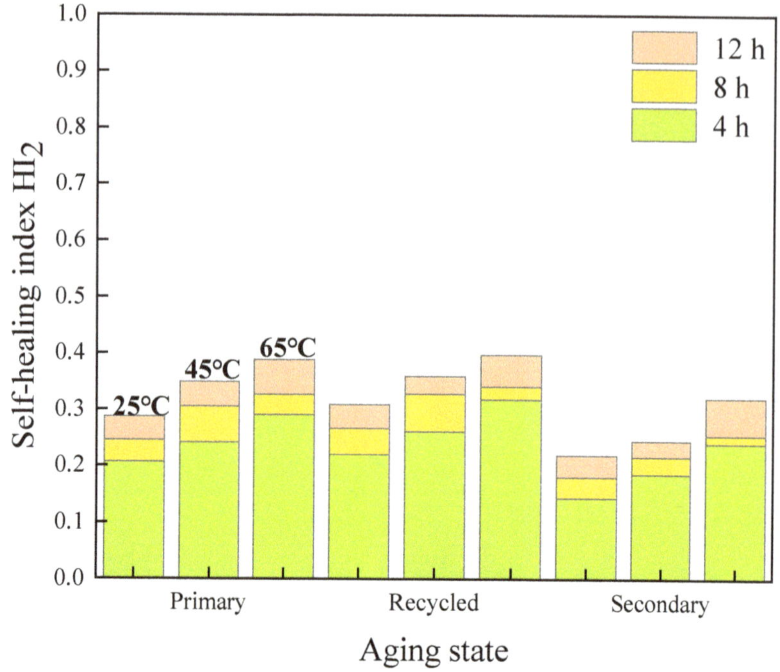

Figure 13. HI_2 of extracted old asphalt mastics.

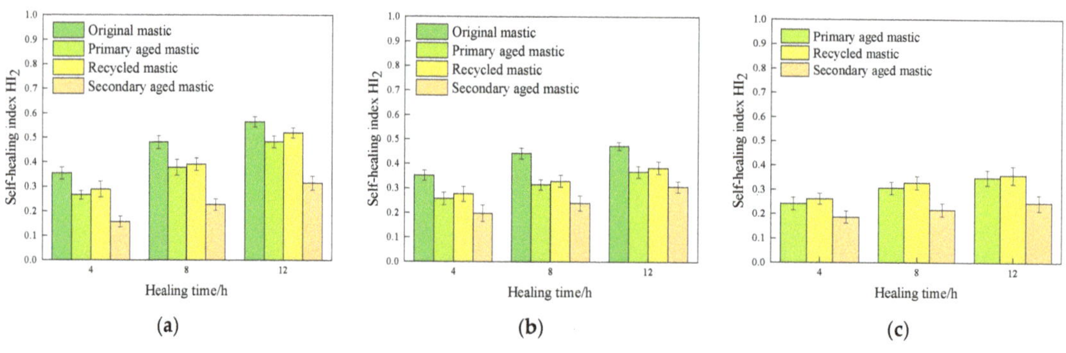

Figure 14. Results of self-healing index HI_2 of asphalt mastics at different healing time. (**a**) HI_2 of 70# asphalt mastics 45 °C; (**b**) HI_2 of SBS-modified asphalt mastics at 45 °C; (**c**) HI_2 of extracted old asphalt mastics at 45 °C.

In terms of the original asphalt mastics, the HI_2 values of 70# asphalt mastics increase to 136% and 159% after curing for 8 h and 12 h, respectively, compared to those of after 4 h healing. Meanwhile, the corresponding HI_2 values of SBS-modified asphalt mastics reach to 126% and 134% of those after 4 h healing, respectively.

After primary aging, the HI_2 of 70# asphalt mastic presents the highest value at different healing time. After curing for 12 h, the fracture healing index HI_2 reach to 182%, 143% and 144% of those after 4 h healing, for 70# asphalt, SBS-modified, and extracted old asphalt mastics, respectively.

After secondary aging, the HI_2 values of all kinds of asphalt mastic decrease significantly, but still increases with an increasing healing time. After curing for 12 h, the HI_2

values reach to 201%, 155%, and 132% of the corresponding asphalt mastics after 4 h healing, for the 70# asphalt, SBS-modified, and extracted old asphalt mastics, respectively. It is worth pointing out that the absolute values of HI_2 still maintain at low levels of less than 0.5 for the primary aged mastics and around 0.3 for the secondary aged ones. In terms of the extracted old asphalt mastic, the HI_2 values range from 0.185 to 0.245 after secondary aging, presenting the weakest fracture healing ability.

3.2.2. Effect of Healing Temperature on Fracture Self-Healing Properties

The results of the tests after 12 h healing are used and illustrated in Figure 15.

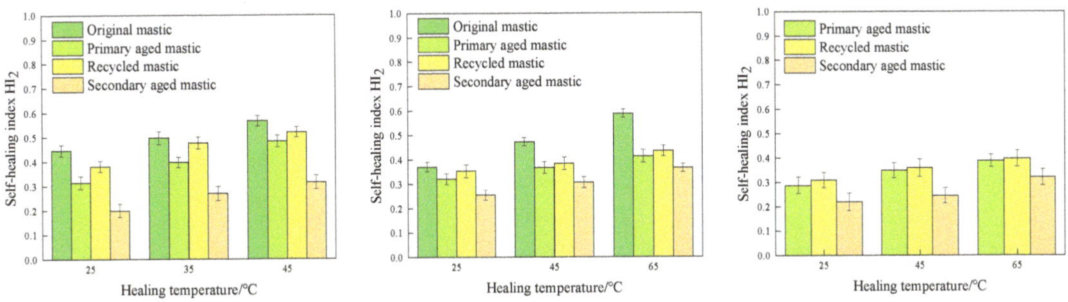

(a) (b) (c)

Figure 15. Results of self-healing index HI_2 of asphalt mastics at different healing temperatures. (**a**) HI_2 of 70# asphalt mastics after 12 h healing; (**b**) HI_2 of SBS-modified asphalt mastics after 12 h healing; (**c**) HI_2 of extracted old asphalt mastics after 12 h healing.

It can be seen from Figure 15 that the HI_2 values of all kinds of asphalt mastics increase with the increase of temperature. As for the unaged original asphalt mastics, after healing 12 h at 35 °C and 45 °C, the HI_2 values of 70# asphalt mastics reached to 111% and 127% of the corresponding mastics healing at 25 °C, respectively. In terms of SBS-modified asphalt mastics, after healing 12 h at 45 °C and 65 °C, the HI_2 values reached to 127% and 158% of the corresponding mastics at 25 °C, respectively. In addition, the HI_2 value of SBS-modified asphalt mastics increases significantly at 65 °C, achieving a higher absolute value than the HI_2 value of 70# asphalt mastics at 45 °C. It indicates that SBS-modified asphalt mastics need higher healing temperature than that of 70# asphalt mastics.

After primary aging, the HI_2 values of 70# asphalt mastics at 25 °C, 35 °C, and 45 °C decrease to 70.6%, 79.9%, and 85.3% of the original unaged asphalt mastics, respectively. At the healing temperatures of 25 °C, 45 °C, and 65 °C, the HI_2 values of SBS-modified asphalt mastic reduce to 86.5%, 77.7%, and 70.4% of the original asphalt mastics, respectively. It can be seen that the primary aged 70# asphalt mastics show superior fracture healing ability than that of SBS-modified asphalt mastics.

After secondary aging, the HI_2 values of all kinds of asphalt mastics decrease significantly, of which the 70# asphalt mastics show the most obvious drop. The HI_2 values of secondary aged 70# asphalt mastics reduce to 44.6%, 54.1%, and 55.7% of the original unaged asphalt mastics, respectively. Meanwhile, the HI_2 values of the secondary aged SBS-modified asphalt mastics decrease to 68.5%, 65.0%, and 62.1% of the original ones, respectively. It could be noticed that the recovery degree of both secondary aged 70# and SBS-modified asphalt mastics can hardly increase with an increasing healing temperature, though the absolute HI_2 value shows a rising trend with temperature.

3.2.3. Effect of Aging State of Asphalt on Fracture Self-Healing Properties

The results of the test at temperature of 45 °C and healing time of 12 h are used and illustrated in Figure 16.

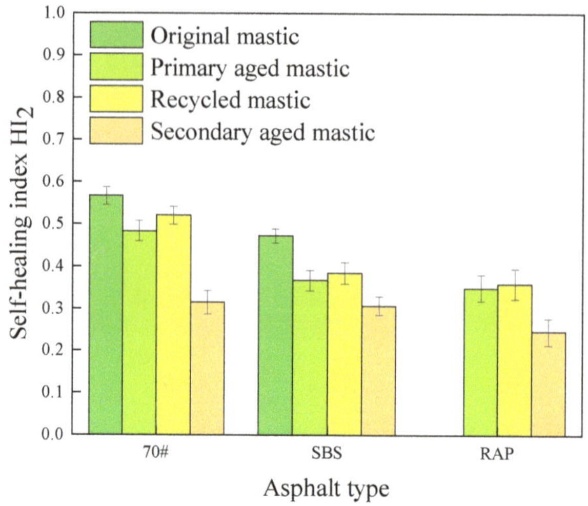

Figure 16. Results of self-healing index HI_2 of asphalt mastics in different aging states.

It can be seen from Figure 16 that as for the original unaged asphalt mastic, the HI_2 value of 70# asphalt mastic is higher than that of SBS-modified asphalt mastic, indicating that 70# asphalt mastics have a better fracture healing ability. After primary aging, the HI_2 value of 70# asphalt mastic decreases from 0.566 to 0.483, which reduced to 85.3% of the original mastic. The HI_2 value of SBS-modified asphalt mastic decreases from 0.472 to 0.367, which reduced to 77.8% of the original one. This indicates that the aging process has a more significant effect on the degradation of self-healing performance of SBS-modified asphalt mastic, which is also related to the slow diffusion behavior of SBS modifier [39].

After adding a rejuvenating agent, the HI_2 of recycled 70# asphalt mastic achieves 91.9% of the original mastic, and the index of SBS-modified asphalt mastic achieved 81.4% of the original mastic. The index of recycled old asphalt mastic is also improved compared to the extracted old asphalt mastic.

After secondary aging, the self-healing ability of all kinds of asphalt mastics deteriorate more seriously than that of the primary aged mastics. The HI_2 values of 70#, SBS, and old asphalt mastics decrease to 60.6%, 79.9%, and 68.2% of the corresponding recycled mastics, respectively.

Overall, during the fracture-healing-fracture test, the asphalt mastics suffered a full fracture. Due to the SBS modifier's interconnecting structure in asphalt, it is difficult for the SBS molecules to diffuse and mix with each other on the two fracture sections during the healing process, resulting in a lower fracture healing index of SBS and old asphalt mastics. Furthermore, the performance of secondary aged SBS-modified asphalt mastics is approaching that of 70# asphalt mastics. It can be partly attributed to the decomposition of SBS modifier [40], resulting in the close fracture healing index values of different mastics.

3.3. Comprehensive Analysis

The highest index values of both fatigue healing and fracture healing of different asphalt mastics were chosen for this comprehensive analysis, and the results are illustrated in Figure 17.

From Figure 17, it can be seen that the fracture healing ability of the three types of asphalt mastics is much inferior to the fatigue healing performance, indicating that low-temperature cracks being hardly cured by self-healing rather than fatigue cracks.

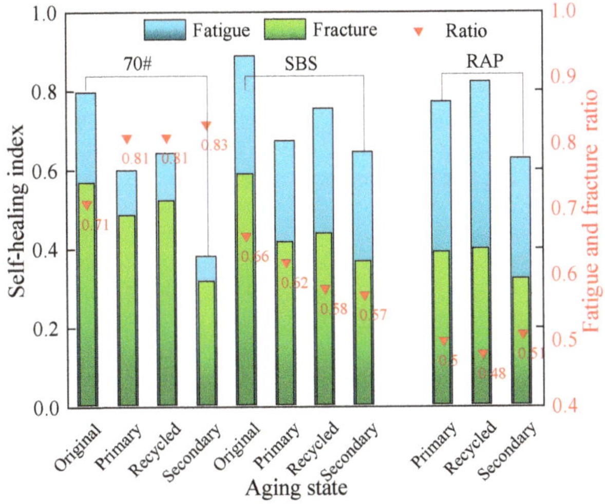

Figure 17. Comparison of fatigue healing and fracture healing indexes.

In terms of 70# asphalt mastics, the ratios of HI_2 over HI_1 reach 71.1%, 80.8%, 81.1% and 83.3% for the original unaged, primary aged, recycled, and secondary aged mastics, respectively. It infers that the fatigue healing ability and fracture healing ability of 70# asphalt mastics are comparable. However, the corresponding ratios only achieve 66.2%, 61.6%, 57.8%, and 56.7% for SBS-modified asphalt mastics. The extracted old asphalt mastics also display small ratio values similar to SBS-modified asphalt mastics. These results indicate that the fatigue and fracture healing properties depend on asphalt type, and 70# asphalt mastics present similar fatigue and fracture healing properties, while SBS-modified asphalt mastics possessing excellent fatigue but insufficient fracture healing properties. The reason can be that SBS modifier can improve the viscoelastic property of asphalt, resulting in excellent fatigue healing ability of SBS-modified asphalt. However, the macromolecule of the SBS modifier can hardly diffuse though the fracture section, leading to inferior fracture healing property.

4. Conclusions

This article used fatigue-healing-fatigue test and fracture-healing-fracture test to investigate the self-healing performance of 70# asphalt, SBS-modified asphalt, and RAP extracted old asphalt mastics under different healing times, healing temperatures, and aging states fabricated by primary and (or) secondary aging processes. The cumulative dissipation energy ratio and tensile fracture stress ratio of asphalt mastics after and before healing was used as fatigue healing index and fracture healing index, respectively. Under the testing conditions mentioned above, the following conclusions can be drawn:

(1) With an increasing healing time and temperature, the fatigue healing index (HI_1) values increase up to around 0.8 and 0.9 for unaged 70# asphalt and SBS-modified asphalt mastics, respectively. The absolute values of fracture healing index (HI_1) values reach just around 0.57 and 0.47. It means both of these two mastics present comparably excellent fatigue healing and inferior fracture healing ability.

(2) Aging state has significant impact on the healing property of asphalt mastics. After primary and secondary aging, the fatigue healing index (HI_1) values reduce from 0.796 of unaged 70# asphalt mastics to 0.598 and 0.378, while the HI_1 values of SBS-modified asphalt mastics reduce from 0.888 of unaged one to 0.672 and 0.674, respectively.

(3) Healing properties are also asphalt-type dependent, and 70# asphalt is more sensitive to secondary aging process. After secondary aging, 70# asphalt mastics show the worst fatigue healing capability and can hardly recover to a satisfactory level, while

SBS-modified asphalt and extracted old asphalt presenting the similar superior healing ability to that of primary aged ones.

(4) Low-temperature cracks of aged asphalts can hardly be cured by self-healing rather than fatigue cracks. After primary and secondary aging, the fracture healing index (HI_2) of all the three types of mastics are only around 0.3, which are much smaller than the HI_1 values under each testing condition, and can hardly increase with an increasing healing temperature.

(5) The recover effect of the rejuvenating agent will be degraded rapidly during the secondary aging process, though both the fatigue and fracture healing performance could be recovered to a certain extent by adding rejuvenating agent. Thus, more effective rejuvenating agents are on demand to be developed.

This study just explored the self-healing performance of asphalt mastics by experiments. There are several points that can be considered for further investigation. Firstly, microstructure and chemical characteristics of asphalt mastics after aging and recycling processes can be investigated by scanning electron microscopy (SEM) and atomic force microscopy (AFM), etc., and the relationships with the self-healing properties can be established. Secondly, only one-time damage was utilized in this study. The effect of multiple damages on the self-healing performance of asphalt mastics also needs further investigation. Thirdly, the effect of secondary aging on the road performance and self-healing properties of recycled asphalt mixtures is also worth studying, in addition to the results of asphalt mastics.

Author Contributions: Conceptualization, B.L. and P.X.; Data curation, B.L. and A.K.; Formal analysis, B.L. and Y.W.; Funding acquisition, P.X. and Y.Z.; Investigation, B.L., Y.W. and Z.W.; Methodology, Z.W., P.X. and Y.Z.; Project administration, A.K. and P.X.; Supervision, Y.Z. and P.X.; Visualization, B.L., Y.W. and Y.Z.; Writing—original draft, B.L. and Y.W.; Writing—review & editing, P.X., Y.Z. and A.K. All authors have read and agreed to the published version of the manuscript.

Funding: This research was funded by the National Natural Science Foundation of China, grant number 52108422; Postgraduate Research & Practice Innovation Program of Jiangsu Province (Yangzhou University) (SJCX23_1936). In addition, this research was funded by the Postgraduate education and teaching reform and practice project of Yangzhou University, grant number JGLX2021_010 and the Yangzhou City & University cooperation program, grant number YZ2021167.

Institutional Review Board Statement: Not applicable.

Informed Consent Statement: Not applicable.

Data Availability Statement: Data are contained within the article.

Conflicts of Interest: The authors declare no conflict of interest.

References

1. Sadeghi, P.; Goli, A. Investigating the Cracking Performance of Asphalt Mixture Containing RAP Using Different Cracking Indices. *Case Stud. Constr. Mater.* **2023**, *19*, e02340. [CrossRef]
2. Alae, M.; Xu, L.; Cao, Z.; Xu, X.; Xiao, F. Fatigue and Intermediate-Temperature Cracking Performance of Rejuvenated Recycled Asphalt Binders and Mixtures: A Review. *J. Clean. Prod.* **2023**, *384*, 135587. [CrossRef]
3. García, A.; Norambuena-Contreras, J.; Bueno, M.; Partl, M.N. Single and Multiple Healing of Porous and Dense Asphalt Concrete. *J. Intell. Mater. Syst. Struct.* **2015**, *26*, 425–433. [CrossRef]
4. Nalbandian, K.M.; Carpio, M.; González, Á. Analysis of the Scientific Evolution of Self-Healing Asphalt Pavements: Toward Sustainable Road Materials. *J. Clean. Prod.* **2021**, *293*, 126107. [CrossRef]
5. Zhuang, C.; Guo, H.; Li, F.; Hao, Y.; Chen, K.; Li, G.; Ye, Y. Study on Adhesion Performance and Aging Strength Degradation Mechanism of SBS Modified Asphalt with Different Anti-Aging Additive. *Materials* **2023**, *16*, 4881. [CrossRef] [PubMed]
6. Sun, Q.; Wang, X.Y.; Wang, S.; Shao, R.Y.; Su, J.F. Investigation of Asphalt Self-Healing Capability Using Microvasculars Containing Rejuvenator: Effects of Microvascular Content, Self-Healing Time and Temperature. *Materials* **2023**, *16*, 4746. [CrossRef] [PubMed]
7. Sousa de Oliveira, L.; Feitosa de Albuquerque Lima Babadopulos, L.; Soares, J.B. Evolution of Asphalt Binder Stiffness during Fatigue Loading and Rest Periods and Its Impact on Fatigue Life. *Int. J. Fatigue* **2021**, *144*, 106041. [CrossRef]
8. Sudarsanan, N.; Kim, Y.R. A Critical Review of the Fatigue Life Prediction of Asphalt Mixtures and Pavements. *J. Traffic Transp. Eng. (Engl. Ed.)* **2022**, *9*, 808–835. [CrossRef]

9. Ani, O.J.; Shafabakhsh, G.A.; Mirabdolazimi, S.M. Presenting the Self-Healing Potential Model by Induction Heating Method in Warm Asphalt Mixtures by Artificial Neural Network. *J. Adv. Res. Fluid Mech. Therm. Sci.* **2023**, *105*, 1–20. [CrossRef]
10. HasaniNasab, S.; Arast, M.; Zahedi, M. Investigating the Healing Capability of Asphalt Modified with Nano-Zycotherm and Forta Fibers. *Case Stud. Constr. Mater.* **2019**, *11*, e00235. [CrossRef]
11. Norambuena-Contreras, J.; Gonzalez-Torre, I. Influence of the Microwave Heating Time on the Self-Healing Properties of Asphalt Mixtures. *Appl. Sci.* **2017**, *7*, 1076. [CrossRef]
12. Zhang, S.; Yue, J.; Wu, J.; Li, R. Influence of Rock Asphalt on Self-Healing Behavior of Asphalt Mastics. *Constr. Build. Mater.* **2021**, *310*, 124851. [CrossRef]
13. Sun, D.; Sun, G.; Zhu, X.; Ye, F.; Xu, J. Intrinsic Temperature Sensitive Self-Healing Character of Asphalt Binders Based on Molecular Dynamics Simulations. *Fuel* **2018**, *211*, 609–620. [CrossRef]
14. Grossegger, D.; Garcia, A. Influence of the Thermal Expansion of Bitumen on Asphalt Self-Healing. *Appl. Therm. Eng.* **2019**, *156*, 23–33. [CrossRef]
15. Fan, S.; Zhu, H.; Yuan, H.; Chen, C. Fracture -Healing Properties of Asphalt Mixtures and Microwave Heating Ther-mo-Sensitivity Analysis of Their Constituent Materials. *J. Clean. Prod.* **2021**, *312*, 127763. [CrossRef]
16. Ding, X.; Rath, P.; Giraldo-Londoño, O.; Buttlar, W.G.; Ma, T. Fracture modeling of rubber-modified binder based on Discrete Element Method. *J. Clean. Prod.* **2022**, *380*, 135017. [CrossRef]
17. Xu, X.; Sreeram, A.; Leng, Z.; Yu, J.; Li, R.; Peng, C. Challenges and Opportunities in the High-Quality Rejuvenation of Unmodified and SBS Modified Asphalt Mixtures: State of the Art. *J. Clean. Prod.* **2022**, *378*, 134634. [CrossRef]
18. Zhou, L.; Huang, W.; Zhang, Y.; Lv, Q.; Yan, C.; Jiao, Y. Evaluation of the Adhesion and Healing Properties of Modified Asphalt Binders. *Constr. Build. Mater.* **2020**, *251*, 119026. [CrossRef]
19. Fakhri, M.; Bahmai, B.B.; Javadi, S.; Sharafi, M. An Evaluation of the Mechanical and Self-Healing Properties of Warm Mix Asphalt Containing Scrap Metal Additives. *J. Clean. Prod.* **2020**, *253*, 119963. [CrossRef]
20. Masri, K.A.; Arshad, A.K.; Samsudin, M.S. Mechanical Properties of Porous Asphalt with Nanosilica Modified Binder. *J. Teknol.* **2016**, *78*, 139–146. [CrossRef]
21. Arshad, A.K.; Samsudin, M.S.; Alam, S.; Khairil, M.; Masri, A.; Arshad, A.K.; Masri, K.A. Engineering Properties of Nanosilica Modified Asphalt Binder. *J. Mech. Eng.* **2019**, *16*, 29–40. [CrossRef]
22. Li, L.; Gao, Y.; Zhang, Y. Crack Length Based Healing Characterisation of Bitumen at Different Levels of Cracking Damage. *J. Clean. Prod.* **2020**, *258*, 120709. [CrossRef]
23. Lv, H.; Zhang, D.; Tan, Y.; Cai, L.; Liu, H. Development of a Complex Index to Characterize the Recovery Evolution of Asphalt Binder after Cyclic Loading. *Int. J. Fatigue* **2022**, *156*, 106664. [CrossRef]
24. Zhao, K.; Wang, W.; Wang, L. Fatigue Damage Evolution and Self-Healing Performance of Asphalt Materials under Dif-ferent Influence Factors and Damage Degrees. *Int. J. Fatigue* **2023**, *171*, 107577. [CrossRef]
25. Wang, L.; Cheng, P.; Zhao, Q. Research on Fatigue–Healing Performance of Asphalt Mixture Based on the Semicircular Bending Test. *Materials* **2023**, *16*, 6382. [CrossRef] [PubMed]
26. Xiang, H.; He, Z.; Tang, H.; Hu, M. Healing Behavior of Thermo-Oxygen Aged Asphalt Based on Molecular Dynamics Simulations. *Constr. Build. Mater.* **2022**, *349*, 128740. [CrossRef]
27. Sun, G.; Hu, M.; Zhu, X.; Ning, W.; Liu, Z.; Sun, D. Decay Pattern of Self-Healing Temperature Susceptibility of Bitumens at Various Ageing States. *Constr. Build. Mater.* **2022**, *324*, 126668. [CrossRef]
28. *JTG F40-2004*; Technical Specifications for Construction of Highway Asphalt Pavements. China Communications Press: Beijing, China, 2004.
29. *JTG E20-2011*; Standard Test Methods of Bitumen and Bituminous Mixtures for Highway Engineering. China Communications Press: Beijing, China, 2011.
30. *JTG/T 5521-2019*; Technical Specification for Highway Asphalt Pavement Recycling. China Communications Press: Beijing, China, 2019.
31. *JTG E42-2005*; Test Methods of Aggregate for Highway Engineering. China Communications Press: Beijing, China, 2005.
32. Leegwater, G.; Scarpas, T.; Erkens, S. Direct tensile test to assess healing in asphalt. *Transp. Res. Rec.* **2016**, *2574*, 124–130. [CrossRef]
33. Qiu, J.; van de Ven, M.; Wu, S.; Yu, J.; Molenaar, A. Evaluating Self-Healing Capability of Bituminous Mastics. *Exp. Mech.* **2012**, *52*, 1163–1171. [CrossRef]
34. Huang, W.; Lv, Q.; Xiao, F. Investigation of Using Binder Bond Strength Test to Evaluate Adhesion and Self-Healing Properties of Modified Asphalt Binders. *Constr. Build. Mater.* **2016**, *113*, 49–56. [CrossRef]
35. Liang, B.; Lan, F.; Shi, K.; Qian, G.; Liu, Z.; Zheng, J. Review on the Self-Healing of Asphalt Materials: Mechanism, Affecting Factors, Assessments and Improvements. *Constr. Build. Mater.* **2021**, *266*, 120453. [CrossRef]
36. Amani, S.; Kavussi, A.; Karimi, M.M. Effects of Aging Level on Induced Heating-Healing Properties of Asphalt Mixes. *Constr. Build. Mater.* **2020**, *263*, 12010. [CrossRef]
37. Xia, W.; Xu, Z.; Xu, T. Self-Healing Behaviors and Its Effectiveness Evaluations of Fiber Reinforced Shape Memory Poly-urethane/SBS Modified Asphalt Mortar. *Case Stud. Constr. Mater.* **2023**, *18*, e01784. [CrossRef]
38. Santagata, E.; Baglieri, O.; Tsantilis, L.; Dalmazzo, D.; Chiappinelli, G. Fatigue and Healing Properties of Bituminous Mastics Reinforced with Nano-Sized Additives. *Mech. Time Depend. Mater.* **2016**, *20*, 367–387. [CrossRef]

39. Zhang, D.; Zhang, H.; Shi, C. Investigation of Aging Performance of SBS Modified Asphalt with Various Aging Methods. *Constr. Build. Mater.* **2017**, *145*, 445–451. [CrossRef]
40. Li, C.; Xie, X.; Wang, L.; Guo, Y.; Zhang, L.; Xue, Z. Evaluation of the Effect of Thermal Oxygen Aging on Base and SBS-Modified Bitumen at Micro and Macroscales. *Constr. Build. Mater.* **2022**, *324*, 126623. [CrossRef]

MDPI
St. Alban-Anlage 66
4052 Basel
Switzerland
www.mdpi.com

Materials Editorial Office
E-mail: materials@mdpi.com
www.mdpi.com/journal/materials

www.ingramcontent.com/pod-product-compliance
Lightning Source LLC
Chambersburg PA
CBHW040840250926
51850CB00007BA/1369